Seminar Nasional Material (SNM 2018)

Science and Materials Technology Innovation to Improve Global Competitiveness

IOP Conference Series: Materials Science and Engineering
Volume 599

Bandung, Indonesia
16 - 17 November 2018

Editors:

Toto Winata
Ferry Iskandar
Akfiny H. Aimon

ISBN: 978-1-5108-9469-3
ISSN: 1757-8981

Printed from e-media with permission by:

Curran Associates, Inc.
57 Morehouse Lane
Red Hook, NY 12571

Some format issues inherent in the e-media version may also appear in this print version.

Additional copies of this publication are available from:

Curran Associates, Inc.
57 Morehouse Lane
Red Hook, NY 12571 USA
Phone: 845-758-0400
Fax: 845-758-2633
Email: curran@proceedings.com
Web: www.proceedings.com

Table of contents

Volume 599

Seminar Nasional Material (SNM 2018)

16–17 November 2018, Bandung, Indonesia

Accepted papers received: 31 July 2019
Published online: 4 September 2019

Preface

Papers

Energy and Electronics Materials

Nanomaterials and Catalysts

Biomaterials

IOP Conf. Series: Materials Science and Engineering **599** (2019) 011001 doi:10.1088/1757-899X/599/1/011001

PREFACE

The Seminar Nasional Material (SNM) is an annual national conference organized by the Physics of Electronic Materials Research Group, Department of Physics, Faculty of Mathematics and Natural Sciences, Institut Teknologi Bandung. The Seminar Nasional Material (SNM) 2018 was held on November 16-17 2018 at Institut Teknologi Bandung with the main theme "Science and Materials Technology Innovation to Improve Global Competitiveness". The aim of this event was to support Material Research in Indonesia which is one of the research with a substantial contribution to the technological progress and sustainable development in Indonesia. In addition, this seminar is also expected to be a place for communication and enhance the creativity of researchers, lecturers, and students involved in the field of materials research.

There were presentations from 1 keynote speaker, 5 guest speakers and 82 participants in the The Seminar Nasional Material 2018. The scopes of the seminar were covered several topics which included 20 presentations in the fields of Energy and Electronics Materials, 18 presentations in the field of Composite Materials, Metals and Ceramics, 23 presentations in the field of Nanomaterials and Catalysts and 21 presentations in the field of Biomaterials. The output of this seminar published in National Proceedings in Bahasa Indonesia and Proceedings of IOP Conference Series: Materials Science and Engineering. After peer-reviewed by experts in related fields, finally, this proceeding contains 31 accepted papers in IOP Conference Series: Materials Science and Engineering.

The Seminar Nasional Material (SNM) 2018 committee would like to acknowledge the enthusiasms of the participants, support from the sponsors, all members of the committees, the advisory board, Faculty of Mathematics and Natural Sciences of ITB, and all those who had contributed to the success of The Seminar Nasional Material (SNM) 2018.

Dr.Eng. Ferry Iskandar
Chairman of the Seminar Nasional Material (SNM) 2018

Prof. Toto Winata (Institut Teknologi Bandung, Indonesia), Dr. Ferry Iskandar (Institut Teknologi Bandung, Indonesia), and Dr.rer.nat. Akfiny H. Aimon (Institut Teknologi Bandung, Indonesia).
Editorial Board

IOP Conf. Series: Materials Science and Engineering **599** (2019) 011002 doi:10.1088/1757-899X/599/1/011002

Peer review statement

All papers published in this volume of *IOP Conference Series: Materials Science and Engineering* have been peer reviewed through processes administered by the proceedings Editors. Reviews were conducted by expert referees to the professional and scientific standards expected of a proceedings journal published by IOP Publishing.

The Modifications of Electronic and Optical Properties of Bulk Molybdenum Disulfide by Oxygen Substitution

Louis Charvia, Shibghatullah Muhammady, Yudi Darma*

Quantum Semiconductor and Devices Lab., Department of Physics, Faculty of Mathematics and Natural Sciences, Institut Teknologi Bandung, Bandung, 40132, Indonesia

*E-mail: yudi@fi.itb.ac.id

Abstract. We study the electronic and optical properties of bulk MoS_2 and MoSO systems using the plane-wave method within the generalized gradient approximation. The structural properties show that the O substitution at S sites tunes the z-axis Wyckoff positions of all atoms and the bond angles. The electronic band structures show that the substitution tunes the conduction band minimum at 0.8 Σ to M and significantly promotes more localized valence states compared to that of MoS_2 system. The localization, mainly applying to Mo $4d$ and S $3p$ states, is more pronounced above −4 eV. Below −4 eV, O $2p$ states are dominant indicating that they are more stable than S $3p$ states. However, the substitution slightly increases the indirect and direct (K → K) bandgap of MoS_2 system. From the optical properties, both systems show the strong optical dichroism. By means of the substitution, the σ_1 width significantly enhances, while the plasmonic-state energy levels decrease. Our result emphasizes that the O substitution significantly tunes the electronic and optical properties of bulk MoS_2 system.

1. Introduction

Molybdenum disulfide (MoS_2) is a semiconducting layered transition-metal dichalcogenide (TMD) which is the most promising two-dimensional (2D) systems for nanoelectronic applications [1-4]. One of the interesting phenomena of the system is that it exhibits the layer-number-dependence of band structure. The bulk and monolayer MoS_2 systems have the indirect bandgap of 1.88 eV [5] and direct bandgap of 1.90 eV [1], respectively. However, the bulk system exhibits the indirect minimum bandgap (E_g) of 1.29 eV [5]. The reduction of the layer thickness from bulk to monolayer systems increases E_g, indicating the tunable band structures [6]. The monolayer system can strongly emit photons. The system exhibits the very high luminescence quantum efficiency of more than 10^4 compared with that of the bulk system. Furthermore, the monolayer system shows the strong photoluminescence (PL) compared to the bilayer and bulk systems [1]. The defect sites in the monolayer MoS_2 can be used to significantly improve its hydrogen evolution activity [7]. Moreover, MoS_2 can also be applied in the supercapacitor by building the MoS_2-H_2O multilayer [8] and for the energy storage by building reduced graphene oxide/MoS_2 hybrids [9].

The doping treatment using the transition metals at Mo site tunes the electronic and magnetic properties of MoS_2 system [10]. The previous experimental report has shown that the n-type substitution increases the local chemical affinity, while the p-type substitution shows the larger mobility when the electron beam exposes the system. The result emphasizes the application possibility in nanoelectronic devices [11]. The other report has shown the possibility of doping at S sites by using nonmetal and transition metal elements. The doping treatment significantly modifies the magnetic behavior of the monolayer system. Notably, H, Br, and Cr dopings induce the half-metallic behavior of the system [12]. However, there is no report of optical properties and plasmonic states in MoS_2 system doped at S sites, as O and S are included in one group in the periodic table.

In this paper, we report electronic and optical properties of bulk MoS_2 and MoSO systems. In the MoSO system, one S atom is substituted by one O atom. We employed the plane-wave method within the generalized gradient approximation (GGA) [13]. Firstly, this paper reveals band structures and density of states from the electronic structure calculation. Secondly, the optical properties are calculated in the term of energy-dependent optical conductivity (σ_1), which is extracted from the imaginary part of dielectric function (ε_2). Lastly, the electron-energy loss function (L) also calculated to indicates the plasmonic-state energy levels.

2. Computational details

Quantum ESPRESSO package code [14, 15] was carried out for calculating the electronic and optical properties of unit cells of bulk MoS_2 and MoSO systems. The crystal structure models of both systems visualized by VESTA [16] are presented in figure 1. The structural parameters in this calculation are directly adopted from the previous experimental report. At room temperature, the bulk MoS_2 exhibits the hexagonal crystal structure with the structural parameters of a = 3.15 Å, c = 12.30 Å, z_S = 0.621 within the space group of $P6_3/mmc$ [17]. Each Mo atom is surrounded by O atoms within the triangular prismatic shape. From top view along c-axis, the hexagonal pattern consists of 3 Mo and 3 S atoms [18]. For MoSO system, two O atom substitute one S atom at atomic positions of (2/3, 1/3, $-z$) and (1/3, 2/3, $-z$ + 1/2) in the unit cell of MoSO system. We employed the Perdew-Burke-Ernzerhof (PBE)-type exchange-correlation functional energy for all atoms within the GGA method [13]. This method was used for various systems in our previous reports [19-21]. All-electron potentials are approached by the norm-conserving pseudopotentials [22-24]. The calculation carried out kinetic cutoff energy of 1.6 keV, a Γ-point-centered k-point mesh of 7 × 7 × 2, and threshold energy of 10 meV. All atomic positions are optimized using the 'relax' calculation with threshold force of 50 meV/Å involving the Broyden–Fletcher–Goldfarb–Shanno (BFGS) algorithm [25-28]. On the other hand, the lattice parameters (a and c) are set to be constant.

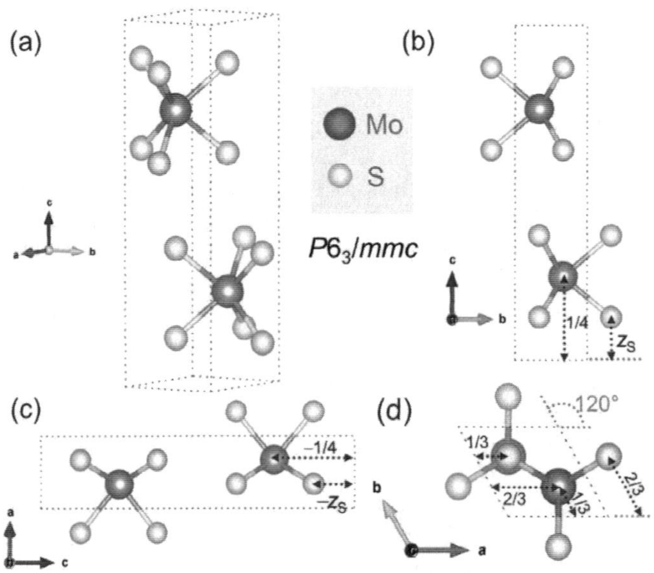

Figure 1. Crystal structure of bulk MoS_2 for views along (a) diagonal, (b) a, (c) b, and (d) c axes. The Wyckoff positions of Mo and S atoms are Mo (2c: 1/3, 2/3, 1/4) and S (4f: 1/3, 2/3, z_S).

The Drude-Lorentz model [29] is used in the optical properties calculation. This model covers the interband and intraband parts of complex dielectric functions (ε) [30]. From both parts, ε is divided into real and imaginary parts of dielectric functions (ε_1, ε_2) which are connected by the K ramers-Kronig relation [31, 32]. Both ε_1, ε_2 will be reported elsewhere in future. In this paper, we show the imaginary-part optical conductivity (σ_1) expressed as

$$\sigma_1 = \frac{\omega \varepsilon_2}{4\pi} \quad (1)$$

where ω is the photon frequency. We also reveal the loss function (L) spectrum calculated by

$$L = -\mathrm{Im}\left\{\frac{1}{\varepsilon}\right\} = \frac{\varepsilon_2}{\varepsilon_1^2 + \varepsilon_2^2} \quad (2)$$

The energy-dependent L can accurately assign plasmonic-state energy levels [32].

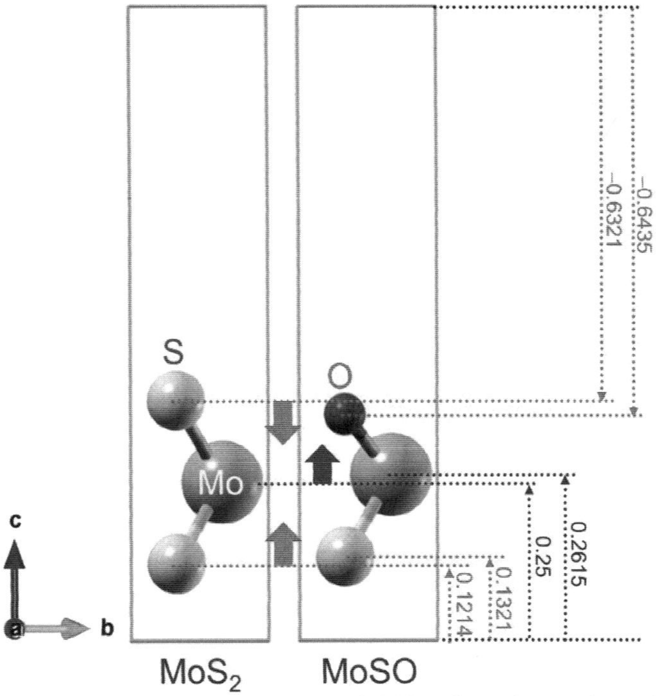

Figure 2. Schematic illustration of shifts of z-axis atomic positions in MoS_2 system induced by the O substitution at S site in MoSO system.

Table 1. Optimized Wyckoff positions of MoS_2 and MoSO systems.

Atom	Atomic positions (x, y, z) ($a = b = 3.15$ Å and $c = 12.30$ Å)		
	MoS_2		MoSO
	Present work	Previous experiment [17]	
Mo (2c)	1/3, 2/3, 1/4	1/3, 2/3, 1/4	1/3, 2/3, 0.2615
S (4f)	1/3, 2/3, z_S ($z_S = 0.6214$)	1/3, 2/3, z_S ($z_S = 0.621$)	1/3, 2/3, z_S ($z_S = 0.6321$)
O (4f)	–	–	1/3, 2/3, z_O ($z_O = 0.6435$)

3. Results and Discussions

3.1. Structural properties

Firstly, we investigate the structural optimization results of both MoS_2 and MoSO systems. Table 1 presents optimized Wyckoff positions of both systems. As the results, the present calculation well reproduces the experimental results [17]. On the other hand, the O substitution at S sites slightly modifies the atomic positions. The *z*-axis Wyckoff position of all atoms increases. The result indicates the atomic bonding modifications. Figure 2 presents the schematic illustration of shifts of the *z*-axis atomic positions, which is visualized by XCrysDEN [33]. We find that the S–Mo–S bond angle of 82.012° decreases to 73.921° as the O–Mo–S bond angle. The previous report shows the S–Mo–S bond angle of 82.207° [17]. Hence, the structural properties of MoS_2 system in the present work is comparable with the previously reported results. We suggest that the structural properties of MoSO system can also be used to predict experimental results.

3.2. Electronic properties

Figure 3 presents the band structures of bulk MoS_2 and MoSO systems. We find the indirect E_g = 0.94 eV from Γ to 0.8Σ points in MoS_2 system, which is lower than that of the previous experimental report, i.e., 1.29 eV from reflectivity [5] and *L* spectra [34]. The underestimated E_g can be caused by the limitation of GGA in describing the exact Kohn-Sham band structure within GGA [35]. We also find the direct transition at K point of 2.66 eV. The band structures show the delocalized states along Γ–M and K–Γ points, indicated by the delocalized valence-band feature. However, the localized states are found along M–K points, which are indicated by the localized valence-band feature.

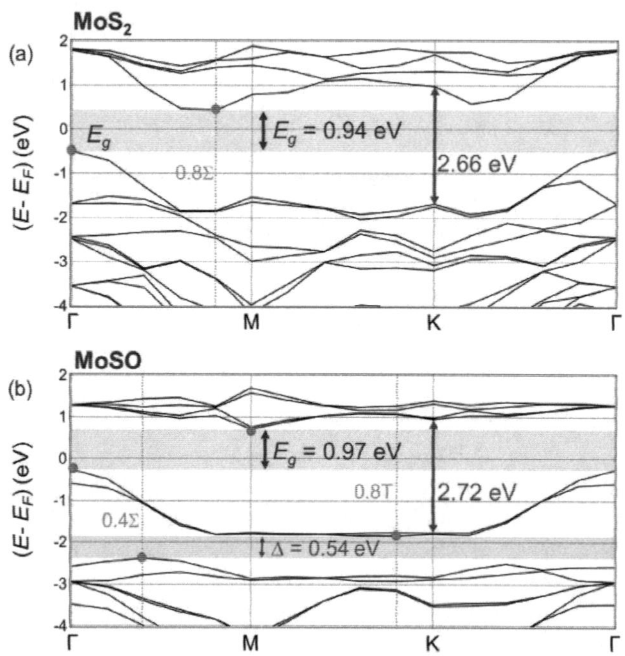

Figure 3. Band structures of bulk (a) MoS_2 and (b) MoSO systems. Indirect and direct bandgaps are assigned. The blue shades denote the minimum bandgap (E_g) of both systems, while the green shade denotes the bandgap (Δ) within the valence band of MoSO system. The red dots denote the points involved in the indirect band transitions. Fermi energy level (E_F) is denoted at 0 eV.

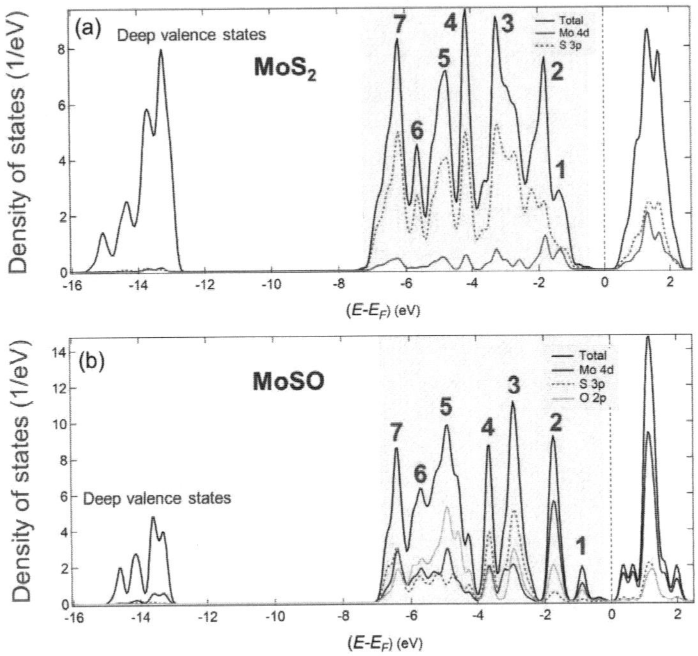

Figure 4. Total and partial density of states of bulk (a) MoS_2 and (b) MoSO systems. Near Fermi energy level (E_F), the width and peaks of valence band are assigned by the soft red shade and numbers (1 to 7), respectively. Fermi energy level (E_F) is denoted at 0 eV.

MoSO system shows the indirect E_g of 0.97 eV from Γ to M points. This result shows that the substitution slightly increases E_g. However, it changes the transition point at the conduction band minimum from 0.8Σ path to M point. This result is different with that of the monolayer MoS_2, as revealed by the previous first-principle calculation. The O substitution at S site in the monolayer MoS_2 does not only change the transition point but also changes E_g type from direct to indirect types [36]. We also find the in-valence-band interband transition (Δ) of 0.54 eV from 0.4Σ to 0.8T paths. The presence of Δ is induced by the localized valence band feature at around –2 eV. The band structure also indicates the localized states for all *k*-points. Hence, the substitution significantly promotes more localized valence states.

Figure 4(a) shows the total and partial density of states (DOS) of bulk MoS_2 system. We find that MoS_2 system shows E_g around Fermi energy level (E_F), which mainly promoted by Mo 4*d*, followed by S 3*p* states. Near E_F, the valence band is dominated by S 3*p*. This result approves the theoretical electron configuration of S^{2-}: [37] $3s^2\ 3p^6$. However, the conduction band is dominated by both Mo 4*d* and S 3*p* states. The presence of Mo 4*d* states in the conduction band approves the partially occupied $Mo^{2+}\ 4d^2$ orbital. The presence of S 3*p* states may be resulted due to the strong hybridization between Mo 4*d* and S 3*p* states. We also assign seven peaks (1-7) in the valence band. All seven peaks overlap each other, indicating the dominant delocalized states, as presented in figure 3(a).

Figure 4(b) presents the total and partial density of states (DOS) of bulk MoSO system. We find that E_g is mainly promoted by Mo 4*d* and O 2*p* states. The conduction band is dominated by Mo 4*d* states approving the valence electron configuration of $Mo^{2+}\ 4d^2$. Interestingly, more complex features in the valence band are observable. Seven peaks in the valence band of MoSO system are assigned. We find that the peaks 1-4 are localized above –4 eV, which are in contrast to that of MoS_2 system. This result confirms the localized bands in figure 3(b). However, the peaks 5-7 below –4 eV are delocalized, as also found in MoS_2 system in figure 4(a). Hence, the substitution significantly modifies the valence band structure alignment of MoS_2 system near E_F.

3.3. Optical properties

Figure 5 presents the energy-dependence of σ_1 bulk MoS_2 and MoSO systems. In MoS_2 system, we find different curves of σ_1 along *xy* plane (σ_{1xy}) and *z* axis (σ_{1z}), as shown in figure 5(a). This result indicates optical dichroism, which has also been found in α-PbO [38] and TiO_2-based systems [39]. The optical dichroism indicates that the bulk MoS_2 is anisotropic, as previously shown by different dielectric constants along different directions (*xy*-plane and *z*-axis) [40]. In region A, two major peaks are observed for both *xy*-plane and *z*-axis directions. The peak shape is similar, indicating the weak dichroism in this region. In region B, we find high peak at around 5.7 eV along *z* axis, while *xy* plane does not show any significant peak. This result indicates the strong dichroism in the region. Between both regions A and B, a major peak at around 4.5 eV is found for *xy* plane. Figure 5(b) also shows the optical dichroism in MoSO system. In contrast to MoS_2 system, both regions A and B show strong optical dichroism along both *xy*-plane and *z*-axis directions, indicated by the significant different shapes between σ_1 of both systems. Notably, region A of MoSO system shows the multiple peaks at which their intensities are lower than that of the double peaks of MoS_2 system. We suggest that the presence of multi-peaks are induced by the transitions between the localized valence states near E_F to conduction states. Furthermore, σ_1 width (~7.16 eV) of MoSO system is wider than that of MoS_2 system (~6.39 eV), indicating the wider absorption range in MoSO system.

Figure 6 presents *L* of bulk MoS_2 and MoSO systems. In the *L* spectrum, the peaks represent plasma energy as the oscillatory response of delocalized valence electrons due to the plasma excitation. This excitation is induced by the inelastic scattering between the atomic electrons in a material and an electron beam. The excitation results when the speed of electron beam is higher than the Fermi velocity [41]. Figure 6(a) shows two *L* peaks at 11.3 and 12.1 eV along *xy*-plane (L_{xy}) in MoS_2 system. At the lower energy level of 10.9 eV along *z* axis, we find a higher peak compared to both other peaks. All these peaks imply the plasmonic states. This result indicates the dichroism of plasmonic state. By considering the difference between energy levels for the peaks at 11.3 and 10.9 eV, the dichroism is 0.4 eV. Figure 6(b) shows two *L* peaks at 10.9 and 11.3 eV along *xy*-plane (L_{xy}) in MoSO system. A peak at 9.5 eV is also found along *z* axis (L_z). By considering the difference between energy levels for the peaks at 10.9 and 9.5 eV, the dichroism of MoSO system is 1.4 eV, which indicates that MoSO system exhibit the higher plasmonic-state dichroism than that of MoS_2 system. Additionally, plasmonic-state width is also increased by the substitution (green shade).

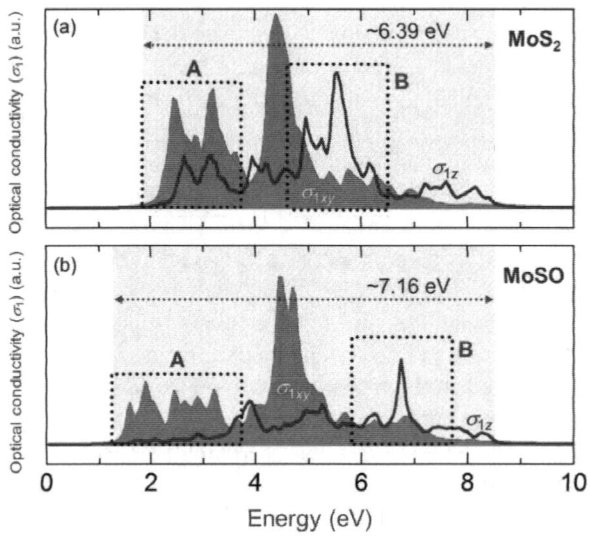

Figure 5. Optical conductivity (σ_1) of bulk (a) MoS_2 and (b) MoSO systems. Red face and blue line denote σ_1 along *xy* plane (σ_{1xy}) and *z* axis (σ_{1z}), respectively. The σ_1 width is assigned by the green shade. The dashed squares (A and B) emphasize the significant σ_1 modification.

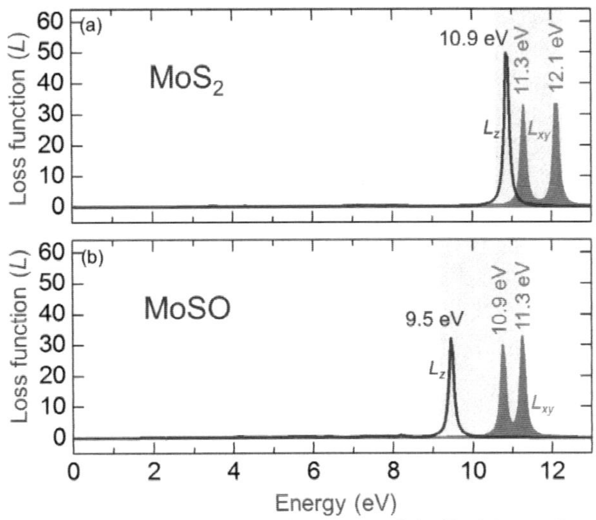

Figure 6. Electron-energy loss function (L) of bulk (a) MoS_2 and (b) MoSO systems. Red face and blue line denote L along xy plane (L_{xy}) and z axis (L_z), respectively. The green shade denotes the plasmonic-state width.

4. Conclusions

We have studied the electronic and optical properties of bulk MoS_2 and MoSO systems using the GGA. As the results, the structural properties show that the O substitution modifies the z-axis Wyckoff positions of atoms as well as the bond angles between cations (Mo) and anions (O or S). The substitution also tunes the conduction band minimum and provokes more localized valence states compared to that of MoS_2 system. We find that the localization is more pronounced above –4 eV, which mainly comes from Mo 4d and S 3p states. On the other hand, O 2p states are dominant below –4 eV. However, there is only a small shift of E_g due to the substitution. Furthermore, the optical properties show that both systems show the strong optical dichroism of σ_1 and L. The substitution significantly increases the σ_1 width and decreases the plasmonic-state energy levels. This study shows the significant tuning of the electronic and optical properties of bulk MoS_2 system by the O substitution at S sites.

5. References

[1] Mak K F, Lee C, Hone J, Shan J and Heinz T F 2010 *Phys. Rev. Lett.* **105** 136805
[2] Radisavljevic B, Radenovic A, Brivio J, Giacometti V and Kis A 2011 *Nat. Nanotechnol.* **6** 147
[3] Lee H S, Min S-W, Chang Y-G, Park M K, Nam T, Kim H, Kim J H, Ryu S and Im S 2012 *Nano Lett.* **12** 3695
[4] Radisavljevic B and Kis A 2013 *Nature Mater.* **12** 815
[5] Beal A R and Hughes H P 1979 *J. Phys. C: Solid State Phys.* **12** 881
[6] Kumar A and Ahluwalia P K 2012 *Eur. Phys. J. B* **85** 186
[7] Ye G, Gong Y, Lin J, Li B, He Y, Pantelides S T, Zhou W, Vajtai R and Ajayan P M 2016 *Nano Lett.* **16** 1097
[8] Geng X, Zhang Y, Han Y, Li J, Yang L, Benamara M, Chen L and Zhu H 2017 *Nano Lett.* **17** 1825
[9] Cheng T, Xu J, Tan Z, Ye J, Tao Z, Du Z, Wu Y, Wu S, Ji H and Yu Y 2018 *Energy Storage Mater.* **10** 282
[10] Cheng Y C, Zhu Z Y, Mi W B, Guo Z B and Schwingenschlögl U 2013 *Phys. Rev. B* **87** 100401
[11] Lin Y C, Dumcenco D O, Komsa H P, Niimi Y, Krasheninnikov A V, Huang Y S and Suenaga K 2014 *Adv. Mater.* **26** 2857
[12] Yue Q, Chang S, Qin S and Li J 2013 *Phys. Lett. A* **377** 1362

[13] Perdew J P, Burke K and Ernzerhof M 1996 *Phys. Rev. Lett.* **77** 3865
[14] Giannozzi P, Baroni S, Bonini N, Calandra M, Car R, Cavazzoni C, Ceresoli D, Chiarotti G L, Cococcioni M, Dabo I, Corso A D, Gironcoli S d, Fabris S, Fratesi G, Gebauer R, Gerstmann U, Gougoussis C, Kokalj A, Lazzeri M, Martin-Samos L, Marzari N, Mauri F, Mazzarello R, Paolini S, Pasquarello A, Paulatto L, Sbraccia C, Scandolo S, Sclauzero G, Seitsonen A P, Smogunov A, Umari P and Wentzcovitch R M 2009 *J. Phys.: Condens. Matter* **21** 395502
[15] Giannozzi P, Andreussi O, Brumme T, Bunau O, Nardelli M B, Calandra M, Car R, Cavazzoni C, Ceresoli D and Cococcioni M 2017 *J. Phys.: Condens. Matter* **29** 465901
[16] Momma K and Izumi F 2011 *J. Appl. Crystallogr.* **44** 1272
[17] Dickinson R G and Pauling L 1923 *J. Am. Chem. Soc.* **45** 1466
[18] Kadantsev E S and Hawrylak P 2012 *Solid State Commun.* **152** 909
[19] Muhammady S, Kurniawan Y, Ishiwata S, Rousuli A, Nagasaki T, Nakamura S, Sato H, Higashiya A, Yamasaki A, Hara Y, Rusydi A, Takase K and Darma Y 2018 *Inorg. Chem.* **57** 10214
[20] Muhammady S, Sutjahja I M, Rusydi A, Winata T, Takase K and Darma Y 2017 *Jpn. J. Appl. Phys.* **56** 121201
[21] Muhammady S, Kurniawan Y, Rusydi A and Darma Y 2019 *J. Phys. Chem. Solids* **125** 16
[22] Troullier N and Martins J L 1990 *Solid State Commun.* **74** 613
[23] Troullier N and Martins J L 1991 *Phys. Rev. B* **43** 1993
[24] Troullier N and Martins J L 1991 *Phys. Rev. B* **43** 8861
[25] Broyden C G 1970 *IMA J. Appl. Math.* **6** 76
[26] Fletcher R 1970 *Comput. J.* **13** 317
[27] Goldfarb D 1970 *Math. Comput.* **24** 23
[28] Shanno D F 1970 *Math. Comput.* **24** 647
[29] Fujiwara H 2007 *Spectroscopic ellipsometry: principles and applications* (West Sussex: John Wiley & Sons Ltd)
[30] Rakić A D, Djurišić A B, Elazar J M and Majewski M L 1998 *Appl. Optics* **37** 5271
[31] Sturm K 1982 *Adv. Phys.* **31** 1
[32] Dressel M and Guner G 2005 *Electrodynamics of solids: Optical Properties of Electrons in Matter* (Cambridge: Cambridge University Press)
[33] Kokalj A 1999 *J. Mol. Graphics Modell.* **17** 176
[34] Hong J, Li K, Jin C, Zhang X, Zhang Z and Yuan J 2016 *Phys. Rev. B* **93** 075440
[35] Perdew J P 1985 *Int. J. Quantum Chem.* **28** 497
[36] Kong L-J, Liu G-H and Qiang L 2016 *Comput. Mater. Sci.* **111** 416
[37] Mak K F and Shan J 2016 *Nature Photonics* **10** 216
[38] Kurniawan Y, Muhammady S, Widita R and Darma Y 2019 *Mater. Res. Express* **6** 055908
[39] Muhammady S, Nurfani E, Kurniawan R, Sutjahja I M, Winata T and Darma Y 2017 *Mater. Res. Express* **4** 024002
[40] Saigal N, Sugunakar V and Ghosh S 2016 *Appl. Phys. Lett.* **108** 132105
[41] Egerton R F 2008 *Rep. Prog. Phys.* **72** 016502

Acknowledgments
Our work was funded by Ministry of Research, Technology, and Higher Education of Indonesia through Hibah Kompetensi Kemenristekdikti 2018, Penelitian Dasar Unggulan Perguruan Tinggi (PDUPT) program 2018 (532w/I1.C01/PL/2018), P3MI and Riset ITB 2019 research program.

Variation of Ammonium Persulfate Concentration Determines Particle Morphology and Electrical Conductivity in HCl Doped Polyaniline

Muhammad Reza[1], Nona Srikandi[1], Auliya Nur Amalina[1], Didi Prasetyo Benu[1], Fry Voni Steky[1], Achmad Rochliadi[1] and Veinardi Suendo[1,2,a]

[1]Division of Inorganic and Physical Chemistry, Faculty of Mathematics and Natural Sciences, Institut Teknologi Bandung, Jl. Ganesha No. 10, Bandung 40132, Indonesia.
[2]Research Center for Nanosciences and Nanotechnology, Institut Teknologi Bandung, Jl. Ganesha No. 10, Bandung 40132, Indonesia.

Corresponding Author: [a]vsuendo@chem.itb.ac.id

Abstract. Polyaniline (PANI) is one of conducting polymers which is widely used in various applications especially energy field, like solar cells and batteries. PANI emeraldine salt (PANI ES) is the only type of conducting PANI which can be easily synthesized using a direct mixing method in an oxidizing solution. Ammonium persulfate (APS) is one example of a strong oxidizing agent which is often used in PANI synthesis. The APS concentrations used in this study ranged from 0.1 to 1 M. Infrared and Raman spectroscopy shows that the addition of APS concentrations above 0.5 M produces the non-conducting fully oxidized PANI and phenazine species. Scanning electron microscope (SEM) showed morphological changes of PANI ES from nanofiber (diameter 80 - 110 nm) to granules (diameter 20 - 70 nm) as APS concentration increases. Based on Electrochemical impedance spectroscopy (EIS), conductivity of PANI increases as APS concentration increases with maximum conductivity of 0.36 S cm^{-1} at 0.5 M.

1. Introduction

The development of conducting polymers has attracted great interest in many applications and different fields. Polyaniline (PANI) is one of the most intensively studied conducting polymers during the last decade due to its ease of synthesis, wide-range conductivity, good environmental stability, and interesting redox properties [1,2]. Many research groups reported that PANI has been used in very broad applications like corrosion protection [3], photovoltaic cells [4], and sensors [5]. PANI has three different forms based on its oxidation state: Leucoemeraldine Base (LEB), Emeraldine Base (EB), and Pernigraniline Base (PEB) (Figure 1). Three forms of PANI differ in the ratio of reduced (benzenoid) and oxidized (quinoid) numbers in the repeating unit [6]. Intrinsicly, all three forms are insulator. EB has a unique structure where it consists of alternating benzenoid and quinonoid rings. EB can be easily transformed into Emeraldine Salt (ES) through acid doping process [7]. This leads to the formation of bipolaron and polaron enhancing the conductivity up to 5 orders of magnitude with respect to PANI EB.

The polymerization method determines the properties of final products, i.e. chemical composition, oxidation state, and average molecular weight. PANI ES can be synthesized using several methods, e.g. direct mixing [8], electrochemical [9], and interfacial/emulsion processes [10]. Direct mixing is the simplest method to obtain high yield of PANI ES [11]. In this method, the choice of oxidant is a crucial factor to obtain PANI with desired oxidation state. Previous works reported many different oxidants were used in PANI synthesis. This group of oxidant constitutes sodium or ammonium persulfate [12], ammonium cerium (IV) nitrate mixture [13], potassium chromate [14], and hydrogen peroxide [10]. Ammonium persulfate (APS), firstly introduced by MacDiarmid (1986), is one example of a strong oxidant leading to the formation of PANI emeraldine. APS dissolves easily in water facilitating redox reaction in aqueous medium. Its by-product is also soluble in water and does not interact with PANI simplifying the purification of polymerization product.

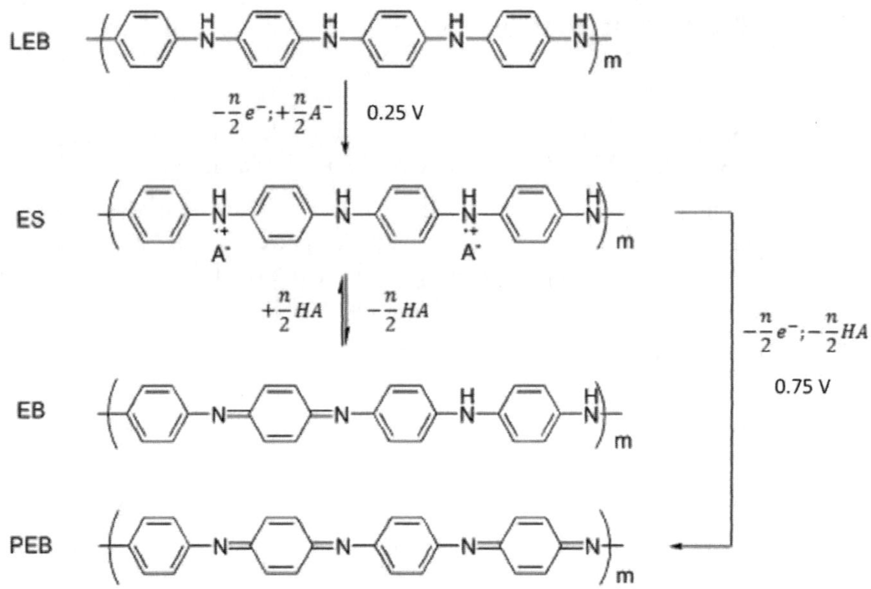

Figure 1. Three isolative species of PANI based on its oxidation state [6].

In this work, we investigate the effect of APS concentration on physicochemical properties of PANI ES. PANI ES synthesis was simply carried out using the direct mixing method with various APS concentration. We expect the optimum APS concentration will produce PANI ES with high conductivity.

2. Experimental Section

2.1. Materials

Aniline and acetone were purchased from Sigma-Aldrich. Hydrochloric acid (HCl) and Ammonium persulfate (APS) were purchased from Merck. All reagents are analytical grade and used without any further purification.

2.2. Synthesis of HCl doped Polyaniline (PANI ES)

PANI was synthesized using a direct mixing method [12]. Two solutions were initially prepared: (1) 3.64 mL of aniline was dissolved in 100 mL of demineralized water and (2) varied amount of APS was dissolved in 100 mL of HCl aqueous solution to obtain different concentration of APS (Table 1). These two solutions were then put at room temperature for 1 hour. Solution (2) was added dropwise to solution (1) under constant stirring for 15 minutes. The mixture was then transferred to a cooling bath (T = -20 °C) and kept for 4 days. Temperature was increased to 8 °C for 18 hours to melt the frozen mixture. The reaction mixture was then filtered, rinsed with HCl and acetone reapeatedly, and dried in vacuum oven for 48 hours.

Table 1. APS concentration in HCl doped polyaniline synthesis using direct mixing method

Sample code	APS Concentration (M)
PANI I	0.1
PANI II	0.2
PANI III	0.4
PANI IV	0.5
PANI V	0.6
PANI VI	0.8
PANI VII	1.0

2.3. Characterizations
The chemical structure of PANI was characterized by a Bruker Alpha FTIR-ATR spectrometer and a Bruker Senterra Raman spectrometer with laser line excitation at 785 nm. LCR meter Agilent E4980A was used to determine the electric conductivity of the samples. Electron micrographs of samples were obtained using a SEM JEOL-JSM-6510LV.

3. Results and Discussion

3.1. Synthesis of HCl doped Polyaniline (PANI ES)
Direct mixing synthesis produces green powder samples with slightly different tone depends on APS concentration (Figure 2). At low APS concentration (PANI I and II), slightly bluish powder samples were obtained. Bluish green color corresponds to the presence of less oxidized species of PANI. As APS concentration increases (PANI III to VI), the color becomes intense green which is typical color of PANI ES. When APS concentration reaches 1 M, the color of PANI VII turns black which corresponds to the non-conducting fully oxidized PANI species, the Pernigraniline salt/PANI PS [15].

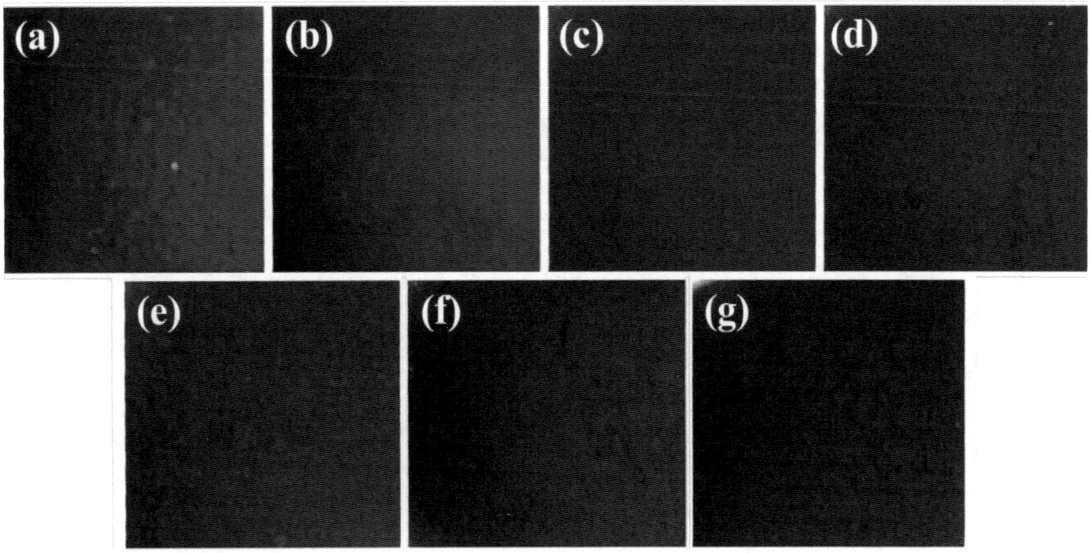

Figure 2. HCl doped PANI samples with different green tones synthesized at different ammonium persulfate concentration: (a) PANI I (b) PANI II (c) PANI III (d) PANI IV (e) PANI V (f) PANI VI and (g) PANI VII.

3.2. Chemical structure investigation
To confirm our previous findings, Infrared and Raman spectroscopy analysis were carried out. In general, Infrared spectra of all samples provide similar features (Figure 3a). The main characteristic bands of PANI are associated with C–C and C=C stretching from quinonoid rings (1609 and 1574 cm^{-1} respectively), C=C from benzenoid rings (1485 cm^{-1}), C–N (1310 cm^{-1}), $C{\sim}N^{\cdot+}$ in polaronic structure

(1249 cm^{-1}), N=Q=N (1177 cm^{-1}), Q=NH^+–B and B–NH^+–B (1152 cm^{-1}) [16]. The last band at 804 cm^{-1} is associated with the out of plane bending vibration mode of C–H [16]. We observed changes in Infrared spectra with the increase of APS concentration. As APS concentration increases, bands at 1609 and 1177 cm^{-1} are also increasing in intensity. This corresponds to the formation of higher oxidation that indicated by the increase of quinonoid ring numbers. Another change is observed at high APS concentration (PANI VI and VII) with the presence of the band at 1689 cm^{-1}. This followed by a significant increase of band intensity at 1485 cm^{-1}. This is in consistent with the presence of non-conducting phenazine species that commonly considered as defect in PANI [17].

Figure 3b shows Raman spectra of PANI samples presenting some characteristic features of PANI ES. Both bands at 419 and 515 cm^{-1} correspond to the in-plane amine deformation. Band at 575 cm^{-1} associates to the ring (**6b**) deformation. Bands at 720 and 807 cm^{-1} correspond to the imine (C–N=C bending) and amine (C–N–C bending) deformations, respectively. The intense single band at 1177 cm^{-1} is the characteristic of C–H bending in the emeraldine (**9a**) species. The presence of polaron species is confirmed by the occurrence of band at 1332 cm^{-1} which correlates to the $C–N^{\cdot+}$ stretching. The band at 1374 cm^{-1} is attributed to the C–N stretching vibration that appeared in the quinonoid ring neighborhood. Band at 1510 cm^{-1} is assigned to the N–H deformation vibrations associated with the semiquinoid structure. The band at 1590 cm^{-1} corresponds to the C=C stretching in Q [18]. Band intensity at 1332 cm^{-1} increases with APS concentration indicating more polaron species in PANI. This change is followed by the increase of band intensity at 575 cm^{-1} that associates with the formation of non-conducting phenazine species.

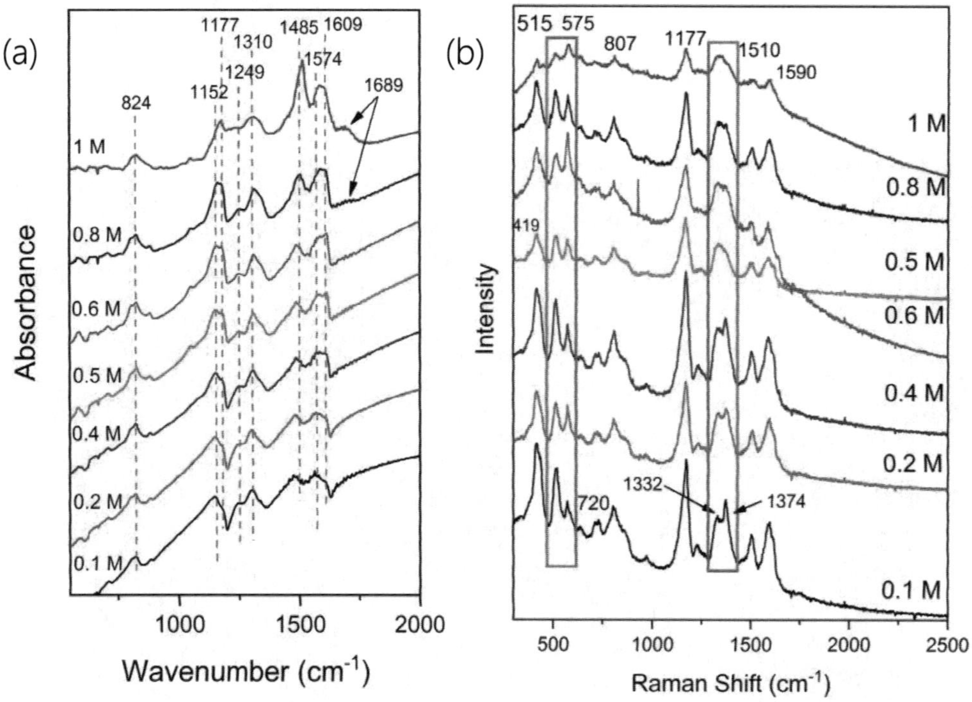

Figure 3. FTIR (a) and Raman spectra (b) of synthesized PANI with different APS concentrations.

3.3. Particle Morphology

SEM images of all synthesized PANI ES are presented in Figure 4. Particle morphology of PANI ES evolve from nanofiber to granules with APS concentration. Average diameter of nanofiber PANI ES varied from 80 to 110 nm (Figure 4h). The nanofiber average diameter increases from PANI I to IV before abruptly decreases at PANI V. Further increase of APS concentration, PANI VII particles tend to form granules. The granules is an agglomerate structure, consist of small particles with diameter 20 – 70 nm. This mechanism is in a good agreement with proposed mechanism in previously reported works how the PANI particles formed and evolved [19-21]. In the reaction mixture, two major species are present, namely the anilinium and radical anilinium cations. The presence of radical anilinium cations has been discovered by Ding et al using the radical trapping method [22]. These two species will react to initiate the polymerization forming active cationic radicals, then followed by propagation steps consuming the anilinium cations [11]. At low APS concentration, the anilinium cations is the major species in the reaction mixture. This will produce longer chains of PANI ES due to the limitation of radical anilinium cations. Those PANI chains will interact each other through intermolecular interactions (Hydrogen bonding) to form nanorod particles.

The diameter of nanorods increases with APS concentration because the cationic radicals concentration increases. The cationic radical concentration determine the reaction kinetics. The reaction rate increases with cationic radical concentration, while the cationic radical still act as the limitting reagent. The low APS concentration is a kinetic controlled reaction condition. In this condition, the monomers are present in abundant. The diameter of nanorods will be determined solely by the synthesis duration. All reactions were carried out with the same reaction time, therefore faster reaction produces larger particles. At high APS concentration, the radical anilinium cations is the major species in the reaction mixture. The competition of propagation and termination process will be severed due to high radical/reactive species concentration. As consequence, shorter PANI ES chains are produced. Those chains interact each other forming small particles that agglomerates resulting granules.

Figure 4. SEM images of PANI synthesized with different ammonium persulfate concentration: (a) PANI I (b) PANI II (c) PANI III (d) PANI IV (e) PANI V (f) PANI VI and (g) PANI VII. Average nanofiber diameter as a function of APS concentration (h).

3.4. Conductivity measurement

Figure 5a presents the frequency response of electrical conductivity for samples synthesized with different APS concentration. The steepness in conductivity decrease profile tends to increase with APS concentration and reached the maximum at PANI IV. This kind of conductivity profile is typical for materials with metallic conduction. Thus, PANI ES samples synthesized at low APS concentration provide metallic conduction. This behavior is less pronounced at higher APS concentration. PANI VI and VII have a semiconductor conduction type. Semiconductor conduction shows an increase of conductivity in their frequency response profile. The conductivity value at 20 Hz tends to decrease as APS concentration increases with the maximum value of 0.36 S cm^{-1} at PANI IV (Figure 5b). This due to the larger nanofiber diameter with APS concentration as observed in SEM images (Figure 4). We also determined the number of charge carrier (n) in each sample by using Drude model approach using the following equation [23]:

$$n(\omega_0) = \frac{2m_e}{\pi e^2} \int_0^{\omega_0} \sigma(\omega)\, d\omega \qquad (1)$$

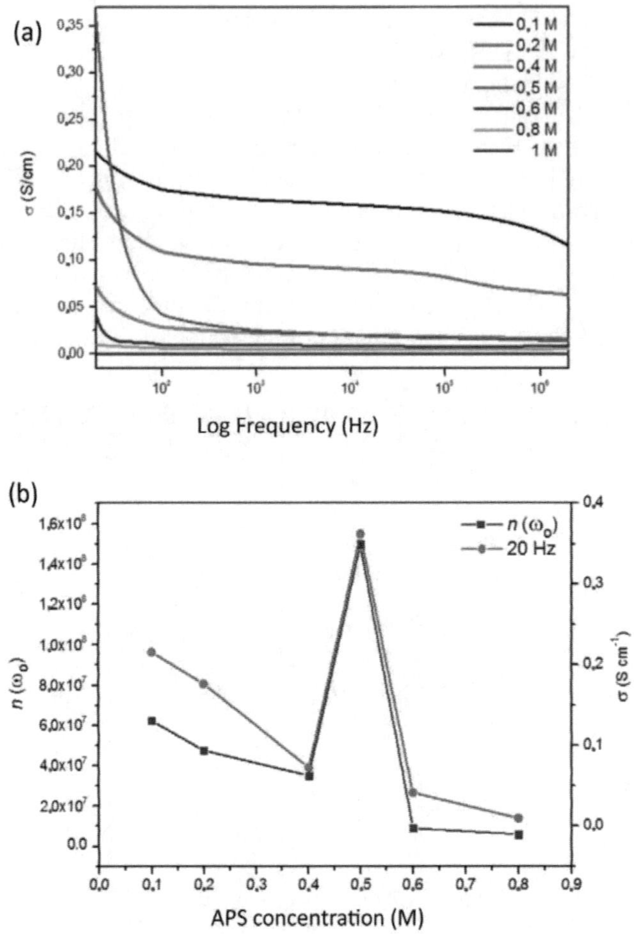

Figure 5. Conductivity profile (a) and value (b) at 20 Hz of PANI ES.

It is obtained that n value is in order $10^7 - 10^8$ and have a similar tendency with conductivity value. We can correlate this phenomenon to our previous findings that PANI ES synthesized at high APS concentration contains phenazine structure and have granular structures. The presence of phenazine structure can either stabilize or trap the charge carriers reducing the electrical conductivity [24]. Nanorod morphology is also well known to facilitate better electron flow than granules in PANI ES due to the lower limit of percolation of rod-like particles with respect to spherical particles [25].

4. Conclusions

PANI IV, which has ammonium persulfate (APS) concentration of 0.5 M, demonstrates the optimum condition for HCl doped PANI synthesis using direct mixing method. PANI ES has nanorod morphology, while granules were obtained at PANI VII. Infrared and Raman spectroscopy revealed that APS concentration usage below or above 0.5 M will lead to the formation of lower or higher oxidation state of PANI ES respectively. PANI ES has metallic conduction at low APS concentration, while semiconductor conduction produced at high APS concentration. The conductivity value decreases with APS concentration. The presence of phenazine species is also responsible decreasing electrical conductivity besides particle morphology and PANI oxidation state.

5. References

[1] Huang J 2006 *pure Appl. Chem.* **78** 15
[2] Wang R, Huang L and Tian X 2012 *J. Phys. Chem. C* **116** 13120 – 13126
[3] Chen Y, Wang X H, Li J, Lu J L and Wang F S 2007 *Electrochim. Acta* **52** 5392–5399
[4] Brabec C J, Sariciftci N S and Hummelen J C 2001 *Adv. Funct. Mater.* **11** 15–26
[5] Yan X B, Han Z J, Yang Y and Tay B K 2007 *Sens. Actuators B* **123** 107–113
[6] Gospodinova N and Terlemezyan L 1998 *Prog. Pol. Sci.* **23** 1443 – 1484
[7] Stejskal J 2002 *Pure Appl. Chem.* **74(5)** 857 – 867
[8] MacDiarmid A G, Chiang J C, Richter A F, Somasiri N L D and Epstein J A 1987 *Reidel* 105–120
[9] Genies E M and Łapkowski M 1988 *Synth. Met.* **24** 61–68
[10] Gosh P, Siddhanta S K, Haque S R and Chakrabarti A 2001 *Synth. Met.* **123** 83–89
[11] Wei Y, Tang X and Sun Y 1989 *J. Pol. Sci.* **27** 2385 – 2396
[12] MacDiarmid A G and Chiang J C 1986 *Synth. Met.* **13** 193–205
[13] Foke Y and Schlenoff J B 1995 *Polymer* **36** 639–643
[14] Adams P N, Abell L, Middleton A and Monkman A P 1997 *Synth. Met.* **84** 61–62
[15] Detsri E and Dubas S T 2009 *J. Met. Mater. Miner.* **19** 39–44
[16] Trchova M and Stejskal J 2011 *Pure Appl. Chem.* **83** 1803 – 1817
[17] Ciric-Marjanovic G, Trchova M and Stejskal J 2008 *J. Raman Spectrosc.* **39** 1375
[18] Trchová M, Morávková Z, Bláha M and Stejskal J 2014 *Electrochim. Acta* **122** 28–38
[19] Mohilner D M, Adams R N and Argersinger W J 1962 *J. Am. Chem. Soc.* **84** 3618–3622
[20] Stejskal J 2002 *Pure Appl. Chem.* **74 (5)** 857–867
[21] Chiang J C, MacDiarmid A G 1986 *Synth. Met.* **13** 193–205
[22] Ding Y, Padias A B and Hall Jr H K 1999 *J Polym Sci A: Polym Chem* **37** 2569
[23] Hidayat F, Lau Y, Qadafi A, Reza M, Benu D P, Steky F V and Suendo V *in preparation*
[24] Kellenberger A, Dmitrieva E and Dunsch L 2011 *Phys. Chem. Chem. Phys* **13** 3411–3420
[25] Lee K, Cho S, Park S H, Heeger A J, Lee C W and Lee S H 2006 *Nature* **441** 65

Acknowledgements

This work was financially funded by ITB Research Grant 2018 through ITB Research and Innovation Program 2018. Muhammad Reza acknowledges Ministry of Education and Culture of Indonesia for scholarship through Beasiswa Unggulan.

Study on Graphene Oxide (GO) Supernatant Dilution to the Optical and Electrical Properties of TCF Based-reduced Graphene Oxide (RGO) Films

Dina Rahmawati[1], Akfiny Hasdi Aimon[1*], Ferry Iskandar[1,2], Arie Wibowo[2,3], and Khairurrijal Khairurrijal[4]

[1]Energy and Environmental Laboratory, Departement of Physics, Institut teknologi Bandung, Jl. Ganesa 10 Bandung 40132, Indonesia
[2]Research Center for Nanosciences and Nanotechnology (RCNN), Institut Teknologi Bandung, Jl. Ganesa 10 Bandung 40132, Indonesia
[3]Material Science and Engineering Research Group, Faculty of Mechanical and Aerospace Engineering, Institut Teknologi Bandung, Jl. Ganesha 10, Bandung, Indonesia
[4]Research Center for Biosciences and Biotechnology, Institute for Research and Community Services, Institut Teknologi Bandung, Jalan Ganesa 10, Bandung 40132, Indonesia

*E-mail: akfiny@fi.itb.ac.id

Abstract. Graphene is a potential material to replace indium tin oxide (ITO) and fluorine tin oxide (FTO) as a transparent conductive films (TCF) due to its high conductivity and transparency. In this study, fabrication of graphene-based TCF was carried out by means of the drop-casting method, with graphite oxide as the raw material. The graphene oxide (GO) was diluted by adding 1, 3, 5, and 7 ml of distilled (DI) water. The sample was also heat treated to ensure complete reduction process. The deposited films were characterized using UV-Vis spectroscopy and I-V to investigate their optical and electrical properties. Based on the measurement results, sample with 3 ml DI water addition gives transparency and sheet resistance (Rs) of 41.01 % and 9.43 kΩ/sq, respectively. This material has a great potential to be applied as the optoelectronics material.

1. Introduction

Transparent conducting film (TCF) is a thin layer of material that has high electrical conductivity and excellent optical transparency. To this day, ITO and FTO are two types of material that commonly used as TCF [1,2]. However, these two materials are high expensive due to its rare earth raw material and high production cost [3]. Therefore, graphene, which has electrical conductivity of 6000 S/cm and high transparency of 97 %, is a strong candidate as a substitute material ITO and FTO to be applied as TCF [4,5].

Various synthesis methods have been developed to synthesize graphene including the chemical vapor deposition (CVD) [6], epitaxial growth on the Si-C substrate [7], a mechanical exfoliation of graphite (Scotch-tape) [8], and wet chemical [9]. Wet chemical method is a promising method to synthesize graphene for its low-cost process, large-scale fabrication, efficient and effective synthesis process

[10,11]. This method is carried out by synthesizing supernatant of graphene oxide (GO) and followed by reduction process, in order to achieve RGO sample.

It is necessary to study the electrical properties and transparency of RGO-based TCF. Dilution of GO supernatant is one of several aspects that needs to be optimized to improve the conductivity and transparency of the film. In this study, GO supernatant was diluted by addition of 1, 3, 5, and 7 ml of distilled (DI) water. The deposited films were characterized to study sheet resistance (Rs) and transparency (Tr) properties of the samples. This study gives new information on preparation and properties of RGO-based TCF.

2. Experimental Setup

Fabrication of RGO films was initiated by synthesize graphite modified using a Marcano [10] and Hummer [11] methods. The graphite oxide exfoliation process was then carried out using an ultrasonic homogenizer for 30 minutes, followed by centrifugation at 4000 rpm. Afterward, 1 ml GO supernatant was by dissolving it in 1, 3, 5, and 7 ml of DI water. Sample was mixed with LAA in a ratio of 5 : 1, and the solution was stirred at 900 rpm in the ice bath for 30 minutes. Then the deposition process is carried out onto a glass substrate by dripped 50 µl / of supernatant. The evaporation process was carried out slowly at room temperature for 12 hours. Furthermore, the thermal reduction treatment by heating in the furnace at 200 °C for 30 minutes. RGO-LAA films were characterized using I-V and UV-Vis spectroscopy to investigate their sheet resistance (Rs) and transparency (Tr). The synthesized GO was tested by FTIR to investigate its chemical composition.

3. Result and discussion

Figure 1. shows FTIR spectra of the synthesized GO. There is a broad peak at wavenumber of 3000-3700 cm^{-1} which indicates stretching vibrations of hydroxyl group [12]. FTIR spectra in figure 1 also shows several sharp peaks correspond to vibration of C-H bond (2917 and 2847 cm^{-1})[10], C=O bond (1739 cm^{-1}) [12,13], C=C bond (1619 cm^{-1}), C-O bond (1224 and 1054 cm^{-1}) [13,14]. The FTIR results indicate that GO supernatant has been successfully synthesized.

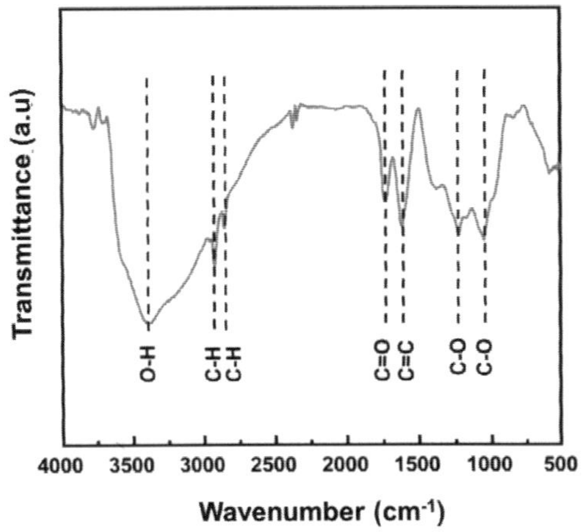

Figure 1. FTIR spectra of the synthesis of GO.

Figure 2 shows appearance of GO supernatant without (a) and with (b) dilution using DI water. It can be seen that the GO supernatant before dilution has a dark brown color. However, after dilution with 1, 3, 5, and 7 ml in the DI water, the supernatant solution turned into brighter color. The results of UV-Vis characterization of GO films is shown in figure 2 (c). Transparency of GO films increases as higher amount of DI water is added. RGO film that was prepared without addition of DI water has Tr value of 1.24 %. Meanwhile, RGO film that were prepared by adding 1, 3, 5, and 7 ml DI water show higher Tr value of 26.07, 41.01, 53.84, and 61.12 % respectively.

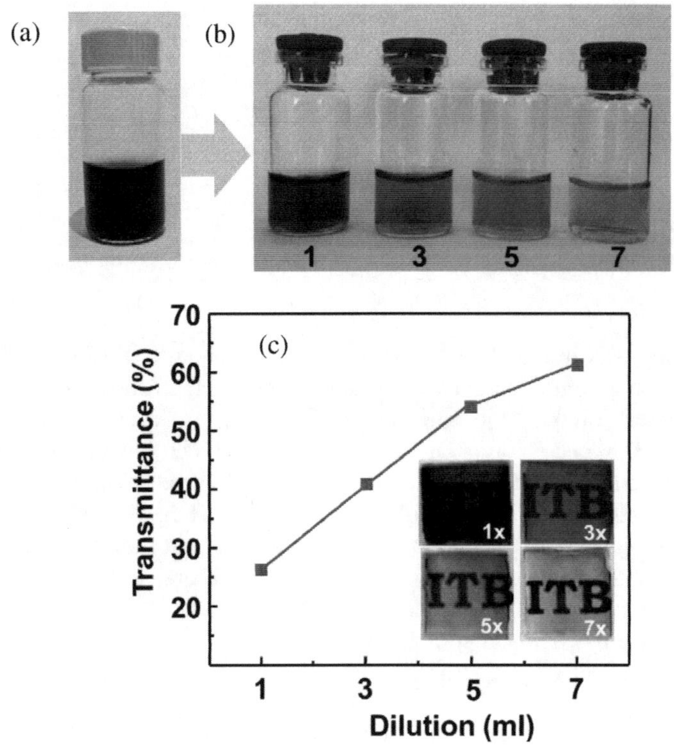

Figure 2. Appearance of GO supernatant (a) without and (b) with addition of 1, 3, 5, and 7 ml of DI water (left to right) (c) Transmittance of GO films prepared by diluting GO supernatant.

Figure 3 shows Rs value of the RGO films using I-V characterization. RGO films that were prepared with addition extra DI water show Rs of 3.42 kΩ/sq. Samples with 1, 3, 5, and 7 ml of DI water dilution give of 8.06; 9.43; 45.36 and 89.88 kΩ/sq, repectively. Based on UV-Vis and I-V characterization, it can be concluded that DI water dilution provides higher Tr and Rs values. This result is reasonable as more dilution will lower concentration of GO in solution. However, in order to be able to apply the sample as a TCF, material with high Tr and low Rs values is favorable. From I-V and UV-Vis characterization results, it can be concluded that the sample with 3 ml of DI water addition has the optimum Tr and Rs values. It is suitable enough to be applied as TCF materials.

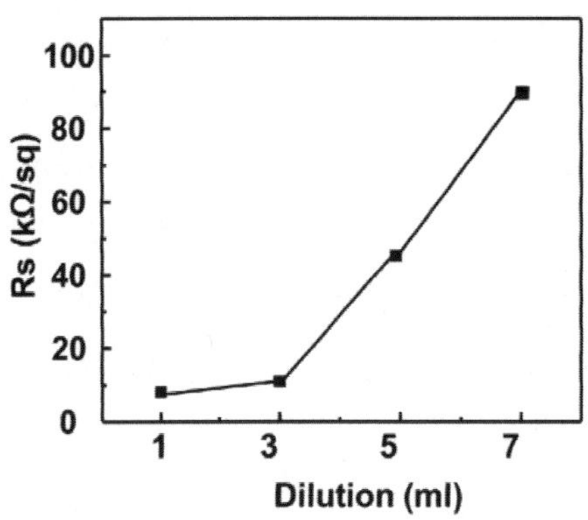

Figure 3. RGO-based TCF sheet resistance (Rs) value of RGO films prepared by diluting GO supernatant.

4. Conclusion

RGO-based TCF was successfully deposited by means of drop casting method. The effect of GO supernatant dilution with 1, 3, 5, and 7 ml DI water were studied based on optical and electrical characterization of RGO film. Sample with 3 ml dilution in DI water gives the best result with Rs 9.43 kΩ/sq and Tr 41.01 %, respectively. This result has potential to be applicated as TCF material.

5. References

[1] Zheng Q, Li. Z, Yang. J, & Kim. J. K 2014 Graphene oxide-based transparent conductive films *Progress in Materials Science* **64** 200-247
[2] Sima C, Grigoriu. C, and Antohe. S 2010 *Thin Solid Films* **519**(2) 595-597
[3] Guo H, Lin. N, Chen. Y, Wang. Z, Xie. Q, Zheng. T, and Peng. D. L 2013 *Scientific reports* **3** 2323
[4] Huang X, Zeng. Z, Fan. Z, Liu. J, and Zhang. H 2012 *Advanced Materials* **24**(45) 5979-6004
[5] Peng S, Han. X, Li. L, Zhu. Z, Cheng. F, Srinivansan. M, Adams. S, And Ramakrishna. S 2016 *Small* **12**(10) 1359-1368
[6] Sutter P. W, Flege. J.-I, and Sutter. E. A 2008 *Nature materials* **7**(5) 406
[7] Whitener Jr. K. E and Sheehan. P. E 2014 *Diamond and related materials* **46** 25-34
[8] Novoselov K. S, Geim. A. K, Morozov. S. V, Jiang. D. A, Zhang. Y, Dubonos. S. V and Firsov. A. A 2004 *science* **306**(5696) 666-669
[9] Hummers Jr. W. S and Offeman. R. E. 1958 *Journal of the american chemical society* **80**(6) 1339-1339
[10] Husnah M, Fakhri. H. A, Rohman. F, Aimon. A. H, and Iskandar. F 2017 *Materials Research Express* **4**(6) 064001
[11] Iskandar F, Hikmah. U, Stavila. E and Aimon. A. H 2017 *RSC Advances* **7**(83) 52391-52397
[12] Alam S. N, Sharma. N & Kumar. L 2017 Synthesis of graphene oxide (GO) by modified hummers method and its thermal reduction to obtain reduced graphene oxide (rGO) *Graphene* **6**(01) 1-18
[13] Zhu. C, Guo. S, Fang. Y & Dong. S 2010 Reducing sugar: new functional molecules for the green synthesis of graphene nanosheets *ACS nano* **4**(4) 2429-2437
[14] Xu. C, Shi. X, Ji. A, Shi. L, Zhou. C & Cui. Y 2015 Fabrication and characteristics of reduced graphene oxide produced with different green reductants *PloS one* **10**(12) e0144842

Acknowledgments

The author would like to DIKTI for research funding 2018. In addition, the author would also like to thank LPDP as an institution that has provided scholarship during master program.

IOP Conf. Series: Materials Science and Engineering **599** (2019) 012004 doi:10.1088/1757-899X/599/1/012004

Fabrication of Flexible Graphite Film Using a Mechanical Method and Its Application in Tio_2/Graphite Solar Cell as a Flexible Counter Electrode

Dui Yanto Rahman, Fisca Dian Utami, Desyana Olenka Margaretta, Euis Sustini and Mikrajuddin Abdullah

Department of Physics, Faculty of Mathematical and Natural Sciences, Bandung Institute of Technology, Jalan Ganeca 10 Bandung 40132, Indonesia

E-mail: duiyantorahmanssi@gmail.com and mikrajuddin@gmail.com

Abstract. We have successfully fabricated solar cell using flexible counter electrode of graphite powder. TiO_2 /graphite suspension was simply poured onto heated Fluorine Tin Oxide (FTO) substrate and kept heated at 200°C for 2 hours. PVA.LiOH polymer electrolyte was employed as a positive charge transfer medium. The counter electrode was made of graphite powder which is manually deposited on transparent plastic using tube-shaped metal which its surface covered with tissue of 2 mm thick. Graphite powder was deposited by straight motion of 20 movements for one coating or one layer. The same activity was repeated on the stacked film to increase the conductivity of the film. Resistivity and transmittance measurement of the film was performed by four-point probe method and UV-Vis equipment respectively. The film thickness and 2-D porosity were measured using an electronic microscope and ImageJ software orderly. The film resistivity and transmittance gradually decreased with the increase of coating number and reached the resistivity of 0.35 Ω.cm and transmittance of 5% for 40 times of coating. At the same time, the film thickness was saturated at the 30 times coating and reached the thickness of 9.4 μm. The film 2-D porosity which indicated how well the surface of the film covered the graphite also decreased to reach the film porosity of 18%. The most conductive film (40 times coating) was used as a flexible counter electrode in TiO_2/graphite solar cell. The efficiency of this structure was 0.4 %, a reasonable efficiency achievement for a cheap material and easy method used.

1. Introduction

The dependence of worldwide people on the polluting energy source such as gas, coal, natural oil has motivated the search of non-polluting energy source. In addition, the availability of this kind of energy was limited and not abundant. Solar cell as a clean energy source was one of the most promising energy source to provide the need on energy in the future due its wide availability source. Various types of solar cell have been produced since it was firstly found such as bulk silicon solar cells [1-3], thin-film solar cells [4-6] and organic solar cells [7-9]. Each development has been carried out mainly on the concern of much easier method of fabrication, much cheaper materials used, much comparable efficiency to the present applied solar cell especially silicon based solar cell.

The development of organic solar cells was mainly dominated by the dye-sensitized solar cell (DSSC) which used the dye as a photon absorbing material and titanium dioxide (TiO_2) as the medium for the electron injected by the dye to flow towards the front electrode, iodide/triiodide redox couple electrolyte as hole transport medium and Pt-counter electrode [10, 11]. However, the solar cell efficiency was relatively small to compete with most commercially silicon-based solar cells. In addition, the cost of producing such a solar cell which consisted mainly of material and process was relatively expensive. It made this type of solar cell still in the distance to its realization and commercialization.

Having participated in developing the solar cell using much less expensive material and a simpler method, we have developed a new type of solar that uses TiO_2 as a photon absorbing material, metal particles as a means of transporting the electron out to the front electrode (FTO), PVA.LiOH polymer electrolyte as the medium for transporting positively charge and Aluminium as counter electrode [12-17].

Further improvement is required to increase the performance of the solar cell. Due to the limited range absorption of TiO_2 in the ultraviolet range, it is essential to have a material assisting TiO_2 in absorbing wider range absorption to get a higher current. Besides, the development of using flexible counter electrode should be an important concern to realize inexpensive plastic solar cell. The attempt of using flexible graphite sheets has been demonstrated [18], but meet the problem of sophisticated method.

In the present research, we proposed a new kind of graphite based counter electrode which is deposited on the plastic using a manual method. The method was easier compared to the previous method [19]. Previoulsy, we have fabricated flexible graphite-based film using manual circle motion but produced low conductivity and has not been employed as flexible counter electrode [20,21]. Another manual straight motion needs to be employed to increase its conductivity. Deposition other materials such as TiO_2 on the transparent plastic using spray method has been successfully conducted and employed in photocatalyst application [22-24]. The flexible graphite based counter electrode was then applied to a solar cell using graphite/TiO_2 as photon absorbing material. The use of flexible counter electrode was aimed at preparing the future whole flexible solar cell using cheap materials and easy method of fabrication.

2. Material and Experiment

2.1 Preparation of flexible graphite counter electrode.

1 g of Graphite powder (Giva utama, Indonesia) was manually lubricated on the surface of the tube-shaped metal covered with tissue of 2 mm thickness. Manual straight motion was employed to attach the graphite powder on the surface of transparent plastic (Yashica, Indonesia) with the size of 2 x 2 cm^2 using 20 movements for one coating time. To increase the conductivity of the film, repeated process was conducted on the previous attached graphite film.

Keithley 224 current source and Keithley 182 voltmeter were employed to measure film resistivity by using Four-Point Probe method. To investigate the transmittance of the film, we used UV-Vis electrometer (Mikropack Brands, NanoCalc 2000, Florida, USA). Scanning Electron Microscope (SEM) (JEOL JSM-6360LA, operated at 20 kV) was to observe film morphology. The film thickness was characterized by using an electronic microscope. 2D-porosity characterization was conducted to know how well the graphite powder covered the surface of the plastic homogeneously. The image was observed using an electronic microscope with 40 times of magnification.

ImageJ software was employed to measure the 2-D porosity of graphite film. The procedure of porosity measurement was easily conducted by opening the image of the film followed by adjusting the threshold of the film on the image feature. The porosity was automatically measured through the information of film uncovered by the graphite powder (white color) in percent unit (%).

2.2 Preparation of solar cell.

5 g of graphite powder (technical grade, Giva Utama, Indonesia) was stirred homogeneously in 12 ml of mineral water (Aqua, Indonesia) for 15 minutes. TiO_2 powder (technical grade, Bratachem, Indonesia) was then added to the suspension with the amount of 10% o the amount of graphite powder.

The suspension of TiO_2 and graphite was stirred for 30 minutes. In a separate place, the FTO glass was heated on a hot plate at 100 ° C for 30 minutes. Using a small spatula, the last suspension was then poured onto the FTO surface. The suspension was then stuck evenly on the FTO surface. To increase the good contacts between the FTO surface and the TiO_2 /graphite particles, the formed film was then held at 200 °C for 2 hours. Separately, 0.38 g of LiOH (Kanto, Japan) was dissolved in 20 ml of mineral water placed in a beaker for 15 minutes. 1.8 g of polyvinyl alcohol (PVA) (Bratachem, Indonesia) was then added to the solution and heated at 100 °C for 60 minutes to form a gel polymer electrolyte. The solar cell module was constructed by manually spreading the polymer electrolyte on the surface of the TiO_2/graphite film and fixing the flexible graphite based counter electrode on the opposite side using two clampers on both sides of the cell.

3. Result and Discussion

The success of graphite powder deposition on the plastic was seen at figure 1. The graphite powder was homogeneously and strongly stuck on the plastic. It is confirmed by touching activity of hands and the investigation of its conductivity property for more than a month which still has the same value of resistivity.

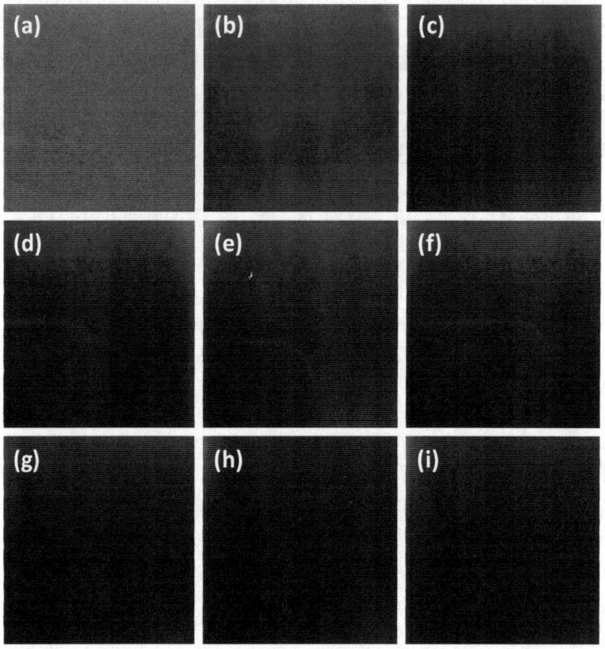

Figure 1. Photo of graphite-based flexible film with size of 2 cm x 2 cm after (a) 1 times (b) 5 times (c) 10 times (d) 15 times (e) 20 times (f) 25 times (g) 30 times (h) 35 times and (i) 40 times of coating.

Figure 2 (a) was the SEM image of flexible graphite film confirming that graphite powder was homogeneously deposited on the plastic. Figure 2(b) describes the cross section of graphite film informing that graphite film was well deposited on the surface of the plastic.

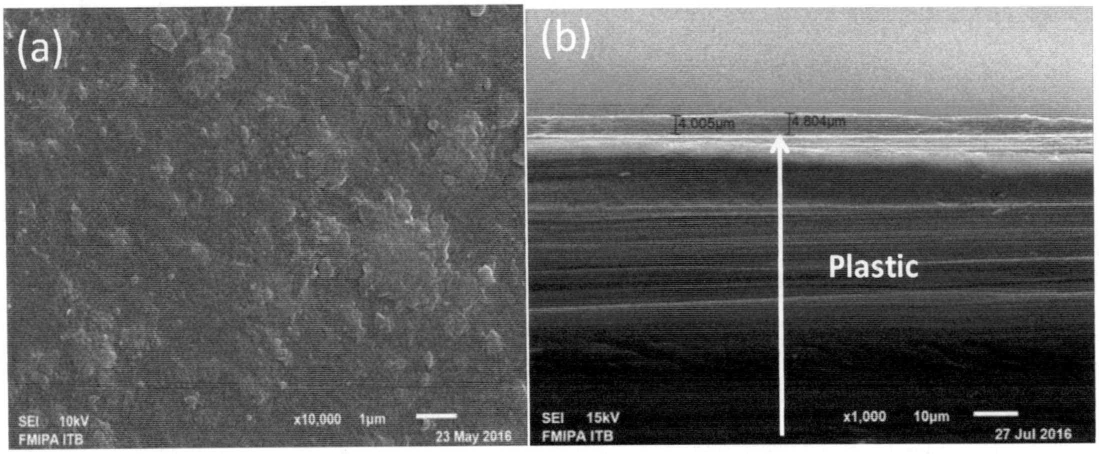

Figure 2. SEM image of (a) surface morphology of graphite-based flexible film. (b) Cross section of graphite-based flexible film after 5 times of coating.

The investigation of the influence of number of coating to the properties of the film was fully described at Table 1. Table 1 informed us that in general that the film conductivity was increased with the increase of number of coatings. This is mostly caused by the increase of film thickness as we see at Table 1. The increase of the film conductivity affected by the film thickness was also reported by Liu et al. who used copper materials [25], Dayal et al. who used silver materials [26], Camacho and Oliva who used Aluminium, copper, and gold materials [27] Those experimental data was theoretically calculated and well fitted by Fred [28].

Table 1. Relationship between number of coatings and resistivity, thickness, transmittance, 2-D porosity of graphite-based film

Number of coatings (times)	Resistivity (Ω.cm)	Thickness (µm)	Transmittance (%)	2-D Porosity (%)
1	1.18	1.3	53	34
5	6.5	4.5	47	30
10	2.1	6.6	36	21
15	1.9	7.5	14	25
20	1.8	9	13	24
25	2.1	9.5	11	24
30	1.9	10.4	8	20
35	1.35	9	7	17
40	0.35	9.4	5	18

Furthermore, we investigated the influence of coating number on the film properties more completely. The increase of coating number influenced not only the conductivity and thickness of the film but also its transmittance. As the number of coating increased, the transmittance of the film decreased. This is also due to the increase of the film thickness. We can see it clearly in figure 3(c) that the transmittance of the film decreased from 53 % to 5 % with the increase in the number of coatings. The increase of film thickness caused the increase of photon absorption by the film as the amount of graphite powder attached on the surface of the plastic increased. The investigation of the effect of the film thickness on the light absorbance was also reported by Islam et al.who found that the absorption of the blend of poly (3-hexylthiophene) (P3HT) and Phenyl C61 butyric acid methylester (PCBM) film increased with the increase of the film thickness [27]. This is then strongly confirmed by Yu Xie et al. who found that the light absorbance of P3HT:PCBM thin film increased with increase of film thickness from 60 nm to 80 nm [29].

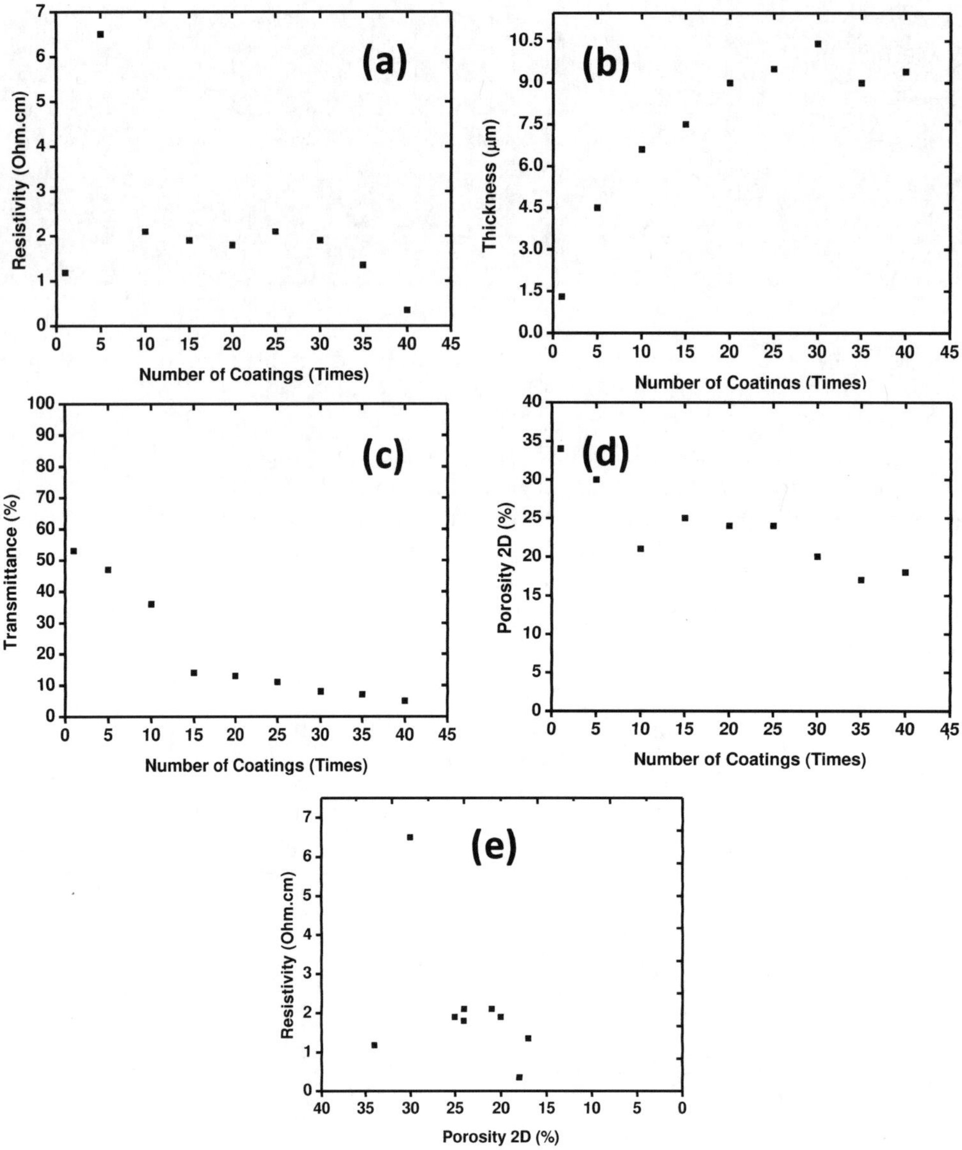

Figure 3. Graph of relationship between (a) number of coating and film resistivity (b) number of coatings and film thickness (c) number of coating and film resistivity (d) number of coating and film 2-d porosity and (e) 2-D porosity and film resistivity.

The film transmittance decrease was supported by the decrease of 2-D film porosity which tends to decrease with the increase of the coating number. It means that the film was increasingly covered by the graphite powder deposition on the surface of the film with the increase of coating number (figure 3(d)). The film 2-D porosity also revealed that the film conductivity was affected not only by the increase of the film thickness, but also how homogeneously the surface of the plastic was covered by the graphite powder. It is clearly observed from table 1 that the thickness of the film was not increased any more after 30 times of coating but the conductivity of the film continuously increased. The reason for this phenomenon was clearly obtained by observing the 2-D porosity behavior that still continuously decreased from 20% to 17% with the increase of coating after 30 times of coating. Figure 4 is the 2-D porosity of the film describing clearly the covering of the graphite powder to the surface of the plastic. From figure 4, we can find that the increase of coating number influenced the covering of graphite powder to the surface of the plastic. It is obviously understood that the more homogeneously the film is covered by the graphite powder, the more conductive the film is. The film 2-D porosity can likely be compared to the surface roughness of the film. The increase of 2-D porosity indicates the increase of the surface roughness of the film. The resistivity of the film highly depends on the surface roughness of the film [27]. The increase in the surface roughness makes the film resistivity increase. The film resistivity is caused by electron scattering from the surface [30, 31]

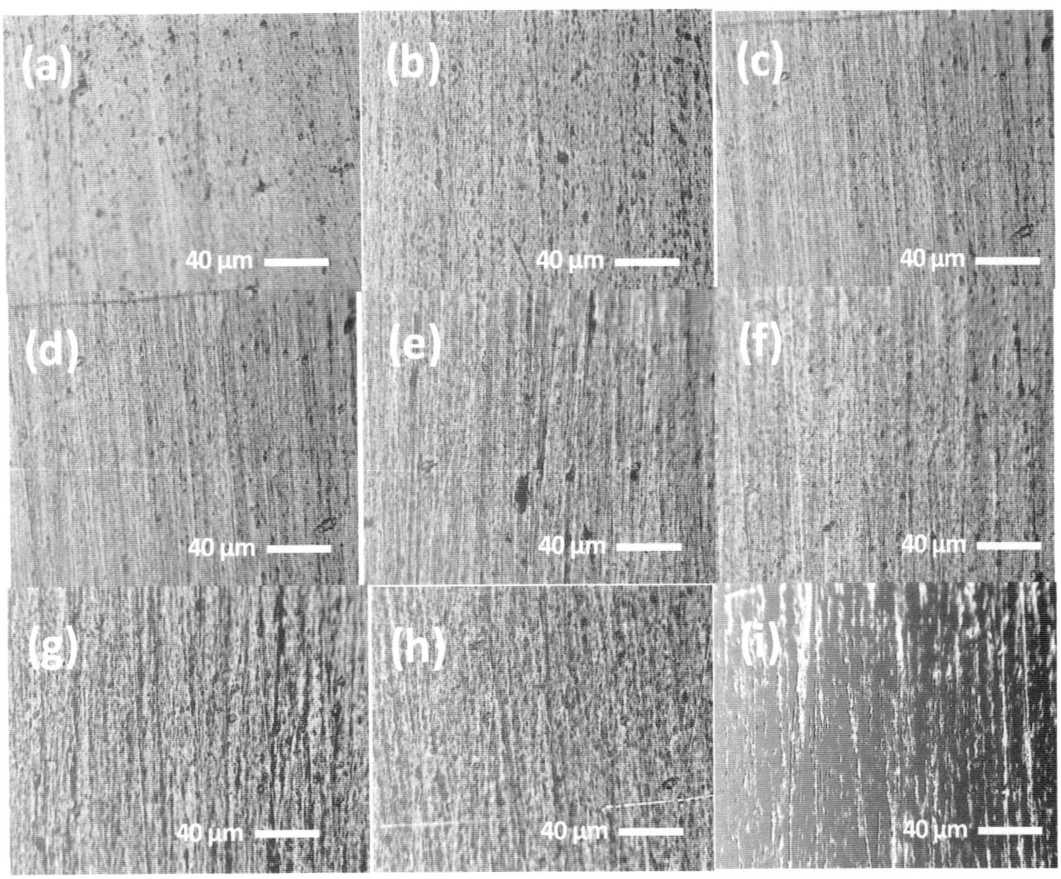

Figure 4. Image of 2-D porosity of graphite-based flexible film (a) one coating (b) 5 times (c) 10 times (d) 15 times (e) 20 times (f) 25 times (g) 30 times (h) 35 times and (i) 40 times, of coating.

After successfully coating graphite powder on the surface of the plastic, we tried to apply the flexible graphite film of 40 times coating as flexible counter electrode of solar cell containing FTO as front electrode, graphite/TiO_2 as photon absorbing materials, PVA polymer electrolyte as a positive charge transfer medium and flexible graphite film as counter electrode. The performance of this structure was compared to the same solar cell structure, but with a different counter electrode (Aluminium). Figure 5

(a) describe the morphology of graphite/TiO_2 anode film while figure 5(b) describes I-V characteristic of solar cell using flexible graphite counter electrode and Aluminium counter electrode.

Figure 5. (a) SEM image of graphite/TiO_2 film photo anode (b) I-V characteristic of solar cell using aluminium and flexible graphite film.

From figure5 (b) we can see that solar cell using flexible graphite film work well with current-voltage value when applied to I-V characterization measurement. This solar cell structure achieved the efficiency of 0.4 % with current of 1,1 mA and voltage of 0.43 V. The efficiency was lower compared to the efficiency of the structure solar cell, but using Aluminium counter electrode which achieved the efficiency of 2.14 % with current of 3.7 mA and Voltage of 0.68 V. This is due to the low conductivity of flexible graphite counter electrode which have resistivity of 0.35 Ω.cm while Aluminium counter electrode has a resistivity of 2.3 x 10^{-4} Ω.cm. It is clearly understood since the solar cell efficiency strongly depends on the conductivity of the counter electrode [19]. The performance of this solar cell still has a chance to be improved when the resistivity of the graphite counter electrode can reach the value of 10^{-4} cm. The way of improving the conductivity of the film is to increase the number of coating to more than 40-80 times. It is requested that coating more than 40 times will increase the covered area of graphite film. It will be clearly observed from 2-D porosity characterization. Another way to improve the conductivity of the film is by combining straight and circle motion when depositing the graphite powder on the surface of the plastic.

4. Conclusion

We have successfully fabricated the flexible graphite based counter electrode using a manual method. The film reached the resistivity value of 0.3 Ω.cm for 40 times of coating. The most conductive flexible film has been applied as counter electrode of solar cell using graphite/TiO_2 composite as absorbing material. This solar structure achieved the efficiency of 0.4%. Although the efficiency of the solar cell was below 1%, but it still has a room of improvement by increasing the conductivity of the graphite-based film through increasing the number of coating and combining the straight and circle motion when depositing the graphite powder aimed at homogeneously covering the surface of the plastic with graphite powder.

5. References

[1] Chapin, D. M., Fuller, C. S., & Pearson, G. L. 1954 A new silicon p-n junction photocell for converting solar radiation into electrical power *Journal of Applied Physics* **25**(5) 676-677

[2] Zhao, J., Wang, A., Green, M. A., & Ferrazza, F. 1998 19.8% efficient "honeycomb" textured multicrystalline and 24.4% monocrystalline silicon solar cells. *Applied Physics Letters*, *73*(14)

1991-1993

[3] Green, M. A., Emery, K., Hishikawa, Y., dan Warta, W. 2010 Solar cellefficiency tables (version 36). *Progress in Photovoltaics: Research andApplications* **18** 346-352

[4] Chopra, K. L., Paulson, P. D., & Dutta, V 2004 Thin-film solar cells: an overview. *Progress in Photovoltaics: Research and Applications* **12**(2-3) 69-92

[5] Contreras, M. A., Egaas, B., Ramanathan, K., Hiltner, J., Swartzlander, A., Hasoon, F., & Noufi, R. 1999 Progress toward 20% efficiency in Cu (In, Ga) Se_2 polycrystalline thin-film solar cells. *Progress in Photovoltaics: Research and applications* **7**(4) 311-316

[6] Green, M. A 2007 Thin-film solar cells: review of materials, technologies and commercial status. *Journal of Materials Science: Materials in Electronics* **18**(1) 15-19

[7] Chamberlain, G. A. 1983 Organic solar cells: A review. *Solar cells* **8**(1) 47-83

[8] Ameri, T., Dennler, G., Lungenschmied, C., & Brabec, C. J 2009 Organic tandem solar cells: A review. *Energy & Environmental Science* **2**(4) 347-363

[9] Hoppe, H., & Sariciftci, N. S. 2004 Organic solar cells: An overview. *Journal of materials research* **19**(7) 1924-1945

[10] Glunz, S. W 2007 High-efficiency crystalline silicon solar cells. *Advances in OptoElectronics*, ***2007***

[11] Green, M. A., Emery, K., Hishikawa, Y., dan Warta, W 2010 Solar cellefficiency tables (version 36). *Progress in Photovoltaics: Research andApplications* **18** 346-352

[12] Saehana, S., Arifin, P., Khairurrijal, Abdullah, M 2012 A new architecture for solar cells involving a metal bridge deposited between active TiO_2 particles, *J. Appl. Phys.* **111**

[13] Saehana, S., Prasetyowati, R., Hidayat, M.I., Arifin, P 2011 Efficiency Improvement in TiO_2-Particle Based Solar Cells after deposition of metal in spaces between particles, *Int. J. Basic Appl. Sci.* **1** 15–28

[14] Saehana, S., Darsikin, Yuliza, E., Arifin, P., Khairurrijal, Abdullah, M 2014 A new approach for fabricating low cost DSSC by using carbon-ink from inkjet printer and its improvement efficiency by depositing metal bridge between titanium dioxide particles, *J. Sol. Energy Eng.* **136**(4)

[15] Rokhmat, M., Wibowo, E., Sutisna, Yuliza, E., Khairurrijal, Abdullah, M. 2015 Enhancement of TiO_2 Particles Based-Solar Cells Efficiency by Addition of Copper(II) Nitrate and Post-Treatment with Sodium Hydroxyde, *Adv. Mater. Res.* **1112** 245–250

[16] Yuliza, E., Saehana, S., Rahman, D. Y., Rosi, M., Khairurrijal, Abdullah, M 2013 Enhancement performance of dye-sensitized solar cells from black rice as dye and black ink as counter electrode with inserting copper on the space between TiO_2 particles by using electroplating method, *Materials Science Forum* **737** 85-92

[17] Rahman, D. Y., Rokhmat, M., Yuliza, E., Sustini, E., Abdullah, M 2016 New design of potentially low-cost solar cells using TiO2/graphite composite as photon absorber, *Int. J. Energy Environ. Eng.* **7** (2016) 289–296

[18] Chen, J., Li, K., Luo, Y., Guo, X., Li, D., Deng, M., Meng, Q 2009 A flexible carbon counter electrode for dye-sensitized solar cells. *Carbon* **47**(11) 2704-2708

[19] Veerappan, G., Bojan, K., & Rhee, S. W 2011 Sub-micrometer-sized graphite as a conducting and catalytic counter electrode for dye-sensitized solar cells. *ACS applied materials & interfaces* **3**(3) 857-862

[20] Rahman, D. Y., Sustini, E., Abdullah, M 2017 Fabrikasi Transparant Conducting Film Berbahan Dasar Grafit Pada Substrat Plastik dengan Proses Mekanik. *Jurnal Matematika dan Sains,* **22**(1) 33-36

[21] Rahman, D. Y., Sutisna, Utami, F. D., Margaretta, D. O., Sustini, E., Abdullah, M. (2019). Highly easy and low cost fabrication of graphite-based flexible transparent conducting film. *J Phys. Conf. Ser.* **1204** (2019) 012084

[22] Utami, F. D., Rahman, D. Y., Sustini, E., Abdullah, M. 2019. Immobilization of TiO_2 on transparent plastic and its application in photocatalytic wastewater treatment. *J Phys. Conf. Ser.* **1171**(1) 012030

[23] Utami, F. D., Rahman, D. Y., Sutisna, Kamirul, Margaretta, D. O., Abdullah, M. 2019 Photocatalyst based on TiO_2 and its application in organic wastewater treatment using simple spray method *J Phys. Conf. Ser.* **1204** (2019) 012086
[24] Margaretta, D. O., Permadi, K. W., Rahman, D. Y., Utami, F. D., Viridi, S., Abdullah, M. 2019 Antibacterial Investigation Activity of Titania Anatase technical grade on polypropylene sheet *J Phys. Conf. Ser.* **1204**(2019) 012051
[25] Rossnagel, S. M., Kuan, T. S. 2004 Alteration of Cu conductivity in the size effect regime. *Journal of Vacuum Science & Technology B: Microelectronics and Nanometer Structures Processing, Measurement, and Phenomena* **22**(1) 240-247
[26] Dayal, D., Rudolf, P., Wiβmann, P. 1981 Thickness dependence of the electrical resistivity of epitaxially grown silver films. *Thin Solid Films* **79**(2) 193-199
[27] Camacho, J. M., Oliva, A. I. 2005 Morphology and electrical resistivity of metallic nanostructures. *Microelectronics Journal* **36**(3-6) 555-558
[28] Lacy, F. 2011 Developing a theoretical relationship between electrical resistivity, temperature, and film thickness for conductors. *Nanoscale research letters* **6**(1) 636
[29] Meyerovich, A. E., Ponomarev, I. V. 2002 Surface roughness and size effects in quantized films. *Physical Review B*, **65**(15), 155413.
[30] Fuchs, K. 1938 The conductivity of thin metallic films according to the electron theory of metals. In *Mathematical Proceedings of the Cambridge Philosophical Society* (Vol. **34**, No. 1, pp. 100-108). Cambridge University Press
[31] Sondheimer, E. H. 1952 The mean free path of electrons in metals. *Advances in physics* **1**(1) 1-42

Acknowledgments

This work was supported by fellowship grant from Indonesia Endowment Fund for Education (LPDP). The PMDSU fellowships from the Ministry of Research, Technology and Higher Education, Republic of Indonesia F.D.U. is highly acknowledged. This work was also supported by KK Research Grant from Bandung Institute of Technology 2019.

The Effect of PVA and PEG Infusion on Titania Antibacterial Coating

D O Margaretta[1,2], K W Permadi [1], D Y Rahman[1], F D Utami [1], S Viridi[1], and M Abdullah[1*]

[1]Department of Physics, Bandung Institute of Technology, Jalan Ganeca 10, Bandung 40132 Indonesia
[2] State Islamic Institut of Tulungagung, Jalan Mayor Sujadi Timur no.46 Tulungagung Indonesia

*e-mail: desymargaretta@gmail.com

Abstract. Indonesia has great potential of the Solar energy, irradiated by the Sun throughout the years. Cotton textiles antibacterial has been prepared by coating Titania (TiO_2) with Polyethylene glycol (PEG) and Polyvinyl alcohol (PVA) as adhesives. The method of depositing cotton using a simple dip coating. Cotton dry for 15 minutes with a halogen lamp, after the deposition process. Then, we do the antibacterial activity test using Plate Count Method. As a result, the coated cotton textiles with Polyvinyl alcohol (PVA), Polyethylene glycol (PEG) has antibacterial properties. Characterization by the Attenuated Total Reflectance (ATR) obtained that Titania (TiO_2) surely adhere on the sample. From Attenuated Total Reflectance (ATR) characterization we conclude that Polyvinyl alcohol (PVA) and Polyethylene glycol (PEG) is not affecting the antibacterial ability of Titania (TiO_2). But the infusion of Polyvinyl alcohol (PVA) and Polyethylene glycol (PEG) on Cotton increases the stiffness value of Cotton.

1. Introduction

In this globalization era, there are many problems caused by bacteria. In the public hospital facility, textiles such as clothes, bedsheet, curtain or the wall can be bacteria contaminated. Virus and fungus from blood, feces, urine, vomit and other body secretion that caused infection [1]. Modification textile surface with nanoparticle and inorganic nanocomposite prospect to be antimicrobial [2]. Titania (TiO_2) has unique characteristics in terms of high stability, shelf life, safety, and large antibacterial microorganisms. The photocatalyst is the activated catalyst when illuminated by the light and has a simple mechanism [3]:

$$\text{Photocatalyst material} + h\nu \rightarrow e_{cb}^- + h_{vb}^+$$

Titanium (TiO_2) in nature generally has three phases consisting of anatase and rutile with a tetragonal structure. Another phase is brookite with an orthorhombic structure.

The anatase phase were used as an electron collecting layer in organic photovoltaic and were also used as a catalyst for nanotube and nanoribbon. Explanation the photocatalyst mechanism in this experiment are: UV Light induce the formation of electron pairs, which are the charge carriers that react with H_2O, OH- and O_2 to produce hydroxyl radicals (•OH) and radical superoxide anions ($O_2^{\bullet -}$), which causes decomposition of almost all organic molecules on TiO_2 surfaces[4,5]. The bacteria is one of organic molecule. Bacteria are groups of microorganisms with prokaryotic cells (do not have a membrane-bound nucleus). The bacteria we used in this experiment were Gram-negative bacteria and Gram-positive bacteria. Escherichia Coli (E. Coli) is Gram-negative bacteria, (2 μm length, 0.7 μm diameter and 0.4-0.7 μm width) [4]. Staphylococcus Aureus (S. Aureus) is the Gram-positive (diameter is between 0.7 and 1.2 nm) [6]. After we successfully prove the ability of Technical Grade Titania (TiO_2) Anatase to degrade bacteria. Now we try to coat the dispersion into cotton with Polyvinyl alcohol (PVA) and Polyethylene glycol PEG as crosslinker. In this study, we want to identify the effect of infusion Polyvinyl alcohol (PVA) and Polyethylene glycol (PEG) in dispersions coated on cotton. We do characterization by plate count method and Attenuated Total Reflectance (ATR) to identify the chemical bound and Stiffness tester to count stiffness value of cotton after coating.

2. Materials and Method

2.1. Sample Preparation

We use material consist of Technical Grade Titania (TiO_2) anatase, PVA (Polivynyl Alcohol), PEG Polyethylene glycol (PEG-6000) (pH 5,5-7,0), 1 gram Polyvinyl alcohol (PVA) (pH 5-7) and deionized water. The Comercial Cotton purchased from mart for substrate. For bacteria we use bacteria gram negative and positif i.e Staphylococus Aureus and Eschercia Coli. Both of them, easily found in human area and cause many infections. Experiment method performed on cotton fabric as organic substrate. Polyvinyl alcohol (PVA) and Polyethylene glycol (PEG) both of similar variation dissolved in 50 ml of deionized water. at 70 ° C, then add TiO2 (1g and 10g) with stirring for 60 minutes, then cover the dispersion by immersion coating as shown in the illustration below.

Figure 1. Illustration of experiment.

2.2. Sample Characterization

We do sample characterization to discover the same antibacterial ability of the same method as previous experiment by margaretta et al [7]. To discover the effect of Polyvinyl alcohol (PVA) and Polyethylene glycol (PEG) infusion we use Atomic Transmittance Reflectance to Investigate chemical bond between different chains of atoms in a polymer or other in sample. We also do Stiffness measurement in Balai Tekstil Bandung.

3. Results and Discussion

3.1. Plate Count Method

The antibacterial activity is identified by the plate count method, focusing on test cotton antibacterial abilities. Quantitative results from manual calculations are shown in table 1. The data consist of three types of samples, the first was a sample control, the second was a sample 1 (coated with 1 gram of Titania) and sample 2 (coated with 10 grams of Titania). The results show the percentage of antibacterial in CFU / ml. try to use antibacterial simply tested method with the manual calculation of the percentage of the antibacterial ability.

Table 1. Results of Colony Counts and Percentage Inhibition Tested by Plate Count Method

Treatment	**E. Coli (colony) [CFU/ml]**				**S. Aureus (colony) [CFU/ml]**			
Plate	**1**	**%**	**2**	**%**	**1**	**%**	**2**	**%**
Control	276×10^6	97 %	263×10^6	97%	220×10^6	96%	212×10^6	96%
Sample 1	2×10^6	99 %	3×10^6	98.8 %	2×10^6	99%	2×10^6	99%
Sample 2	0	100 %	0	100%	0	100%	0	100%

TiO2 technical grade show its efficacy inhibits the growth of Bacteria. The table shows that the number of bacteria in the sample is reduced compared to the control. The difference in the photocatalytic activity of photocatalyst with the different structures closely related to surface area and particle size. Particles bulk structure anatase more reactive than rutile and brookite. This caused by anatase has band gap energy higher 3,2 eV than Rutile in 3,1 eV. So anatase has definitely photocatalytic activity better [8]. Figure below shown us the plate containing the bacteria growth control and growth of bacteria inhibited by antibacterial photocatalyst Titania (TiO_2) in every variation.

Table 2. Plate Count Method picture after incubation (a. sample control [7], b sample 1, c sample 2)

Plate *S. aureus*		**Plate** *E.Coli*	
1	**2**	**1**	**2**

a.Control

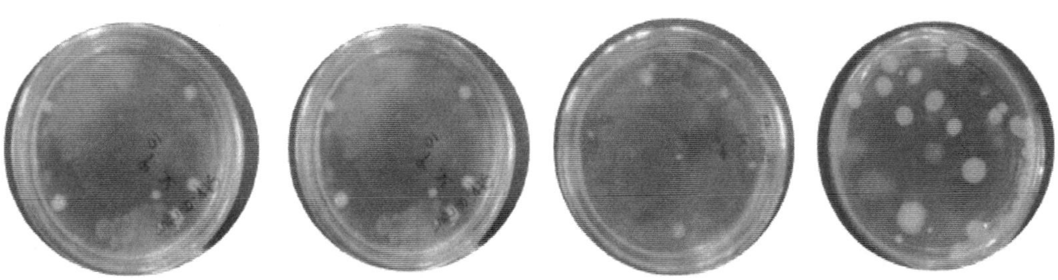

b.Sample 1 (TiO_2 :1 gr)

IOP Conf. Series: Materials Science and Engineering **599** (2019) 012005 doi:10.1088/1757-899X/599/1/012005

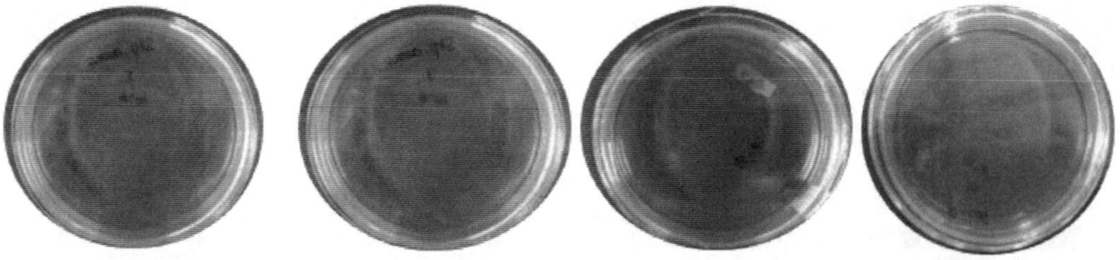

c.Sample 2 (TiO_2 :10 gr)

From the manual colony counting from the table 1 above. We can conclude that both of the sample cotton after coating has the antibacterial tendency. The variables that can kill bacteria, such as temperature and pH, are still within the range of bacteria that can live. E.Coli bacteria can live in range temperature between 10°C- 40°C with pH above 5. And S. aureus can live in temperature between 4°C- 48°C [9] dan pada pH 4.5-9.3[10]. Whereas E.Coli bacteria die at pH below 5 and S. Aureus pH below 4.2. The test temperature conditions is 32°C, it is still in the condition of E. coli and S. aureus bacteria can live [10].

3.2 Effect of Polyvinyl alcohol (PVA) and Polyethylene glycol (PEG) on Cotton

Because TiO_2 irradiated by UV-Light, band gap produced electron and hole pair then caused redox reaction in surface of TiO_2. Effect from electron excitation from valence band to conduction band, the negatively charged electron and oxygen banded be O_2^- Whereas positively charged *hole* and water will produce Radical hydroxyl [11]. Radical hydroxyl caused pressure oxidation and reduced cytoplasmic eukaryotic and prokaryotic cells, causing cell damage on bacteria [12]. We do ATR characterization and we get the result showed in Figure 3a.

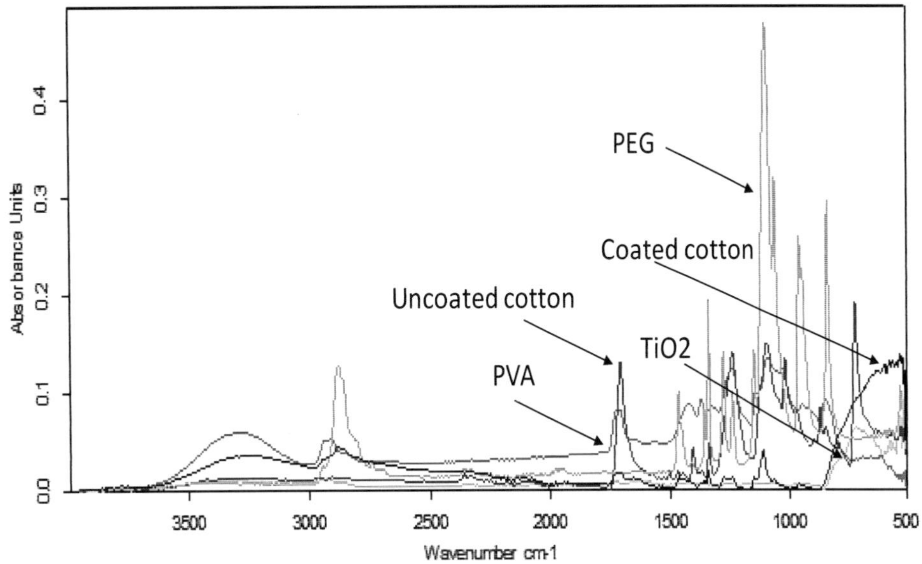

Figure 2 a. ATR chemical bound.

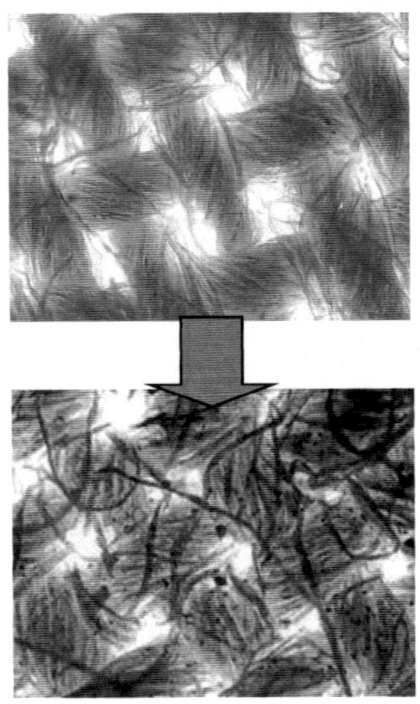

Figure 2b. Cotton picture from light microscope.

FTIR characterization results on samples of cotton shows that peaks 1006.73 C = O [11] is the culmination of ketones which indicates the presence of crosslinking Polyethylene glycol (PEG) and cotton. 720.93 peak is a characteristic possessed by Ti - O anatase TiO_2 where vibration characteristics are at 500-800 cm-1. Peaks 1714.29 C = O is the culmination of ketones which shows crosslinking reaction of Polyvinyl alcohol (PVA) [13]. From figure above showed the morphology of cotton taken with microscope. Before coating and after coating, shown different morphology. After coating the fiber become more rigid, the space between cotton fiber filled by TiO_2 dispersion. The fiber arrangement also become unorganized, different from initial conditions. In this figure we can see that there are thicker coated parts. The rigid condition of cotton did not same if we measured it from different position. In this case, for rigidity measurements we use repetitive measurement averages.

3.3 Stiffness Test on Cotton

The stiffness value test on Balai Besar Textile Bandung with Shirley Stifness Tester. The fiber arrangement of cotton consisting of two lanes, vertical threads are called warp and horizontal threads are called woof, horizontal layer coating on this thread cause the stiffness of cotton changes [14].

Table 3. Stiffness Value

	Before Coating (gr/cm)		After Coating (gr/cm)	
	woof	warp	woof	warp
Stiffness value	165.0	107.2	1357.0	1576.4

The stiffness value of coated fabric after coating process increases drastically. The value of initial cotton stiffness was 165.0 g/cm for woof lane and 107.2 g/cm for warp lane. It increases to 1357 g/cm for woof line and 1576.4 g/cm for warp lane. The value increases drastically because of uneven coating deposition on cotton. The solvent does not adhere well in textile. The textile from this experiment can't be used for clothing because it doesn't fulfil with the textile standard for clothing, but it could be useful for other purpose such as rain coat, curtain. The heating process increases the rigidity of cotton warp and woof directions. Mixed polymer Polyvinyl alcohol (PVA) and Polyethylene glycol (PEG) causes the fibers compact and more rigid.

4. Conclusion

In this study it is concluded that infusion of Polyvinyl alcohol (PVA) and Polyethylene glycol (PEG) do not affect the performance of coated TiO_2 in cotton as an antibacterial. Anatase TiO_2 can be a good antibacterial agent that can be activated by sunlight and halogen lights at night. Due to the temperature and pH used during synthesis also testing and incubation do not exceed the optimum of bacteria. Then it is confirmed that the degradation of bacteria is due to the TiO_2 photocatalyst process that produces hydroxyl radicals and destroys bacterial cells.

5. References

[1] Serpone N and Emeline A V 2002 *Int J Photoenergy* **4** 91-131
[2] Dastjerdi R, Montazer M 2010 *Colloids and Surfaces B: Biointerfaces* **79** 5-18
[3] Dizaj S M, Lotfipour F, Barzegar-Jalali M, Zarrintan M H, and Adibkia K 2014 *Materials Science and Engineering: C* **44** 278-284
[4] Fujishima A, Rao T N and Tryk D A 2000 *J. Photochem. Photobiol. C: Photochem. Rev.* **1** 1
[5] Utami F D, Rahman D Y, Sustini E, and Abdullah M 2019 *J Phys. Conf. Ser.* **1171(1)** 012030
[6] Jawetz E, Melnick J L, Adelberg E A, Brooks G F, Butel J S, and Ornston L N 1995 *Mikrobiologi Kedokteran.* (Alih bahasa: Nugroho & R.F.Maulany). Jakarta: Penerbit Buku Kedokteran EGC
[7] D O Margaretta, K W Permadi, D Y Rahman, and F D Utami 2019. Antibacterial Investigation Activity of Titania Anatase Technical grade on polypropylene sheet. *J Phys.Cof.Ser* 1204 012051
[8] Velasco E, Baldovino-Medrano V G, Eric M G, Sonia A G 2015 *Springer* **59** 378–386
[9] Adams M R, Moss M O 1999 *Food Microbilogy* Royal society of chemistry, Oxford: UK. p. 349
[10] Bennett R W, Monday S R 2003 *"Staphylococus aureus, Chapter 4*. In: MD Miliotis, JW Bier (Eds), *International Handbook of Foodborne Pathogens"*. Marcel Dekker, Inc., New York
[11] Coleman H M, Marquis C P, Scott J A, Chin S S, Amal R 2005 *J.Chem Eng* **113:**55-63
[12] Suketa N, Sawase T, Kitaura H, et al. *An antibacterial surface on dental implants, based on the photocatalytic bactericidal effect*. Clin Implant Dent Relat Res. 2005;**7**(2):105-111
[13] Silverstein R M, Bassler G C, Morrill T C 1991 *Spectrometric identification of organic compounds* (New York: Wiley)

IOP Conf. Series: Materials Science and Engineering **599** (2019) 012005 doi:10.1088/1757-899X/599/1/012005

[14] Hill H 1992 *Indonesia's Textiles and Garment Industries*. Singapore: ISEAS

Acknowledgments
Authors are very grateful to Mr.Teddy and Mrs. Fitri from Bioscience and Biotechnology research center ITB for all contribution, and a lot of help support facilities for bacterial testing. And this work is supported by Grant PMDSU 314e/I1.C01/PL/2015 from Ministry of Research, Technology and Higher Education.

The Effect of Pore Geometry on the Flow Unit and Its Impact to Permeability of the Reservoir Rock

G Yasmaniar*, S Prakoso and R Sitaresmi

Petroleum Engineering Department, Faculty of Earth Technology and Energy, Universitas Trisakti
Jalan Kyai Tapa No.1 Grogol, Jakarta Barat 11440, Indonesia

*E-mail: ghanimayasmaniar@gmail.com

Abstract. Rock type is defined as a rock interval with a unique pore geometry and it is related to the characteristics of the hydraulic flow unit. The quality of hydraulic unit of the rock is controlled by pore geometry which is a function of texture and mineralogy of the rock. In this research, it is conducted the study of the effect of pore geometry on the flow unit owned by the reservoir and its impact to the permeability value. This study used 191 of sandstone core samples from Upper Cibulakan Formation in the North West Java Basin. The hydraulic flow unit concept developed by Kozeny-Carman is used to determine the rock type in the reservoir. Afterward, microscopic geology features are identified based on rock texture in the form of grain size, grain sorting, and angularity as part of the pore geometry that controls the hydraulic quality of the rock. This identification was carried out to determine the pore geometry which generally composes each rock type. Based on the results of core data analysis, the reservoir under this study consists of eight flow units. Based on microscopic geology features identification, it is generally found that the better flow unit, the simpler texture it has, which is seen from the better of grain size and grain sorting. On the contrary, the lower quality of the flow unit, the texture tends to be more complex. From this study, it has also known that the better flow unit has a higher permeability value compared to the lower quality of flow unit, which causes the fluid flowing faster and the rock has the better reservoir quality index.

1. Introduction

Permeability is one of the important of rock properties and it is related to the fluid flow from the reservoir. Permeability values in the reservoir are heterogeneously distributed which are influenced by the geological and petrophysical properties of the rock, and it is also associated with rock types that are scattered in the reservoir. Rock type is defined as the interval of rocks with a unique pore geometry and is related to the characteristics of its hydraulic flow unit [1]. Rock typing is the process of classifying rocks into separate units, which is deposited in the same geological conditions and undergoes the same diagenetic process [2-6].

Amaefule introduced the HFU (Hydraulic Flow Unit) concept developed by Kozeny-Carman to classify rocks based on their rock type [7]. Amaefule have shown that the relationship between porosity and permeability in each rock type is an evidence of different hydraulic units [7]. The hydraulic quality of a rock is determined by the geometry of the rock pores [8],[9]. Pore geometry is a function of mineralogy (type, number, morphology, and relative position of pore throat) and texture (grain size, grain shape, grain size uniformity, and grain structure) [10,11].

The objective of this study is to find out the effect of pore geometry to the flow unit of the reservoir rock, then it can also conducted the impact of those to the permeability value of the reservoir. The HFU concept developed by Kozeny-Carman was used to determine the rock type of the reservoir, so that the relationship between porosity and permeability of each rock type can be obtained. After that, microscopic geology features were identified based on rock texture in the form of grain size, grain sorting, and angularity as part of the pore geometry that controls the quality of rock hydraulics. This identification was carried out to determine the pore geometry which generally composes each rock type and how it can impact the permeability value.

2. Methodology

In this study, 191 of sandstone core samples from Upper Cibulakan Formation in the North West Java Basin were used. These core samples had had porosity and permeability value from core analysis in the laboratory. The core samples were used to determine the HFU class owned by the reservoir which was analyzed using the Kozeny-Carman approach [7],[10],[12]. Kozeny-Carman considers that the reservoir pores are a collection of capillary tubes. Kozeny-Carman uses Poisseuille's Law (for flow in capillary tubes) and Darcy's Law (for flow in porous media), to get the relationship between porosity and permeability [13]. In general the Kozeny-Carman equation can be written as:

$$k = \frac{\emptyset_e^3}{(1-\emptyset_e)^2}\left[\frac{1}{F_s \tau^2 S_{gv}^2}\right] \quad (1)$$

Then, new parameter is generated that expressed in the following equation:

$$RQI = \sqrt{\frac{k}{\emptyset_e}} \quad (2)$$

$$\emptyset_z = \left(\frac{\emptyset_e}{1-\emptyset_e}\right) \quad (3)$$

From Equations (2) and (3) then a new parameter is generated, namely FZI (Flow Zone Indicator) which can be expressed in the following equation:

$$FZI = \left[\frac{1}{\sqrt{F_s}\tau S_{gv}}\right] = \frac{RQI}{\emptyset_z} \quad (4)$$

Then, to grouping the data based on the flow unit, the following equation is used:

$$FZI_{discrete} = \text{ROUND}\,(2 \text{ x } \ln(FZI) + 10{,}6) \quad (5)$$

After determining the rock type using the HFU concept developed by Kozeny-Carman, the relationship between porosity and permeability was obtained for each rock type. It was also carried out the identification of microscopic geology features based on available core data. The identified microscopic geology features are grain size, grain sorting, and angularity as part of pore geometry that controls the quality of rock hydraulics. This is obtained to find out the pore geometry which generally compiles each HFU and distinguishes it from other HFUs and how it relates to the permeability value.

3. Results and Discussion

Determination of HFU was carried out by applying Kozeny-Carman's approach using 191 of available sandstone core samples. From the results of this calculation, it is known that the reservoir consists of eight HFUs and it was also found the relationship between porosity and permeability of each HFU which can be seen in Figure 1.

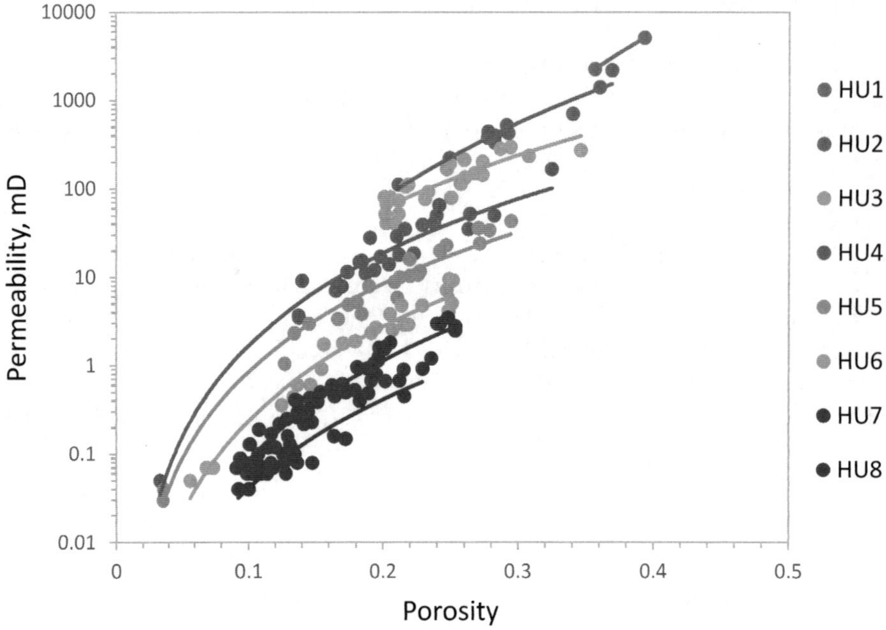

Figure 1. Porosity vs. Permeability Relationship for Each HFU.

From the relationship of porosity and permeability, the empirical equations generated by each HFU can be seen in Table 1, where x is porosity and y is permeability. Using this empirical equation, the permeability value can be calculated by entering the porosity value to the empirical equation according to the HFU group.

Table 1. Empirical Equations of the Relationship of Porosity and Permeability for Each HFU

HFU	Empirical Equation
1	$y = 1E{+}07x^{8.394}$
2	$y = 19{,}8170\ x^{4.879}$
3	$y = 15{,}918x^{3.481}$
4	$y = 5{,}064.7\ x^{3.479}$
5	$y = 1{,}759.9\ x^{3.316}$
6	$y = 840.51\ x^{3.552}$
7	$y = 450.54\ x^{3.718}$
8	$y = 87.244\ x^{3.325}$

Meanwhile, the hydraulic quality of a rock is controlled by the geometry of the rock pores which is a function of the texture and mineralogy of the rock. Thus, eight types of HFU obtained from this study have certain geological features that distinguish one HFU from the other HFU. In this study, microscopic geology features were identified based on rock texture which included grain size, grain sorting, and angularity as part of the pore geometry that controls the quality of rock hydraulics, so the data that generally compile each HFU can be seen in Table 2.

Table 2. Microscopic Geology Features at Each HFU

HFU	Grain Size	Angularity	Grain Sorting
1	Fine Grain	Sub Angular-Sub Rounded	Well Sorted
2	Fine Grain	Sub Rounded	Moderate-Well Sorted
3	Very Fine-Fine Grain	Sub Rounded	Moderate Sorted
4	Very Fine-Fine Grain	Sub Rounded	Moderate Sorted
5	Very Fine-Fine Grain	Sub Rounded	Poor- Moderate Sorted
6	Very Fine-Slightly Fine Grain	Sub Rounded	Poor- Moderate Sorted
7	Very Fine-Fine Grain	Sub Angular-Sub Rounded	Poor- Moderate Sorted
8	Very Fine-Fine Grain	Sub Angular-Sub Rounded	Poor- Moderate Sorted

From Table 2, it can be seen that in general, the better HFU, which is indicated by a small HFU number, the simpler texture it has, that can be seen from the better grain size and grain sorting. Conversely, the lower quality of HFU, which is indicated by the greater number of HFU, the texture tends to be more complex. To be clearer, then the relationship between grain size and permeability was made from the available core data, which can be seen in Figure 2.

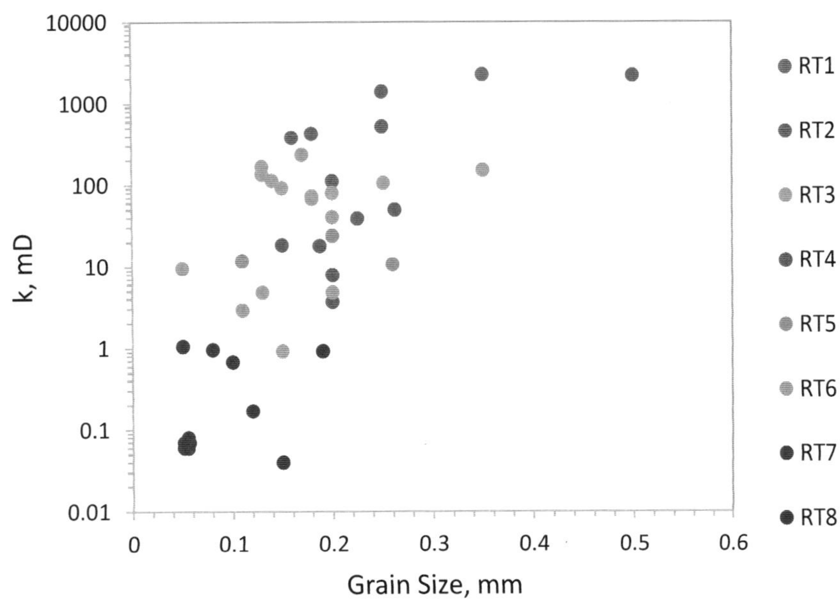

Figure 2. Grain Size vs. Permeability.

From Figure 2, it can be seen that in general, a better HFU has a larger grain size than a lower quality of HFU. In addition, it is also seen that pore size is not only related to the rock volume, but also relates to the ability of rocks to flow the fluid, that named permeability. In general, the increasing grain size will have an effect on increasing permeability value which is also has the better HFU class. For more details, the tabulation of the smallest, the largest, and average permeability value for each HFU can be seen in Table 2.

Table 3. The Smallest, the Largest, and Average Permeability for Each HFU

HFU	Permeability, mD		
	Min	Max	Mean
1	2258.000	5153.000	3705.500
2	112.000	2191.000	650.182
3	40.500	298.000	134.656
4	0.050	168.000	26.499
5	0.030	43.000	12.412
6	0.050	9.550	2.946
7	0.060	3.500	0.763
8	0.040	0.920	0.191

From the results of the tabulation, it is known that the good HFU, which is a small HFU number, has a higher permeability value than the low quality of HFU, which results in faster fluid flow and better reservoir quality index.

4. Conclusions

The results of this study show that the pore geometry composing the rock affects the flow units of the rock and influences the permeability value as well. According to the results of the analysis on 191 of core data from Upper Cibulakan Formation in the North West Java Basin, eight different HFUs were obtained with different equations to estimate permeability value for each HFU. It is also known that the better HFU has the simpler texture of pore geometry that can be seen from the better grain size and grain sorting. Conversely, in the lower quality of HFU, the texture tends to be more complex. Furthermore, the better HFU has a higher permeability value than the low quality of HFU, so that the fluid flows faster and the reservoir quality index gets better.

5. References

[1] Turkey, S., Alkanderi, J., Kumar, P., Alalawi, G., Alhashmi, S., Alharthy, A., and Alraisi, M 2012 Rock Typing and Characterization of Carbonate Reservoirs: A Case Study From South East Kuwait, SPE 163294

[2] Guo, G., Diaz, M.A., Paz, F., Smalley, J., and Waninger, E.A 2005 Rock Typing as an Effective Tool for Permeability and Water Saturation Modeling: A Case Study in Clastic Reservoir in the Oriente Basin, SPE 97033

[3] Chekani, M. 2009 Reservoir Rock Typing in Carbonate Reservoir-Cooperation of Core and Log Data: Case Study, SPE 123703

[4] Gomes, J.S., Ribeiro, M.T., Strohmenger, C.J., Negahban, S., and Kalam, M.Z. 2008 Carbonate Reservoir Rock Typing – The Link between Geology and SCAL, SPE 118284

[5] Michel, R., and Bruno, L. 2014 Rock Typing in Carbonates: A Critical Review of Clustering Methods, SPE-171759-MS

[6] Yasmaniar, G., Sitaresmi, R. and Prakoso, S. 2018 Determination of Rock Type Using Hydraulic Flow Unit Concept to Predict Permeability with Artificial Neural Network, Journal of Earth Science, Engineering, and Technology, 1 2

[7] Amaefule, J.O. and Altunbay, M. 1993 Enhanched Reservoir Description: Using Core and Log Data to Identify Hydraulic (Flow) Units and Predict Permeability in Uncored Intervals/Wells, SPE 26436

[8] Permadi, P., and Susilo, A. 2009 Permeability Prediction and Characteristics of Pore Structure and Geometry as Inferred From Core Data, SPE 125350

[9] Wibowo, A.S., and Permadi, P. 2013 A Type Curve for Carbonates Rock Typing, IPTC 16663

[10] Abbaszadeh, M., Fuji, H., and Fujimoto, F. 1996 Permeability Prediction by Hydraulic Flow Units – Theory and Applications, SPE 30158

IOP Conf. Series: Materials Science and Engineering **599** (2019) 012006 doi:10.1088/1757-899X/599/1/012006

[11] Yasmaniar, G., Sitaresmi, R. dan Prakoso, S., 2018 Pengaruh Atribut Pori Terhadap Rock Type dan Penggunaannya pada Prediksi Permeabilitas, Thesis, Magister Teknik Perminyakan Universitas Trisakti
[12] Mortensen, J., Engstrom, F., and Lind, I. 1998 The Relation Among Porosity, Permeability, and Specific Surface of Chalk From the Gorm Fiels, Danish North Sea, SPE 31062
[13] Tiab, D., and Donaldson, E.C. 2004 Petrophysics – Theory and Practice of Measuring Reservoir Rock and Fluid Transport Properties, Gulf Professional Publishing, Oxford

Effect of Formation Voltage on the Pore Size of Porous Anodic Aluminum Oxide

Manogari Sianturi[1,2], Ariadne L. Juwono[1], Anawati Anawati*[1]

[1]Department of Physics, Faculty of Mathematics and Natural Sciences, Universitas Indonesia, Depok 16424, Indonesia
[2]Universitas Kristen Indonesia, Cawang, Jakarta, 13630, Indonesia

Corresponding author: anawati@sci.ui.ac.id

Abstract. This work is aimed to clarify the effect of formation voltage on the pore size of porous anodic aluminum oxide (PAAO) layer formed on commercially pure aluminum. The PAAO layers were obtained by anodization process in 0.3 M sulfuric acid solution at constant voltages of 10, 15, 20, and 25 V at 10°C. The structure and morphology of PAAO layers were characterized by using FE-SEM. The pore diameters which were estimated by using ImageJ software were 19.61 $\pm$ 15.35 nm, 20.03 $\pm$ 13.59 nm, 20.31 $\pm$ 12.36 nm, 25.06 $\pm$ 12.10 nm, and the wall thicknesses were 33 nm, 49.99 nm, 74.97 nm, 83.33 nm for the PAAO formed at 10, 15, 20 and 25V, respectively. The structure of the porous oxide layer became more uniform and organized, and the diameter of the pores increased linearly with applied voltage. High anodization voltage is known to cause Joule heating because of the fast movement of electrons and ions. It is believed that the Joule heat was transferred to the bulk electrolyte which results in larger pore diameter and interpore distance. The optimum condition to obtain high order PAAO is at 25 V.

Keywords: Porous Anodic Aluminum Oxide; Anodization; SEM

1. Introduction

In the past decade, there has been a renewed interest for researchers to fabricate porous anodic aluminum oxide (PAAO) layer due to the use of PAAO in various nanotechnology applications such as chemo-and biosensing applications [1], molecular separation, catalysis [2], application of biomedical drug delivery, implants and cell growth, electronic applications [3], hemodialysis, [4]. The unique properties of PAAO membranes such as an ordered pore structure, uniform pore size, controlled pore geometry, surface chemistry, and a large surface area make this material used as a template [5].

The process parameters have been extensively studied to achieve highly ordered pores. The Anodic voltage, temperature, type and concentration of electrolyte have been recognized as the most critical process parameters to control the order and geometry of the pores. The solution of aqueous sulfuric acid (H_2SO_4) at 25 V, oxalic acid ($H_2C_2O_4$) at 40 V and phosphoric acid (H_3PO_4) at 195 V are the most common electrolytes used to grow PAAO layer. In addition, other acid electrolytes which can also be used such as selenik, chromate, malonate, tartaric, citrate, malic acid, in which anodic oxide is slightly soluble[6].

The PAAO structure can be defined by the structural parameters namely pore diameter, pore distance, pore length, wall thickness and thickness of the oxide barrier layer. By controlling the process parameter, the structural parameters can be adjusted in the range of 10-400 nm pore diameter, 50-600 nm interpore distance, 10-5000 pore aspect ratio, 10 nm - 150 μm barrier layer thickness, 10^9-10^{11} cm^{-2} pore density and porosity from 5% - 50% [6]. Electrolyte concentrations can directly affect pore size and cell diameter; whereas the two of the pore diameters and interpore distances are directly proportional to the anodization voltage [7]. Furthermore, a linear correlation between pore diameter, cell diameter, and voltage is formed in citric acid, malonic, oxalate, phosphate, sulphuric and tartrate. Earlier researchers had used sulphuric acid at voltages of 19-25 volts, oxalic acid at voltages of 40-70 volts, phosphoric acid at voltages of 160 - 195 volts, at a concentration of 0.3 M to obtain optimum structural parameters. The electrolyte temperature was kept as low as 0°C [4][7][8][9]. The low temperature of anodization was not beneficial for mass fabrication of PAAO layer. In this work, the anodization was performed at a higher temperature 10°C, and the formation voltage at the corresponding temperature was optimized in the range 10-25 V.

2. Methods

The sample was a commercially pure aluminum plate with 0.1 cm thickness cut into 1.5 cm × 3.0 cm. The working area was 5 cm^2. Before anodization, the sample was degreased with acetone and in ethanol in an ultrasonic bath for 3 min. The Al plate was then electropolished under a constant DC current of 500 mA for 4 minutes in a perchloric acid ($HClO_4$) solution (Merck) with ratio 1: 4. The sample was then washed in ethanol and then immediately placed in the anodization cell. Anodization was conducted in 0.3M H_2SO_4(Merck) at 10, 15, 20, and 25 V for 4 h. The sample was placed as anode and platinum, Pt, was placed as a cathode. The distance between the two electrodes was 2,5 cm. The temperature during anodization was maintained at 10 °C, and the electrolyte solution was constantly stirred at 300 rpm. The current density was recorded by a Digital Multimeter (DMM, Rigol). After anodization, the sample was immediately rinsed under running water and finally sprayed with DI water before drying with a hair dryer. The porous AAO layer produced from the above anodization process was characterized by using the Field Emission Microscope Electron (FESEM) JEOL type JIB4610F. To determine the average pore diameter, ImageJ software was used to calculate the average area of pore from FE-SEM image.

3. Results and Discussion

3.1. Current density vs time

The graph of current density output that arises during the anodization process is shown in Figure 1. The graph characteristic was similar at all voltages where a current transient occurred during the first 30 min before stabilized at prolonging anodization time. The formation of porous layers can be divided into two stages: stage I is a linear growth barrier layer, and stage II is the steady-state growth of porous layer. The first stage is characterized by the slope of the current density which was more swooping. At the beginning of the anodization process, the entire metal surface is covered with a compact and uniform oxide layer. The slope of this current density illustrates the mobilization of Al ions and the movement of electrons in the solution to start forming an oxide layer on aluminum metal. The second stage is the distribution of the electric field (E) strongly dependent on inhomogeneities and surface roughness of the oxide layer, which leads to a locally focused electric field and depression at the interface of the electrolyte-oxide. As a result, dissolution assisted by the electric field from the oxide layer occurs, and the first localized pores are formed and become the site for further pore growth. Consequently, an increase in ionic migration will lead to an increase in current density until pores completely cover the entire surface. The anodization process further increases the distribution of individual pores through the oxide layer where the maximum electric field is concentrated at the pore end and decreases towards the pore wall. Prolonged anodization increases the pore order to some extent. The last stage is a competition between oxide growth at the pore bottom and at the same time oxide dissolution processes at barrier oxide layer interfaces. Finally, the pore growth process reaches a steady state, and uniformly distributed pores are formed. The analysis of this study was similar to earlier report [10].

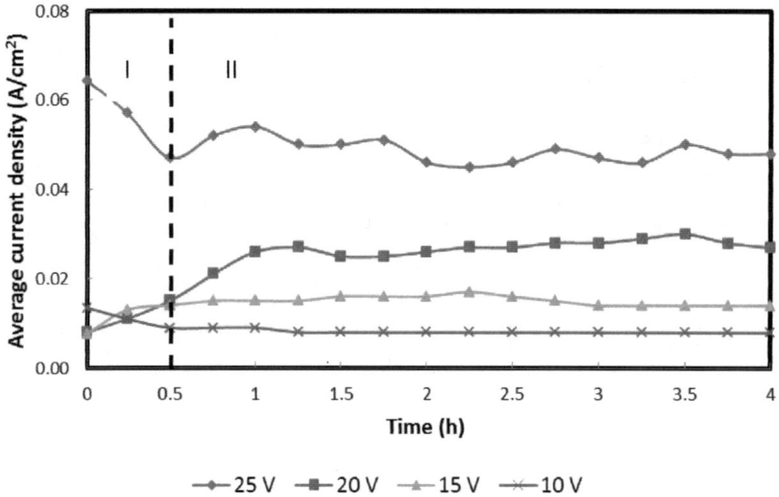

Figure 1. Current density versus time plot at formation voltages of 10, 15, 20, and 25 V in 0.3M H_2SO_4 solution at 10°C.

An interesting phenomenon from this study was that when the formation voltage increased, the current density output increased gradually. The sample anodized at 25 V generated a much greater current density than at lower voltage. The Joule heating which probably occurred at high voltage contributed to increasing the current density. The anodization was conducted at a low temperature of 10°C which slower down the ionic mobility in the electrolyte. The local Joule heating at the metal-electrolyte interface enhanced the ionic mobility which increased the reaction rate for the formation of PAAO layer. The average high current density during anodization gave a more uniform and regular porous oxide growth which also showed greater pore size and interpore distance. The higher regularity of the PAAO pores is required for the application in particular when used as a membrane. A membrane filtered between coarse and fine particles. The higher the regularity of the membrane's pore the better it filtered coarse particles. Figure 2 shows that when the sample was anodized at an increasingly high voltage, the average current density generated increases linearly with slope 0.0027. The average of the current density obtained in this study at various voltages was carried out at a constant temperature of 10 °C. Therefore, the current density generated during the anodization process was strongly influenced by the anodic voltage used.

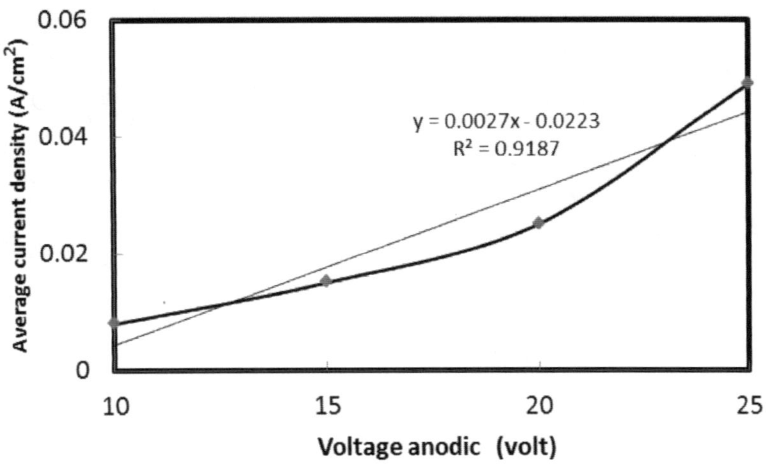

Figure 2. Graph of average current density versus anodization voltage.

3.2. Morphology of porous AAO layers

Figure 3 shows the morphology of PAAO layers produced from the anodization process at 10, 15, 20, and 25 V in 0.3 M H_2SO_4 solution. The average pore diameter sizes were 19.61 $\pm$ 15.35 nm, 20.03 $\pm$ 13.59 nm, 20.31 $\pm$ 12.36 nm and 25.06 $\pm$ 12.10 nm for anodization process voltage 10, 15, 20 and 25V, respectively and the interpore distance was 33 nm, 49.99 nm, 74.97 and 83.33 nm respectively, as measured by using ImageJ software. FE-SEM results revealed that the porous oxide layer morphology grew more regularly and uniformly with increasing formation voltage.

In this study, it was proved that the size of the pore diameter and interpore distance is greater with increasing applied voltage was confirmed in previous studies [11]. The effects of changes in anodization parameters directly affect the characteristics of the pore structure. The pore diameter and interpore distance are directly proportional to the applied voltage [7, 12]. Although other researchers also reported oppositely that anodization in oxalic acid the pore diameter and density increased with decreasing applied voltage[13]. Figure 3 shows that the size of the pore diameter and interpore distance tended to increase with applied voltage.

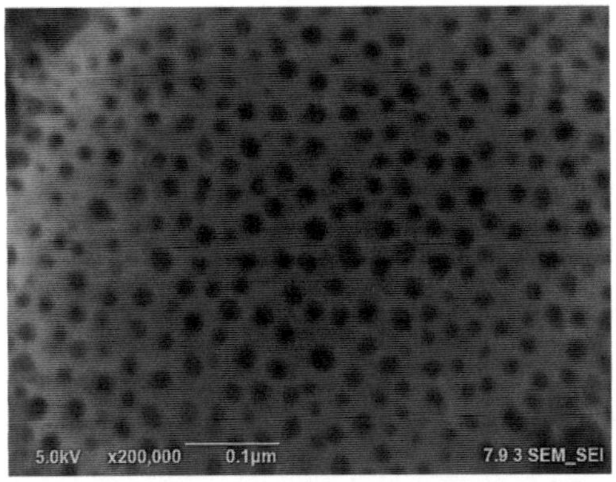

A

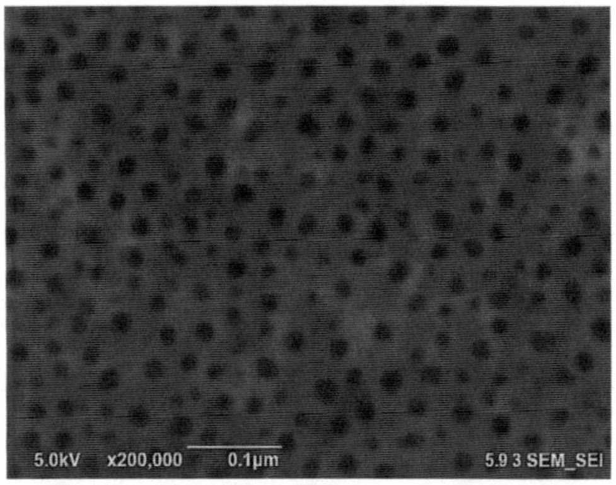

B

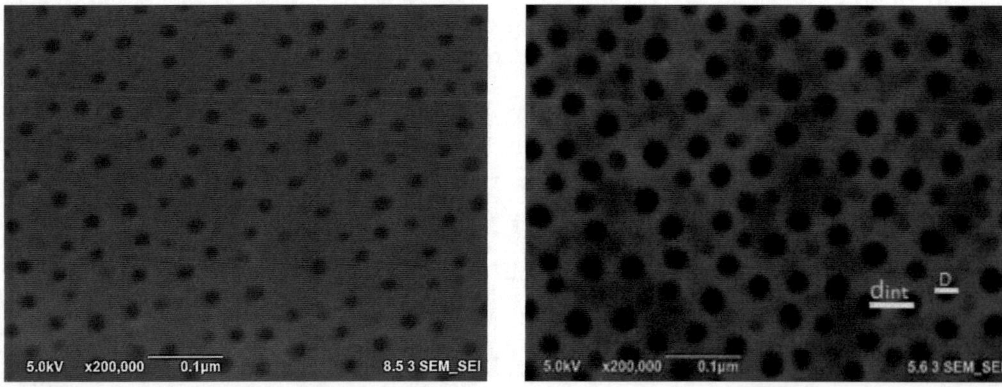

C D

Figure 3. FE-SEM image of the surface morphology of anodic aluminum oxide at A. 10, B. 15, C. 20 and D. 25 V in 0.3 M sulfuric acid.

Table 1 shows the PAAO structure parameters, namely average pore area, average pore diameter (D), and pore distance (d_{int}) obtained from Figure 3. Anodization at an anodic voltage range of 10-20 volts increased the average pore diameter linearly. When the sample was anodized at an anodic voltage greater than 20 volts the average diameter of pore increased rapidly, the phenomenon shows that the voltage anodic used has approached the breakdown voltage of the aluminum sample in the sulfuric acid solution. The interpore size also increased with increasing applied voltage.

Table 1. PAAO parameters of image A, B, C and D

PAAO parameters	Image A	Image B	Image C	Image D
Average pore area (μm^2)	3.02 x 10^{-4} $\pm$ 1..85 x10^{-4}	3.15 x 10^{-4} $\pm$ 1.45 x 10^{-4}	3.24 x 10^{-4} $\pm$ 1.2 x 10^{-4}	4.93 x 110^{-4} $\pm$ 1.15 x 10^{-4}
D (nm)	19.61 $\pm$ 15.35	20.03 $\pm$ 13.59	20.31 $\pm$ 12.36	25.06 $\pm$ 12.10
d_{int} (nm)	33	49.99	74.97	83.33

This study reveals that the morphology of the pores from the results of the characterization of FE-SEM that the average pore diameter, the inter pore distance, and the uniformity of pore increases with the increasing the anodic voltage used [14]. Figure 4 shows that the obtained inter-pore distance increases linearly with a slope of 3.5194. The image D from Fig. 3 shows that the average diameter of the pore and interpore distance were the largest, and most uniform when the sample was anodized at a voltage of 25V.

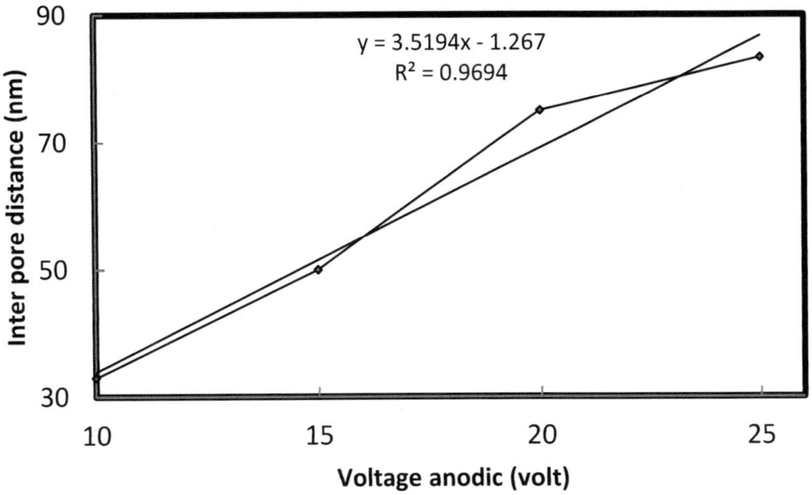

Figure 4. Graph of inter pore distance versus anodization voltage.

4. Conclusions

We have successfully fabricated porous anodic aluminum oxide (PAAO) layers at various formation voltages of 10, 15, 20, and 25 V in a 0.3 M sulfuric acid electrolyte solution at a 10°C. Surface characterization by FE-SEM and further analysis with ImageJ software indicated structural parameter of the PAAO layer increased with applied voltage. The pore diameter, interpore distance, and pore regularity increased with increasing applied voltage. Therefore, the optimum voltage to grow a regular PAAO layer at 10°C in 0.3 M H_2SO_4 solution is at 25 V.

5. References

[1] Santos A, Kumeria T, and Losic D, 2013 Nanoporous anodic aluminum oxide for chemical sensing and biosensors *Trends in Analytical Chemistry* **44** 25–38

[2] Belwalkar A, Grasing E, Van Geertruyden W, Huang Z, and Misiolek W.Z 2009 NIH Public Access *Journal of Membrane Science* **319** 192–198

[3] Mutalib A, Losic D, and Voelcker N.H 2013 Progress in Materials Science Nanoporous anodic aluminium oxide : Advances in surface engineering and emerging applications," *Progress in Materials Science* **58** 636–704

[4] Gujela O.P 2016 Anodic Aluminum Oxide (AAO) Nano Membrane Fabrication under Different Conditions, *in ICRAIE 2016- IEEE International Conference on Recent Advances and Innovations in Engineering*., pp. 0–5

[5] Ingham C.J, ter Maat J, and de Vos W.M 2012 Where bio meets nano: The many uses for nanoporous aluminum oxide in biotechnology *Biotechnol. Adv* **30**(5) 1089-1099

[6] Santos A, Kumeria T, and Losic D 2013 Nanoporous anodic aluminum oxide for chemical sensing and biosensors *TrAC - Trends Anal. Chem* **44** 25–38

[7] Belwalkar A, Grasing E, Van Geertruyden W, Huang Z, and Misiolek W.Z 2008 Effect of processing parameters on pore structure and thickness of anodic aluminum oxide (AAO) tubular membranes," *J. Memb. Sci.* **319**(1–2) 192–198

[8] Stępniowski W.J, Durejko T, Michalska-Domańska M, Łazińska M, and Aniszewska J, 2016 Characterization of nanoporous anodic aluminum oxide formed on laser pre-treated aluminum, *Mater. Charact.*, **122** 130–136

[9] Kasi A.K, Kasi J.K, Afzulpurkar N, Bohez E, Tuantranont A, and Mahaisavariya B, "RETRACTED ARTICLE: Fabrication of anodic aluminum oxide (AAO) nano-porous membrane on both sides of aluminum sheet," in *ICMEE 2010 - 2010 2nd International Conference on Mechanical and Electronics Engineering, Proceedings*, 2010.

[10] Knörnschild G, Poznyak A.A, Karoza A.G, and Mozalev A 2015 Surface & Coatings Technology Effect of the anodization conditions on the growth and volume expansion of porous alumina films in malonic acid electrolyte *Surf. Coat. Technol.*, **275** 17–25
[11] Md Jani A.M, Losic D, and Voelcker N.H, 2013 Nanoporous anodic aluminium oxide: Advances in surface engineering and emerging applications *Prog. Mater. Sci.*, **58**(5) 636–704
[12] Zaraska L, Brudzisz A, Wierzbicka E, and Sulka G.D, 2016 Electrochimica Acta The effect of electrolyte change on the morphology and degree of nanopore order of porous alumina formed by two-step anodization, *Electrochim. Acta*, **198** 259–267
[13] Rizkia V, Soedarsono J.W, Munir B, Suharno B, 2017 An Analysis of The Electrolyte Resistivity Effect on The Pore, *International Journal of Technology (IJTech-2017)* **8**(8) 1479–1488
[14] Lee, W. and Park, S.-J. 2014 Porous Anodic Aluminum Oxide: Anodization and Templated Synthesis of Functional Nanostructures, *Chemical Reviews* **114**(15) 7487–7556

Acknowledgement
This work was fully funded by Hibah Penelitian Dasar Unggulan Perguruan Tinggi Kementrian Riset Teknologi and Pendidikan Tinggi Republik Indonesia under the contract no. 364/UN2.R3.1/HKP05.00/2018.

Agent-based Model and its Potential in Simulating Some Physical Systems

S Viridi* and F Haryanto

Department of Physics, Institut Teknologi Bandung, Jalan Ganesha 10, Bandung 40132, Indonesia

*E-mail: dudung@fi.itb.ac.id

Abstract. Agent-based model (ABM) is very flexible in defining types of interaction between agents including non-physical ones, which can be used to simulate social interaction, economy, until city development. Only our abstraction in defining the interactions is the limitation. For a physical system, an agent can be considered as a particle and interaction between them is simply the interaction force, e.g. nuclear force between nucleons, electrostatic force between ions, gravitation force between asteroids, and contact force between granular particles. In this work, simple rule for the interaction between particle is proposed and several materials phases such as solid, liquid, gas, and granular can be obtained, but only qualitatively.

Keywords: simulation, agent-based model, physical system, system of particles.

1. Introduction

Since its birth agent-based model (ABM) has already been used for various types of system, such as impact of out-of-stocks and supply chain design in manufactures [1], multi-lane traffic behavior [2], profit and risk for offshore hake trawl industry [3], the impact of antibiotic use on transmission of resistant bacteria in hospitals [4], and prediction of a bioeconomy system [5]. One of the advantages of ABM is its flexibility in defining interactions between agents, which is used in this work to obtained various materials phases.

2. Model

Agents or particles are placed in a rectangular grid, where one particle can only occupy one cell in the grid. Visual representation of the grid is shown in figure 1 (left), where its matrix representation is given in figure 1 (right). An empty cell will have the value of 0, while an occupied cell has the value of 1, and the wall is with the value of –1. There area eights agents or $N = 8$ in figure 1.

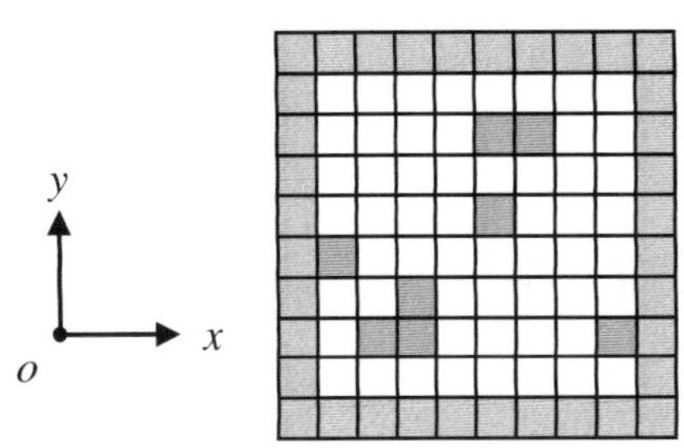

$$\begin{bmatrix} -1 & -1 & -1 & -1 & -1 & -1 & -1 & -1 & -1 & -1 \\ -1 & 0 & 0 & 0 & 0 & 0 & 0 & 0 & 0 & -1 \\ -1 & 0 & 0 & 0 & 0 & 1 & 1 & 0 & 0 & -1 \\ -1 & 0 & 0 & 0 & 0 & 0 & 0 & 0 & 0 & -1 \\ -1 & 0 & 0 & 0 & 0 & 1 & 0 & 0 & 0 & -1 \\ -1 & 1 & 0 & 0 & 0 & 0 & 0 & 0 & 0 & -1 \\ -1 & 0 & 0 & 1 & 0 & 0 & 0 & 0 & 0 & -1 \\ -1 & 0 & 1 & 1 & 0 & 0 & 0 & 0 & 1 & -1 \\ -1 & 0 & 0 & 0 & 0 & 0 & 0 & 0 & 0 & -1 \\ -1 & -1 & -1 & -1 & -1 & -1 & -1 & -1 & -1 & -1 \end{bmatrix}$$

Figure 1. A system of particles in a 10×10 grid constrained by a surrounding wall and consisted of eight agents: a visual representation (left) and its matrix representation (right).

An agent is allowed to move in eight directions as shown in table 1, where each direction is numbered as shown.

Table 1. Allowed moving direction for agents.

Direction	Abbreviation	Number
↑	NO (north)	0
↗	NE (northeast)	1
→	EA (east)	2
↘	SE (southeast)	3
↓	SO (south)	4
↙	SW (southwest)	5
←	WE (west)	6
↖	NW (northwest)	7

At time t an agent is currently staying at cell (i, j) and will move to another cell $(i + di, j + dj)$, if the destination cell is empty or not a wall cell, otherwise, it will remain at the same position. A random number z is generated from a uniform random function f_U, which has a probability density function of

$$f_U(z) = \begin{cases} \dfrac{1}{b-a}, & a \le z < b, \\ 0, & z < a \cup z \ge b, \end{cases} \tag{1}$$

where, in this work, it is set that $a = 0$ and $b = 1$. Since there are eight directions, eight probability values are required, one for each direction. Using table 1 we can define p_0 until p_7, that should meet the condition of normalization

$$\sum_{k=0}^{7} p_k = 1 . \tag{2}$$

Value of p_k means the probability of an agent to take the direction k. According to equation (1), generated value z will be $0 \le z < 1$, the value of l will be produced through

$$l = \begin{cases} 0 \le z < \sum_{k=0}^{l} p_k, & l = 0, \\ \sum_{k=0}^{l-1} p_k \le z < \sum_{k=0}^{l} p_k, & 1 \le l \le 7, \end{cases} \tag{3}$$

where l is simply the direction number in table 1. In region consisted of some cells, we can define a probability vector P

$$P = \begin{bmatrix} p_0 & p_1 & p_2 & p_3 & p_4 & p_5 & p_6 & p_7 \end{bmatrix} \tag{4}$$

which represents the probability of an agent to move in the eight allowed directions. In the system of $G_y \times G_x$ grid (figure 1 is or $G_y = 10$ and $G_x = 10$) we can have several regions with different probability vector. Assume that $G_y = G_x = G$ then there will be $H_y \times H_x$ region with $H_y = H_x = H$, where

$$G < H \tag{5}$$

and

$$\frac{G}{H} = \left\lfloor \frac{G}{H} \right\rfloor. \tag{6}$$

System direction probability will be in the form of a probability matrix

$$\mathbf{P}_{1\times1} = [P] \tag{7}$$

for $G = 1$,

$$\mathbf{P}_{2\times2} = \begin{bmatrix} P_0 & P_1 \\ P_2 & P_3 \end{bmatrix} \tag{8}$$

for $G = 2$, and until

$$\mathbf{P}_{5\times5} = \begin{bmatrix} P_0 & P_1 & P_2 & P_3 & P_4 \\ P_5 & P_6 & P_7 & P_8 & P_9 \\ P_{10} & P_{11} & P_{12} & P_{13} & P_{14} \\ P_{15} & P_{16} & P_{17} & P_{18} & P_{19} \\ P_{20} & P_{21} & P_{22} & P_{23} & P_{24} \end{bmatrix} \tag{9}$$

for $G = 5$. Equation (7) if for the homogeneous region where the system has only one type of direction probability and equation (8) can accommodate the inhomogeneous region with various types of direction probability.

3. Results and discussion

Several types of direction probability are tested and the results are discussed. The system is specified with $G = 100$ and $N \approx 1681$ or less (since it is randomly generated in 41×41 cells with $N_0 = 10000$ as the initial state).

3.1. Single phase system (G = 1)

There are four types of system phase that can be obtained using $\mathbf{P}_{1\times1}$, which are solid, liquid, gas, and granular

$$\mathbf{P}_{1\times1} = [P_{sp}], \tag{10}$$

with probability vector single phase P_{sp} are

$$P_{\text{Sol}} = [0.000 \quad 0.000 \quad 0.000 \quad 0.000 \quad 1.000 \quad 0.000 \quad 0.000 \quad 0.000], \tag{11}$$

$$P_{\text{Liq}} = [0.060 \quad 0.110 \quad 0.110 \quad 0.125 \quad 0.250 \quad 0.125 \quad 0.110 \quad 0.110], \tag{12}$$

$$P_{\text{Gas}} = [0.125 \quad 0.125 \quad 0.125 \quad 0.125 \quad 0.125 \quad 0.125 \quad 0.125 \quad 0.125], \tag{13}$$

$$P_{\text{Gra}} = [0.000 \quad 0.000 \quad 0.000 \quad 0.250 \quad 0.500 \quad 0.250 \quad 0.000 \quad 0.000]. \tag{14}$$

Implementation of equations (11) – (14) and the results are given in figure 2.

IOP Conf. Series: Materials Science and Engineering **599** (2019) 012008 doi:10.1088/1757-899X/599/1/012008

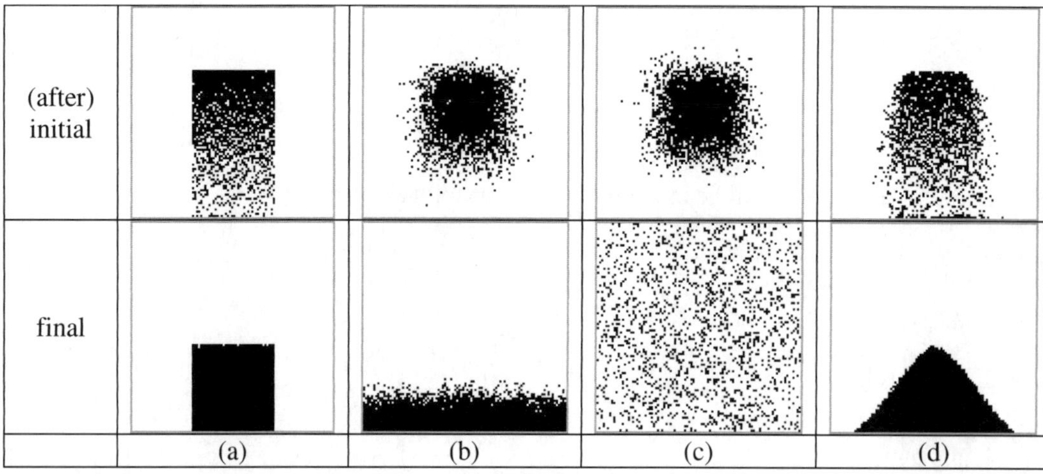

Figure 2. Single phase system (after) initial and final state for various types of materials (a) solid, (b) liquid, (c) gas, and (d) granular.

Different probability vector will give different final state as shown in figure 2. P_{Sol} can only allow agent to move according to the direction of gravitation acceleration $\vec{g}$, P_{Liq} has most direction probability to the direction of gravitation acceleration $\vec{g}$ and still has for the other directions, P_{Gas} has an isotropic direction probability to the eight directions, while P_{Gra} has only three direction probabilities, which are NE, NO, and NW.

3.2. System with G = 2

In this part, a system consists of four regions, where each region can have its own probability vector P_{dir} which could be in the form of P_{NO}, P_{NE}, P_{EA}, P_{SE}, P_{SO}, P_{SW}, P_{WE}, P_{NW}, whose value are given in following table 2. In the table probability vector for isotropic direction, P_{IS} is the same as P_{Gas} in the previous part of this work

Table 2. Defined probability vector P_{dir} for eight directions

Dominant direction	Symbol	Value
↑	P_{NO}	[0.650 0.050 0.050 0.050 0.050 0.050 0.050 0.050]
↗	P_{NE}	[0.050 0.650 0.050 0.050 0.050 0.050 0.050 0.050]
→	P_{EA}	[0.050 0.050 0.650 0.050 0.050 0.050 0.050 0.050]
↘	P_{SE}	[0.050 0.050 0.050 0.650 0.050 0.050 0.050 0.050]
↓	P_{SO}	[0.050 0.050 0.050 0.050 0.650 0.050 0.050 0.050]
↙	P_{SW}	[0.050 0.050 0.050 0.050 0.050 0.650 0.050 0.050]
←	P_{WE}	[0.050 0.050 0.050 0.050 0.050 0.050 0.650 0.050]
↖	P_{NW}	[0.050 0.050 0.050 0.050 0.050 0.050 0.050 0.650]
✳	P_{IS}	[0.125 0.125 0.125 0.125 0.125 0.125 0.125 0125]

There are to examples of implementation of probability vectors P_{dir} from table 2 as shown in figure 3.

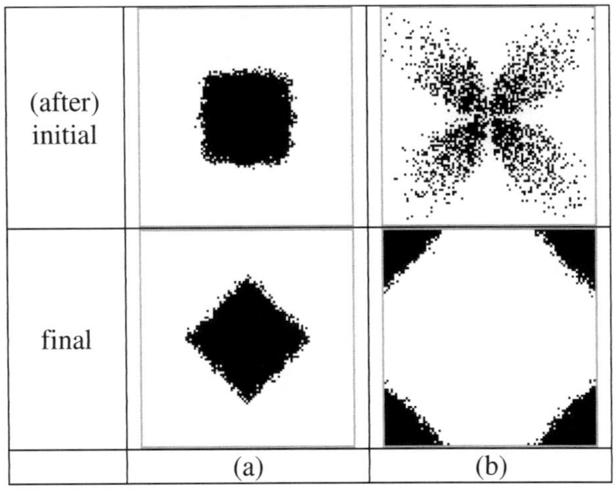

Figure 3. A system with $G = 2$ (after) initial and final state for fluid-like materials: (a) to center and (b) to corner flow.

Probability matrices for figure 3 are

$$\mathbf{P}_{2\times 2}^{\text{to center}} = \begin{bmatrix} P_{\text{SE}} & P_{\text{SW}} \\ P_{\text{NE}} & P_{\text{NW}} \end{bmatrix} \tag{15}$$

and

$$\mathbf{P}_{2\times 2}^{\text{to corner}} = \begin{bmatrix} P_{\text{NW}} & P_{\text{NE}} \\ P_{\text{SW}} & P_{\text{SE}} \end{bmatrix} \tag{16}$$

To center flow will constraint agents in the center region, while to corner flow will bring about a quarter of agents to each corner. Agents can be trapped in the corner since there are two walls constructing a corner and probability vector in the region is toward the corner.

3.3. System with G = 3

There are three systems with $G = 3$, where their probability matrices are given as follow

$$\mathbf{P}_{2\times 2}^{\text{two center}\backslash} = \begin{bmatrix} P_{\text{SE}} & P_{\text{WE}} & P_{\text{SW}} \\ P_{\text{NO}} & P_{\text{IS}} & P_{\text{SO}} \\ P_{\text{NE}} & P_{\text{EA}} & P_{\text{NW}} \end{bmatrix}, \tag{17}$$

$$\mathbf{P}_{2\times 2}^{\text{two center}/} = \begin{bmatrix} P_{\text{SE}} & P_{\text{EA}} & P_{\text{SW}} \\ P_{\text{SO}} & P_{\text{IS}} & P_{\text{NO}} \\ P_{\text{NE}} & P_{\text{WE}} & P_{\text{NW}} \end{bmatrix}, \tag{18}$$

and

$$\mathbf{P}_{2\times 2}^{\text{one center}} = \begin{bmatrix} P_{\text{SW}} & P_{\text{SO}} & P_{\text{SW}} \\ P_{\text{EA}} & P_{\text{IS}} & P_{\text{WE}} \\ P_{\text{NE}} & P_{\text{NO}} & P_{\text{NW}} \end{bmatrix}. \tag{19}$$

Results from equations (17) – (19) are given in figure 4.

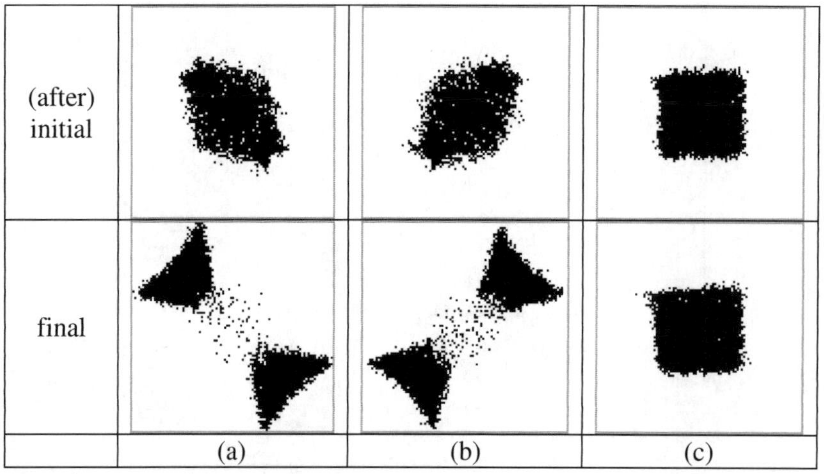

Figure 4. System with G = 3 (after) initial and final state for fluid-like materials: (a) two center \, (b) two center / and (c) one center.

Figure 4(a) and 4(b) are mirrored results to each other, where agents are segregated to opposite corners, while in figure 4(c) agents are confined in the center of the system.

3.4. System with G = 4

Four systems with $G = 4$ are observed, whose probability matrices are given as follow

$$\mathbf{P}_{4\times4}^{(a)} = \begin{bmatrix} P_{NW} & P_{NO} & P_{NO} & P_{NE} \\ P_{WE} & P_{IS} & P_{IS} & P_{EA} \\ P_{WE} & P_{IS} & P_{IS} & P_{EA} \\ P_{SW} & P_{SO} & P_{SO} & P_{SE} \end{bmatrix}, \tag{20}$$

$$\mathbf{P}_{4\times4}^{(b)} = \begin{bmatrix} P_{SE} & P_{SO} & P_{SO} & P_{SW} \\ P_{WE} & P_{IS} & P_{IS} & P_{EA} \\ P_{WE} & P_{IS} & P_{IS} & P_{EA} \\ P_{NE} & P_{NO} & P_{NO} & P_{NW} \end{bmatrix}, \tag{21}$$

$$\mathbf{P}_{4\times4}^{(c)} = \begin{bmatrix} P_{NE} & P_{EA} & P_{EA} & P_{SE} \\ P_{NO} & P_{NW} & P_{NE} & P_{SO} \\ P_{NO} & P_{SW} & P_{SE} & P_{SO} \\ P_{NW} & P_{WE} & P_{WE} & P_{SW} \end{bmatrix}, \tag{22}$$

$$\mathbf{P}_{4\times4}^{(d)} = \begin{bmatrix} P_{SW} & P_{NW} & P_{NE} & P_{SE} \\ P_{SO} & P_{NO} & P_{NO} & P_{SO} \\ P_{SO} & P_{NO} & P_{NO} & P_{SO} \\ P_{SE} & P_{NE} & P_{NW} & P_{SW} \end{bmatrix}, \tag{23}$$

Results from equations (20) – (23) is given in figure 5.

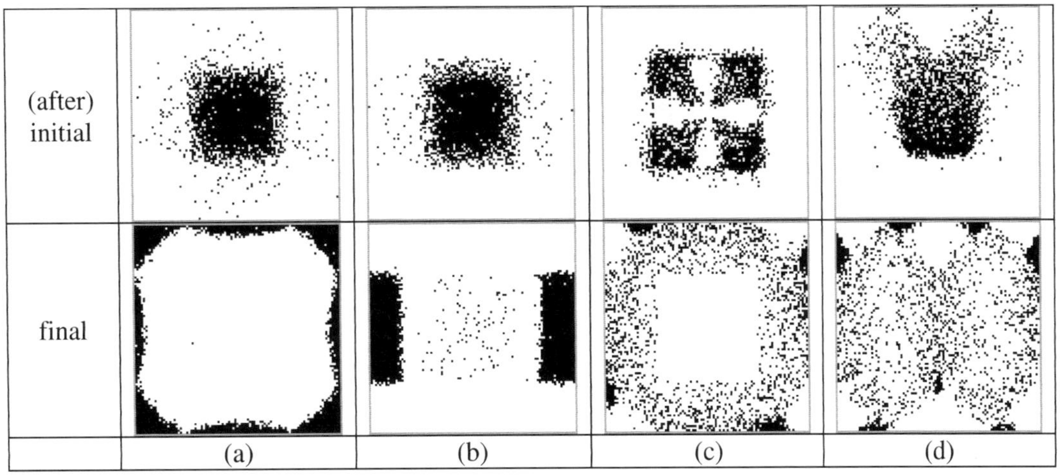

Figure 5. System with G = 4 (after) initial and final state for fluid-like materials: (a) outside deposition, (b) horizontal deposition, (c) single circular flow with center-out, and (d) double circular flow with center-down.

3.5. Discussion

In single phase system (system with $G = 1$), where the probability vector in all cells has the same value, four phases of materials, i.e. solid, liquid, gas, and granular, can be produced qualitatively. Then, by introducing regions with different vector probability ($G > 1$) richer phases of fluid-like materials can be obtained such as center- and corner-deposition ($G = 2$), two- and one-cluster agglomeration ($G = 3$), and outside- and horizontal-deposition, and single- and double-circular flow ($G = 4$). Further investigation with a higher value of G, e.g. $G > 4$, could lead to other interesting phases. From the results it can be said that this method has a very good potential in simulating a physical system, but with the physical reason why a probability vector P should have such value. It would be better when P can be derived from interactions between agents and also between agents and their environment.

4. Summary

Some materials phases, e.g. solid, liquid, gas, granular, deposition, flow, have been simulated using ABM for a single- and multi-region system. This approach can show how initial state evolving to final stable static or dynamics state, but only qualitatively due to the absence of the connection between interaction form and probability vector.

5. References

[1] Rosales C, Whipple JM, Blackhurst J 2018 *Transp. J.* **57** 137
[2] Sugihakim R and Alatas H 2016 *Phys. Lett. A* **380** 147
[3] Cooper R, Jarre A 2017 *Ecol. Econ.* **142** 257
[4] Almagor J, Temkin E, Benenson I, Fallach N, Carmeli Y 2018 *PLoS One* **13** e0197111
[5] Viridi S, Premadi P, Aditiawati P, Maqdir ES, Suheri T, Halid J, Sari KN, Pasaribu US, Sudaryani NM, Latifah N, Rahimah S 2019 IOP *Conf. Ser. Earth Environ. Sci.* **230** 012118

Acknowledgments

Authors wishing to acknowledge to KK Fismatel and SNM committee for invitation and opportunity in presenting this work.

IOP Conf. Series: Materials Science and Engineering **599** (2019) 012009 doi:10.1088/1757-899X/599/1/012009

Analysis of Damage Area of Fiberglass/Polyester Bi-Panel with Rice Husk Ash as a Filler Caused by the Impact of High-Speed Particles

Nazarudin Koto[1,2] and Bambang Soegijono[1*]

[1]Dept. of Physic, Universitas Indonesia, Depok 16424, Indonesia
[2]Research and Development Defence Ministry, Jl Jati No.1st. Pondok Labu, South Jakarta 12450, Indonesia

*Corresponding author: naufal@ui.ac.id

Abstract. Analysis of the damaged area of the composite bi-panel fiberglass/polyester with the addition of rice husk ash as a filler have been presented in this paper. The area of damage is caused by the impact of high-speed particles. Bi-panel composites were made by hand lay-up of S-glass and unsaturated polyester resin BTQN 157 ex with variations of 3, 5, 7, 9 %v rice husk ash as a filler. The impact test were carried out using a 9 mm caliber FN gun in accordance with NIJ level IIIA standards. The results show the area of damage caused by a projectile decrease with the increasing amount of Rice husk %v. There is a significant effect with the addition of rice husk ash filler on the composite of bi-panel fiberglass/polyester to the impact of high-speed particles. The rice husk ash could be used as an alternative for filler in Polyester/Fiberglass composite for body armor application.

Keywords: bi-panel, fiberglass/polyester, rice husk ash, impact,

1. Introduction

Some previous studies have reported the use of metal, ceramic, polymer, natural fiber or composite as body armor materials. But the researches on this material are still ongoing on searching the lightest, the cheapest, environmentally friendly and also the geometry and the number of layer [1]. Tasdemirci et al. (2012) reported the results of research on testing aluminum foam and Teflon composites without inserts and which were inserted by 3 rubber which was carried out using AP Projectile type M61 at a shooting distance of 15 m, experiencing an increase in kinetic energy that was almost the same [2]. Composites given inserts cause delayed kinetic energy that arises and drastically reduces kinetic energy in both models [3-5]. Ritonga (2014) reported that the addition of Hollow Glass Microsphere (HGM) volume fraction of 16% in epoxy showed a maximum toughness value of 21.54 x 10-3J / mm3 [6]. Maples et al. (2014), reported that the compressive and tensile properties that occur in carbon fiber reduced significantly at 120 °C due to the presence of softened polystyrene interleaves. The results of the flexural strength test at 20°C indicate that there is a need for improved adhesion between polystyrene and fiber-carbon with epoxy [7]. It has also been investigated that the average absorption of epoxy matrix kinetic energy with HGM epoxy reinforcement for vehicle front bumper is 86.39% [8]. Meanwhile, the use of ballistic devices with a thickness of 20 mm, capable of absorbing bullet energy is smaller than 170 Joules [9-10].

The existing body armor is still developing to enhance its performance. In addition to the reliability of the body armor such as opaque and capable of providing protection from projectile impact, there are problems that arise when applied to objects that move like a soldier and his combat vehicle. To answer the issue of this body armor, research needs to be done. The material chosen for this composite body armor is a matrix using polyester because it is lightweight, economical and environmentally friendly. Furthermore, the reinforcement uses fiberglass because it increases mechanical strength and is more economical than fiber-kevlar or other aramid fibers which have always been used for making body armor composites.

This objective of this study is to obtain alternative materials other than Kevlar fiber or other aramid fibers which are relatively expensive and imported from abroad. For this reason, it is necessary to try the use of fiberglass as reinforcing fiber and polyester as a matrix and the use of rice husk ash filler to obtain an armored body where all the materials have been produced in Indonesia.

2. Experimental procedures

2.1. Materials

Polyester and fiberglass supplied from PT Justus Sakti in Jakarta Indonesia. Rice husk ash was manually prepared using a table loom. The density of the rice husk ash is 1.40 g/m^3. The parameters of rice husk ash are provided. The resin used in this study is unsaturated polyester BQTN 157 with a density of 1.08 g/m^3.

2.2. Fabrication of composite laminates

The composite was manufactured in the form panel with fiberglass, rice husk ash as a filler and polyester resin as a matrix. The constant fraction volume of fiberglass used in this study was a woven roving S-Glass denier 800 g/cm^2, which was taken from the supplier's data sheet. The resin was cured using methyl ethyl ketone peroxide (MEKP) is an organic peroxide. Rice husk ash supplied from the results of processing themselves. Panel laminated composites of (20 x 20 x 0.4) cm size were used for ballistic testing. The hand-lay-up method was used to fabricate laminates of woven roving fiberglass and rice husk ash as a filler in polyester resin. The samples consisted of 7-layers of woven roving fiberglass in the same direction (0/90°). The woven roving fiberglass fabrics were hand laid-up with the polyester + rice husk ash matrix by mixing resin and MEKP catalyst in the ratio of 5%. Two thick mild steel plates were used as a mold (20 cm x 20 cm) in the fabrication process. The composites were cured by applying compression pressure using load which is equal 150 Kg on the top of the mold and cured at room temperature for 2 days.

2.3. Ballistic testing

The experiment consists of 2 variations where each variation consist of two panels as shown in Figure 1. Each variation consists of 4 composite bi-panel specimens. These bi-panel specimens are stacked together without adhesives. The front panel is panel 1 and the back is panel 2. For variation 1 the front part is panel 1 with the composition without filler, and the rear part is panel 2 with the filler of 3, 5, 7, and 9 %v rice husk ash. Whereas variation 2 is the opposite of variation 1, the front is panel 1 with fillers of 3, 5, 7, and 9 %v rice husk ash and the back is panel 2 without filler. The composite bi-panel the configuration is shown in Figure 1.

The ballistic impact experiment was conducted using 9 calibers (diameter of 9 mm) fragment simulating projectiles (FSPs) with a weight of about 8.1 g. Each sample was impacted according to BA 9000 NIJ standards in 2012 on the requirements of body armor system (Type IIIA in armor level). The ballistic testing setup used is shown in Figure 1. Bullet speed is measured by Chronograph Prochrono Digital brand. The gun used for ballistic testing was the Belgian Fabrique Nationale (FN) 9 mm.

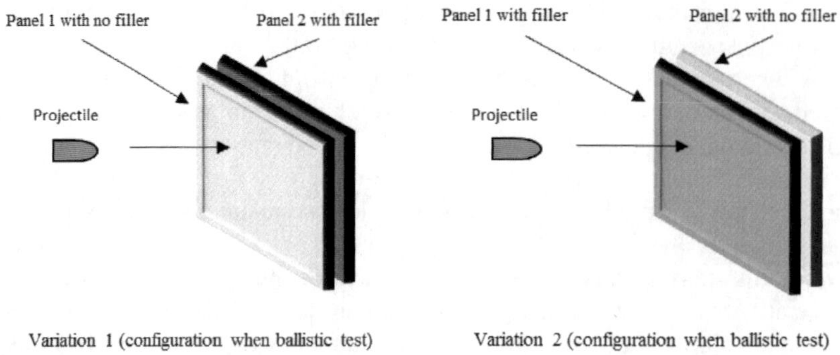

Figure 1. The configuration of variation 1 and 2 when ballistic testing.

3. Results and discussion

Table 1 shows the results of ballistic tests on variations 1 and 2. Variation 1 in specimens 1b and 3b show that the composite bi-panel is not punching until panel 2. However, specimens 2b and 4b penetrate until panel 2. In variation 2 for specimens 1a, 2a, 3a, and 4a show the restrained projectile in panel 1, and panel 2 only shows the matrix crack. So variation 2 shows good results for holding the impact of bullet projectiles. Panel 1 without filler did not show a low performance for the body armor panel because all panels were punching, this was proven in variation 1. Although in variation 2, panel 2 without filler was not punching but panel 1 showed a significant area of damage. The rice husk ash as a filler has an influence on the performance of bi-panels composite on the impact of bullet projectiles.

Table 1. Ballistic test results of composite bi-panel specimens on variations 1 and 2

Variation	Specimen codes	Arrangements of bi-panel with and without filler		Condition after the impact of the projectile	
		Panel 1	**Panel 2**	**Panel 1**	**Panel 2**
1	1b	0 %v	3 %v	punch	restrained
	2b	0 %v	5 %v	punch	punch v_f: 250 m/s
	3b	0 %v	7 %v	restrained	matrix crack
	4b	0 %v	9 %v	punch	punch v_f: 230 m/s
2	1a	3 %v	0 %v	restrained	matrix crack
	2a	5 %v	0 %v	restrained	matrix crack
	3a	7 %v	0 %v	restrained	matrix crack
	4a	9 %v	0 %v	restrained	matrix crack

3.1. The density of the composite

Figure 2 shows the composite with the addition of the fraction of the volume of rice husk ash will reduce the weight of the composite panel. The polyester-fiberglass composite panel shows the density of 1.613 g/cm^3. With the addition of filler 3 %v of rice husk ash, the density of composite panel mass decrease to 1.610 g/cm^3. Likewise, with the addition of 5 %v, the density is 1.609 g/cm^3. Furthermore, the addition of 7 %v. of rice husk ash, the density decreased again to 1.606 gram/cm^3. Then in the volume fraction of 9 %v, the rice husk ash tends to decrease to 1.604 g/cm^3. It turns out that in general the density decreases compared to the panel without filler that is visible. The ballistic test was a double-panel composite with 9 %v. filler shows the most amazing performance. It turns out that by increasing the fraction of the volume of rice husk ash as a filler can improve the higher performance of the double-panel composite for the body armor function.

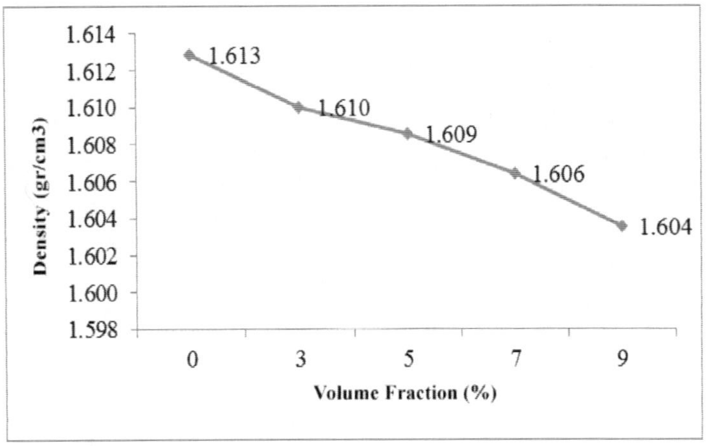

Figure 2. Comparison of bi-panel composite density with variations in volume fraction of rice husk ash filler.

3.2. The ballistic test of variation 1

Figure 3 shows the deformation of the projectile after impact. The projectile with an original diameter of 9 mm became flat after the test (diameter of 20 mm). Based on projectile data it will break and break if it hits a hard surface. Based on the fact that the projectile was held in panel 2 after penetrating panel 1.

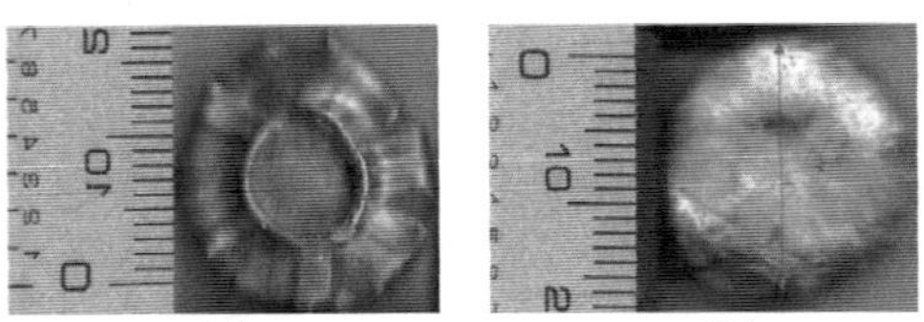

(a) (b)

Figure 3. The projectile after impact was deformed (a) rear view (b) front view.

Figure 4 shows the area of damage that occurs in the bi-panel composite specimen 1b. The specimen 1b arrangement is the first panel without filler and the second panel 3 %v filler. In panel 1 in the front, the former projectile hole is 10 mm wide. The back of the panel 1 break fiber with a hole due to the impact of a 20 mm wide projectile. Then the projectile is held in panel 2 by showing the area of the matrix damage to be wide with a width of 70 mm, and the back of panel 2 is not damaged. The bullet projectile penetrates panel 1 and is then held in panel 2. Panel 1 with filler 3 %v has not been able to hold the projectile. It turns out that panel 2 has a contribution to holding back the rate of projectiles that have experienced a slowdown in panel 1.

Figure 5 shows the results of the ballistic test in specimen 2b with a filler composed of 5 %v of rice husk ash. The specimen 2b arrangement is the first panel without filler and the second panel 5 %v filler. After the ballistic test shows the entire panel on this bi-panel composite were punched by the bullet projectiles. The projectile speed after penetrating both panels is 250 m/Sec from 380 m/Sec. (Table 1). The damage pattern is shown as in Figure 5, with the damaged area of panel 1 in the form of a used projectile hole and there is a small matrix damage. This indicates very little energy distribution to all fiberglass. Likewise, the same thing was shown by fiberglass panel 2 as reinforced which did not respond to kinetic energy. In this case, the filler distributed in the panel is less than optimal for preventing projectile impact.

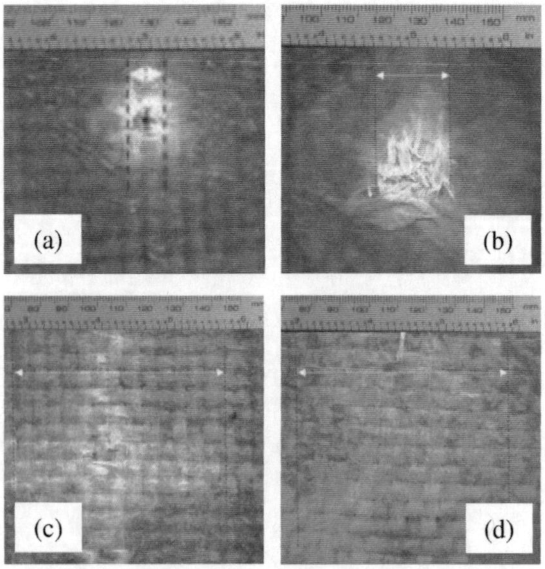

Figure 4 Specimen 1b showing the damaged area in bi-panel composite consists of panel 1 without filler and panel 2 with 3 %v rice husk ash. Front view (a) and back view (b) of panel 1. Front view (c) and back view (d) of panel 2.

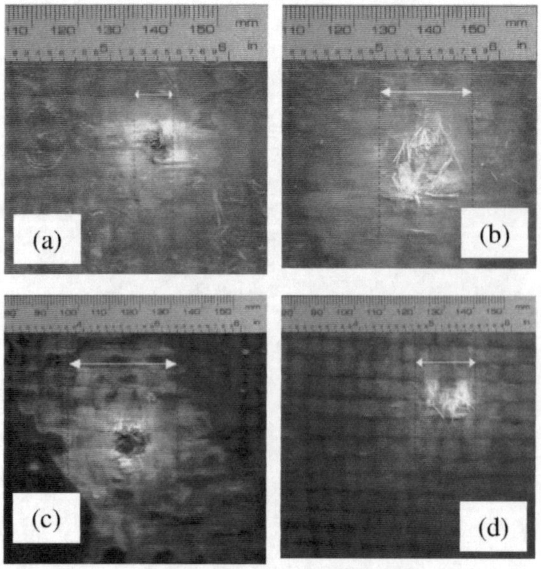

Figure 5 Specimen 2b showing the damaged area in bi-panel composite consists of panel 1 without filler and panel 2 with 5 %v rice husk ash. Front view (a) and back view (b) of panel 1. Front view (c) and back view (d) of panel 2.

Figure 6 shows the damaged area that occurred in bi-panel specimens 3b with a filler composed of 7 %v of rice husk ash. The specimen 3b arrangement is the first panel without filler and the second panel 7 %v filler. In panel 1 in the front, there was a 9 mm wide projectile hole. The back of panel 1 is a damaged matrix of 27 mm wide and punching in Figure 6b. Next panel 2 shows the front the damaged area of the matrix with a width of 18 mm and the rear panel 2 punching with a width of 13 mm in Figure 6d. The ability of panels 1 and 2 to impact bullet projectiles is not good. This is caused by the percentage of filler is smaller and there is a wider distance (space) between panels 1 and 2 than recommended (below 1 mm). It is estimated that there are production defects when making composites. After the ballistic test shows the 1st panel is not penetrated by the bullet projectile and is held in the panel. Next panel 2 shows the basin due to the impact of deformed projectiles as shown in Figure 6.

Figure 7 shows the damaged area that occurs in the bi-panel composite specimen 4a without filler in panel 1 and 9 %v filler in panel 2. In panel 1 in the front, there is a 10 mm wide projectile hole. The back of the panel 1 breaks fiber with a hole due to the impact of a 20 mm wide projectile. Then the projectile is held in panel 2 by showing the area of the matrix damage to be wide with a width of 70 mm, and the back of panel 2 is not damaged. Figure 7 shows the damaged area that occurred in bi-panel 4B specimens of 0 %v filler in panel 1 and 9 %v filler in panel 2. In panel 1 in the front, there is a 9 mm wide projectile hole. The back of the panel 1 is a 27 mm wide and translucent matrix damage as shown in Figure 7b. Next, panel 2 shows the area of front-facing matrix damage with a width of 18 mm. The back panel 2 is punching with broken fiberglass broken matrix with a width of 13 mm in Figure 7d.

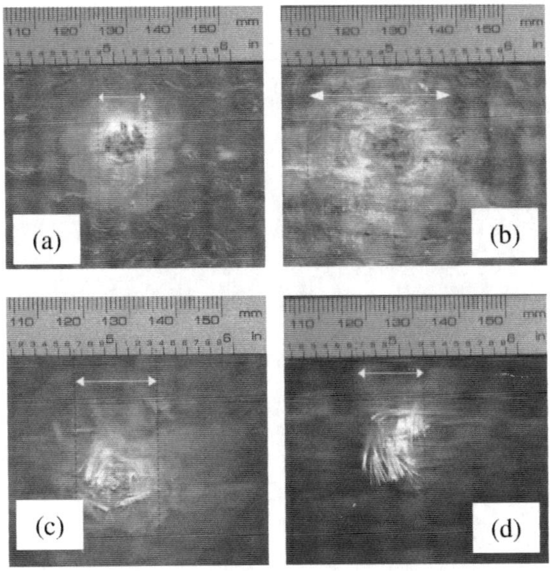

Figure 6 Specimen 3b showing the damaged area in bi-panel composite consist of panel 1 without filler and panel 2 with 7 %v rice husk ash. Front view (a) and back view (b) of panel 1. Front view (c) and back view (d) of panel 2.

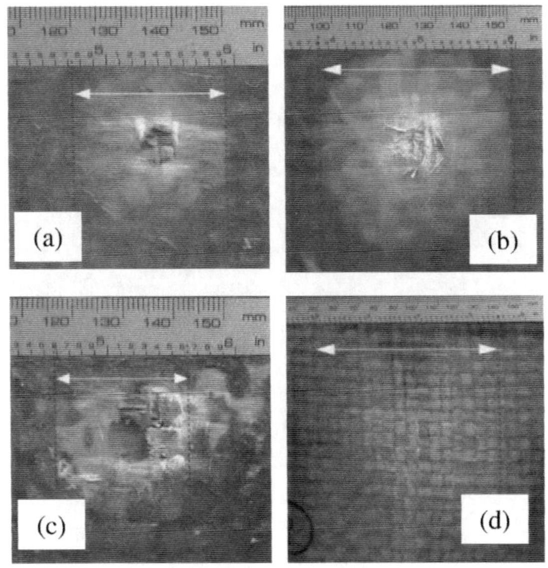

Figure 7 Specimen 4b showing the damaged area in bi-panel composite consists of panel 1 without filler and panel 2 with 9 %v rice husk ash. Front view (a) and back view (b) of panel 1. Front view (c) and back view (d) of panel 2.

This specimen 4b has the ability to resist projectile impact. There is an effect of 9 %v rice husk ash filler on impact ability. The bi-panel composition also affects the panel's impact ability. The wider area of the matrix damage indicates kinetic energy is distributed evenly throughout the direction and layer of the fiber. Specimen 4b shows that the impact ability is not as good as the specimen 3b. This was caused by defects in production when manufacturing composites. There is a larger space than the bi-panel recommendation. It is suspected that there is a defect in the laminate in the form of uneven resin on the entire surface of the fiberglass and the presence of air bubbles trapped inside the panel laminate. The uneven mixture of polyester resin with rice husk ash filler. The overall capability of panel 1 is still below panel 2 because there are still production defects.

3.3. The ballistic test of variation 2

Ballistic test results in specimens 1a in panel 1 composition with 3 %v rice husk ash filler and panel 2 without filler. The test results showed panel 1 succeeded in holding the bullet projectile and was stuck on the panel. Panel 2 does not suffer significant damage, only the matrix breaks. Overall the results of the ballistic test and the damage that occur can be seen in Figure 8. The damage area that occurs in bi-panel specimens 1a 3 %v filler in panel 1 and 0 %v filler in panel 2. In panel 1 the front is visible due to projectile impact 9 mm wide. The back of panel 1 damage the matrix with a width of 17 mm, there are some broken fiberglass parts and the panel has not penetrated as shown in Figure 8b. Furthermore, panel 2 shows the frontal matrix damage area with a width of 18 mm and the rear panel 2 there is no fiberglass damage but there are matrix cracks with a width of 40 mm in Figure 8d. The ability of specimens 1a is better than previous specimens. There is a filler effect of rice husk ash and a bi-panel arrangement of the ability to resist impact. Bi-panel capability increases by changing the position of the panel arrangement as in specimen 1b.

Figure 9 shows the area of damage that occurs in bi-panel specimens 2a 5 %v filler in panel 1 and without filler in panel 2. In panel 1 the front is visible hole due to the impact of projectiles with a width of 9 mm. The back of panel 1 damage the matrix with a width of 60 mm, there are several broken fiberglass parts and the panel has not penetrated. Next panel 2 shows the front area matrix damage area with a slight width of 10 mm and the rear panel 2 there is no fiberglass damage but there are matrix cracks with the pattern as in Figure 9d. The area of damage in panels 1 and 2. Fiberglass is unable to hold projectiles and breaks. The holes that occur to the rear surface of panel 1 with a diameter of 9 mm and the back of 15 mm. In panel 1, the more dominant damage is broken fiberglass breaks and fewer matrix cracks, this indicates the panel does not work better and there is no kinetic energy distribution in this panel. Likewise in panel 2, the same thing is shown as in panel 1 with fewer matrix damage areas and broken fiberglass by forming a hole in panel 2.

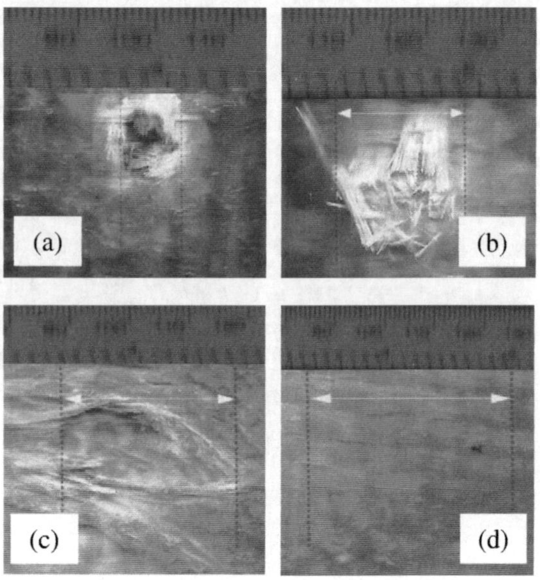

Figure 8 Specimen 1a showing the damaged area in bi-panel composite consists of panel 1 with filler 3% wt rice husk ash and panel 2 without filler Front view (a) and back view (b) of panel 1. Front view (c) and back view (d) of panel 1.

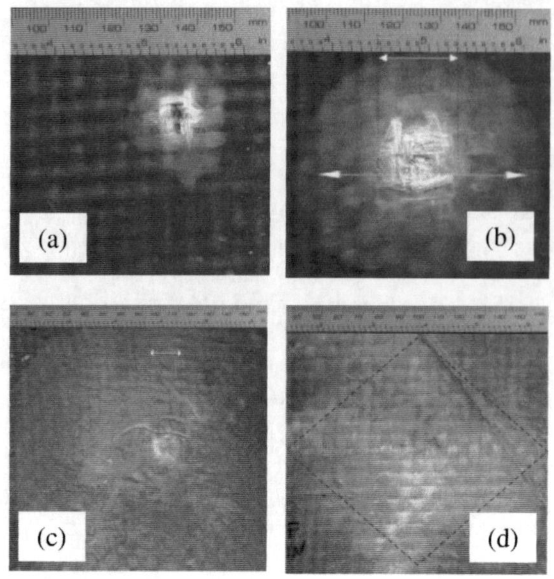

Figure 9 Specimen 2a showing the damaged area in bi-panel composite consists of panel 1 with filler 5% wt rice husk ash and panel 2 without filler. Front view (a) and back view (b) of panel 1. Front view (c) and back view (d) of panel 1.

Figure 10 is the result of ballistic tests on specimens 3a with a filler composition of 7 %v of rice husk ash. The first-panel arrangement with 7 %v filler and second panel without filler. The ballistic test results showed panel 1 was successful in holding the bullet projectile and was stuck on the panel. Panel 2 does not suffer significant damage, only the matrix breaks. The area of damage that occurs in bi-panel specimens 3a in panel 1 of the front is visible holes due to the impact of projectiles with a width of 9 mm. The back of panel 1 damage the matrix with a width of 60 mm, there are several broken fiberglass parts and the panel has not penetrated as shown in Figure 11b. Furthermore, panel 2 shows the front area matrix damage area with a little width of 10 mm and the rear panel 2 does not have fiberglass damage but there is a matrix with the pattern as shown in Figure 11d.

Figure 11 is the result of ballistic tests on specimens 4a with a filler composition of 9 %v of rice husk ash. The first-panel arrangement with 9 %v filler and the second panel without filler. After the ballistic test, the bullet projectiles were embedded in the first panel and the second panel only cracked the matrix. The area of damage that occurs in bi-panel specimens 4a is in panel 1 the front is visible hole due to the impact of projectiles with a width of 9 mm. The back of panel 1 damage the matrix with a width of 60 mm, there are several broken fiberglass parts and the panel has not penetrated as shown in Figure 11b. Next panel 2 shows the front area matrix damage area with a little width of 10 mm and the rear panel 2 there is no fiberglass damage but there are matrix cracks with the pattern as in Figure 11d. The ability of specimens 4a to resist the impact is better than previous specimens. There is a filler effect of rice husk ash and a bi-panel arrangement of the ability to resist impact. Bi-panel capability in resisting the impact increases by changing the position of the panel arrangement as in the specimen 4b.

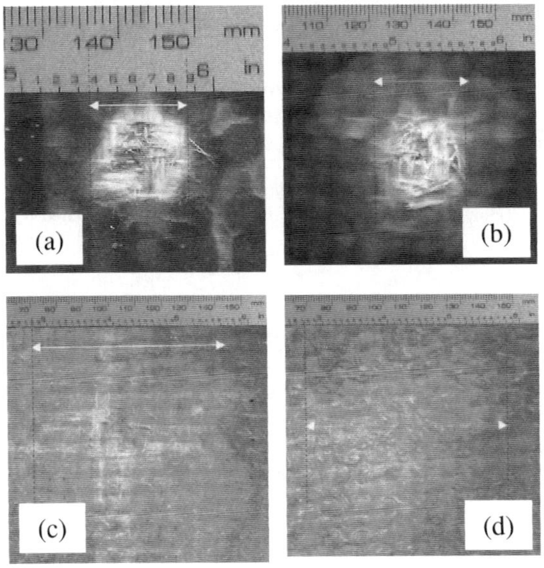

Figure 10 Specimen 3a showing the damaged area in bi-panel composite consists of panel 1 with filler 7 %v rice husk ash and panel 2 without filler. Front view (a) and back view (b) of panel 1. Front view (c) and back view (d) of panel 1.

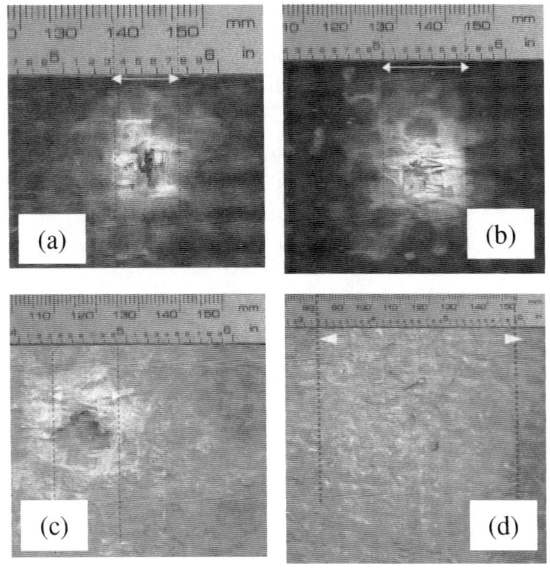

Figure 11 Specimen 4a showing the damaged area in bi-panel composite consists of panel 1 with filler 9 %v rice husk ash and panel 2 without filler. Front view (a) and back view (b) of panel 1. Front view (c) and back view (d) of panel 1.

4. Conclusion

Based on variations in rice husk ash fillers in bi-panel composites, the density becomes decrease with an increase in volume fraction. The damaged area in variation 1 indicates the effect of rice husk ash as a filler can resist the impact of projectiles. The damaged area gets smaller with the increase of filler volume fraction. In variation 2, the filler volume fraction increases become the performance of the composite bi-panel against projectile impact increase. The rice husk ash can affect the performance of bi-panel composites to be better as a body protector.

5. References

[1] Monteiro SN, Milanezi TL, Henrique L, Louro L, Lima ÉP, Braga FO, et al. 2016 Novel Ballistic Ramie Fabric Composite Competing with Kevlar TM Fabric in Multi-Layered Armour *Materials and Design* **96** 263–9

[2] Tasdemirci A, Tunusoglu G, Güden M 2012 The Effect of the Interlayer on the Ballistic Performance of Ceramic/Composite Armour: Experimental And Numerical Study *Int. J Impact Eng* **44** 1–9

[3] Department of Defense. MIL-DTL-46593B: projectile, calibre .22, .30, .50, and 20 mm fragment simulating. Maryland, USA: Department of Defense; 2006

[4] Joshi S.V., Drzal L.T., Mohanty A.K., Arora S. 2004 Are natural fiber composites environmentally superior to glass fiber reinforced composites? *Compos. A: Appl. Sci. Manuf.* **35** 371–376

[5] Anwar M, Miyagawa T, Gaweesh M, Using rice husk ash as a cement replacement material in concrete, The First International Conference on Ecological Building Structure, San Rafael, California 2001, pp. 671–684

[6] Ritonga, W., Pengaruh Variasi Fraksi Volume, Temperatur Curing dan Post-Curing Terhadap Karakteristik Tekan Komposit Epoxy-Hollow Glass Microspheres Impak. Indonesia. (2014)

[7] Habeeb G.A, Fayyadh M.M., 2009 Rice husk ash concrete: the effect of RHA average particle size on mechanical properties and drying shrinkage, *Aust. J. Basic Appl. Sci.* **3** 1616–1622

[8] Hamzeh Y, Ziabari K.., Torkaman J, Ashori A, Jafari M 2013 Study on the effects of white rice husk ash and fibrous materials additions on some properties of fiber-cement composites, *J. Environ. Manag.* **117** 263–267
[9] Torkaman J, Ashori A, Sadr Momtazi A., 2014 Using wood fiber waste, rice husk ash, and limestone powder waste as cement replacement materials for lightweight concrete blocks, *Constr. Build. Mater* **50** 432–436
[10] National Institute of Justice, U.S. Department of Justice 2000 Ballistic Resistance of Person Body Armour. Law Enforce Correct Stand Test Program (Washington: National Institute of Justice, Office of Science and Technology) pp. 1–67

Acknowledgment
The authors would like to show their appreciation Research and Development, Ministry of Defence and University of Indonesia for supporting this research activity.

Study on Heating Profile of Liquid-mixing-synthesized $ZnTiO_3$

F U Ermawati[1, *] and N A Imamah[1]

[1]Physics Department, Faculty of Mathematics and Natural Sciences, Universitas Negeri Surabaya, Surabaya 60231, Indonesia

*The Corresponding Author: frida.ermawati@unesa.ac.id

Abstract. The synthesis of monophasic $ZnTiO_3$ powder has been reported as one of the challenges in material chemistry because under normal condition and as prepared by various synthesis methods, $ZnTiO_3$ phase readily decomposes to Zn_2TiO_4 and TiO_2 rutile phases. This is due to the temperature range of $ZnTiO_3$ phase formation was very narrow, i.e. 600 - 800 °C; above that temperature, the decomposition of $ZnTiO_3$ phase took place. This paper is intended to introduce a facile route, named liquid mixing method, to prepare $ZnTiO_3$ powder from Zn and Ti metal powders as the raw materials and HCl as the solvent. Via this route, $ZnTiO_3$ phase was formed at a temperature as low as 550 °C with particle size distributed homogeneously. The heating profile of $ZnTiO_3$ phase from 440 to 900 °C was studied based on TG/DTA, XRD and FTIR data. However, when a 'normal' stoichiometric ratio of Zn: Ti = 1: 1 was applied, TiO_2 rutile phase becomes dominant, while $ZnTiO_3$ phase is only a secondary phase. This affects to the crystallinity of the $ZnTiO_3$ phase. An effort to overcome the dominance of TiO_2 phase was also demonstrated in this paper.

1. Introduction

The synthesis of ZnO-TiO_2 nano composite usually results in the formation of poliphasic compounds, $ZnTiO_3$ (hexagonal), Zn_2TiO_4 (cubic) and $Zn_2Ti_3O_8$ (cubic) along with other secondary impurity phases, such as TiO_2 rutile. It has been reported that obtaining a single phase of either $ZnTiO_3$ is one of the challenges in material fabrication [1-4]. Various synthesis methods have been reported to produce $ZnTiO_3$, including the conventional solid-state reaction [3], sol-gel processes [2,4], a modified alcoholysis method [1]. However, pure hexagonal ilmenite type $ZnTiO_3$ was not obtained by those methods, due to the reason described above. The product of $ZnTiO_3$ has been useful, such as for catalytic sorbents for the desulfuration of hot coal gases, for ethanol, NO and CO gas sensor materials, for microwave resonator materials, and more preferably for low temperature co-fired ceramic [2].

A relatively new approach in fabricating ceramic powders, named liquid mixing method, has been successfully synthesise nearly single phase of $MgTiO_3$-based ceramic powder at low temperature [5-7] and the fabricated $MgTiO_3$-based ceramics exhibited excellent dielectric characteristics [6,8-9]. The liquid mixing method is to dissolve metal powders, as raw materials, separately in strong acid solvents until consistent aqueous solutions are obtained, mixing all the solutions homogeneously and drying the slurry. The dried powder is calcined to obtain the desired crystalline phase. Metal powder that has a high solubility level in strong acid solvents is very beneficial when combined with this simple synthesis method to accelerate the intimate mixing process between the metalic ions. By this way, the expected crystalline phase can be formed at low calcination temperatures and nano-sized powders was obtained.

However, when multiple raw materials is required, the stoichiometry of the materials tends to be relatively complicated. Occasionally this could interfere the intimate reactions; as a result, impurity phase might also be detected.

In this paper, the liquid mixing method was adopted to synthesise $ZnTiO_3$ powders due to simplicity of the process and the nature of crystalline phase formation at low temperature. The aims was to study the heating profile of $ZnTiO_3$ phase from 440 to 900 °C, to investigate the crystallinity and distribution of particle size in the samples. This paper also presents an effort to overcome the dominance of TiO_2 phase during preparation of $ZnTiO_3$ phase.

2. Experiment

$ZnTiO_3$ powder was prepared following the liquid mixing route given in Reff. [5,6]. Zn and Ti metal powders (Merck) were used as the starting materials and HCl 12 M as the solvent. The metal powders were weighed stoichiometrically (Zn: Ti = 1: 1) and dissolved individually in HCl using a stirring hotplate. The resulting $ZnCl_2$ and $TiCl_4$ liquid solutions were mixed homogeneously and the mixed liquid solution was heated at 110 °C to obtain dried powder. The powder was crushed manually for uniform particle size.

Thermogravimetric/differential thermal analysis (TG/DTA) was carried out to the powder from room temperature to 600 °C at the heating rate of 10 °C/min using a DTG-60 and DSC-60 plus Shimadzu to obtain information on any changes to physical and chemical parameters that occur due to heating and which can be measured as a function of temperature with constant heating rate. The powder was calcined at different temperatures as suggested by the TG/DTA curves using a Nabertherm furnace to study phases evolution during heating. Phase identification was monitored from the XRD spectra measured from $2\theta = 20 - 65°$ using an X'Pert Diffractometer (Philips) with CuK_α radiation at the rate of 0.02 °/min. The crystallinity (%) was analysed from the XRD spectra according to the ratio between the intensity of the $ZnTiO_3$ peaks and that of all peaks, or from the ratio between the area of the $ZnTiO_3$ peaks and that of all peaks. FTIR measurement was performed using a Thermo-Scientific FTIR spectrometer from 4000 to 400 cm^{-1}. The distribution of particle size and the average value were examined using a particle size analyzer (PSA) Zetasizer Nano Series Software Version 7.0 (Malvern Instrument).

3. Results and Discussion

3.1 Heating profile

Figure 1 shows the TG/DTA curves of the dried $ZnTiO_3$ powder measured from room temperature (RT) to 600 °C at the rate of 10 °C/min. As shown, after the powder was heated constantly to 600 °C at the rate of 10 °C/min, the TG curve in Figure 1 recorded three different mass losses, *i.e.* A from RT to 243 °C, B from 243 to 440 °C and C from 440 to 600 °C. Within the A mass loss event, the DTA curve detected some minor thermal events, followed by a broad-endothermic peak. These thermal events indicate that a dynamic reaction took place within that temperature range; and the most probable produced compound is zinc chloride x-hydrates ($ZnCl_2{\cdot}xH_2O$) [10,11] that unfortunately can not be detected by XRD. This hydrate component is presumed originally from the HCl used (*i.e.* 37 % HCl) which means that the rest content (63 %) is H_2O. The $ZnCl_2{\cdot}xH_2O$ compound is then decomposed into $ZnCl_2{\cdot}H_2O$ [11,12], accompanied by a release of the solvent used, in this case, H_2O. The calculated mass loss A in Figure 1 is 5.44 % which is 2.10 % lower than the experimental value, 7.54 %. This difference might be due to the presence of H_2O moisture in the sample which then evaporated with heating, or it is attributable to a compound containing Ti, probably $TiCl_4$ which later due to heating, the compound transformed into TiO_2 and HCl.

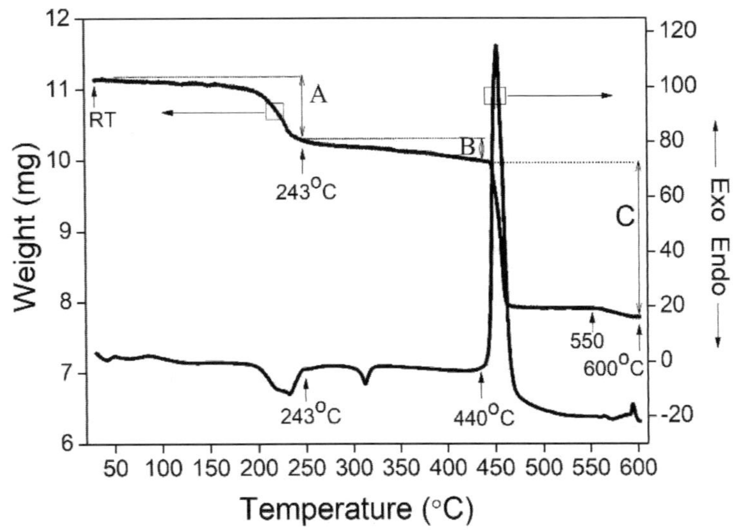

Figure 1. TG/DTA curve of the dried powder measured from RT to 600 °C at the constant rate of 10 °C/min.

Within the temperature range 243 – 440 °C, the mass loss B is about 2.20 %. At the same time, the DTA curve recorded a noticeable endothermic peak at 315 °C which corresponds to the transition of 1 mol $ZnCl_2 \cdot H_2O$ to zinc hydroxide chloride (Zn(OH)Cl) [13] followed by release of 1 mol HCl gas. The TiO_2 compound remains unchanged. At the end of the B mass loss event, a very significant exothermic peak was recorded. Further heating from 440 to 600 °C also generates a massive mass loss C of 22.0 %. This substantial mass loss corresponds to the formation of $ZnTiO_3$ phase and the release of 1 mol HCl gas as a result of the reaction between Zn(OH)Cl and TiO_2 [14]. Figure 2 depicts the FTIR data of the powders calcined from 440 to 900 °C, while Figures 3 and 4 show the associated XRD spectra.

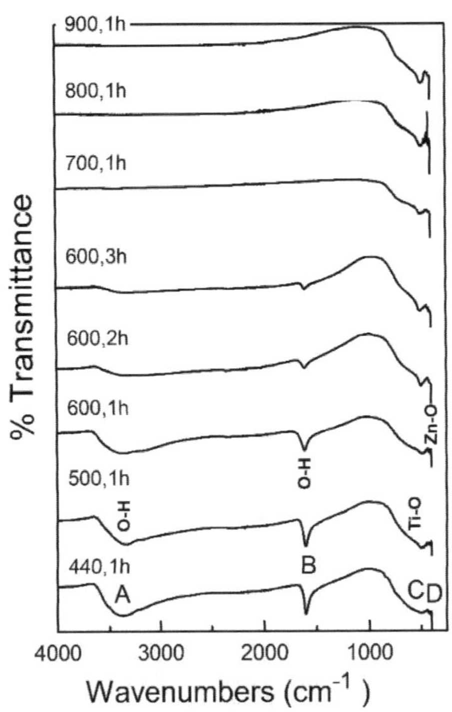

Figure 2. FTIR of the dried powder after being calcined from 440 to 900 °C, each for 1 hour (h) and at 600 °C for 1, 2 and 3 h. The absorption peak A = 3446.18 cm^{-1}, B = 1600.90 cm^{-1}, C = 494.09 cm^{-1} and D = 418.02 cm^{-1}. Details are given in the text.

Within the 4000 to 400 cm^{-1} range, the FTIR curves only detect four different absorption bands, namely a broad band A at the position of 3446.18 cm^{-1}, B at 1600.90 cm^{-1}, C at 494.09 cm^{-1} and D at 418.02 cm^{-1}. The broad band A relates to an O-H stretching band in a lattice water [15]. The intended lattice water is most likely $ZnCl_2 \cdot xH_2O$ as analysed on the TG curve in Figure 1; this broad band corresponds to the adsorbed H_2O [15]. This specific absorption peak was recorded on the FTIR curves labeled 440 1h, 500 1h, 600 1h, 600 2h and 600 3h in Figure 2. The peak intensities are gradually weakens with heating temperature; and this peak totally disappeared on the curves labeled 700 1h, 800 1h and 900 1h. This fact implies that the x-value in the $ZnCl_2 \cdot xH_2O$ compound was quite big and therefore a high temperature was required to completely remove the hydrate content in the compound.

The sharp B band at 1600.90 cm^{-1} in Figure 2 was identified as an O-H deformation or bending vibration band [15]; in this work, this particular band corresponds to Zn(OH)Cl, the hydrolysis product of $ZnCl_2 \cdot H_2O$. Similar to the previous A band, this B band was also traced on the FTIR curves labeled 440 1h, 500 1h, 600 1h, 600 2h and 600 3h and the peak disappeared afterward. The C band at 494.09 cm^{-1} and the D band at 418.02 cm^{-1} are respectively identified as a Ti-O stretching vibration band and a Zn-O vibration band [15]. Unlike the previous A and B absorption peaks, the peaks of Ti-O and Zn-O remain detected on all FTIR curves, starting from the curves labeled 440 1h to 900 1h. These Ti-O and Zn-O bands show the fact that two independent phases containing Ti and Zn are present in the samples, presumably TiO_2 and $ZnTiO_3$. The XRD spectra in Figures 3 and 4 clarify this argument.

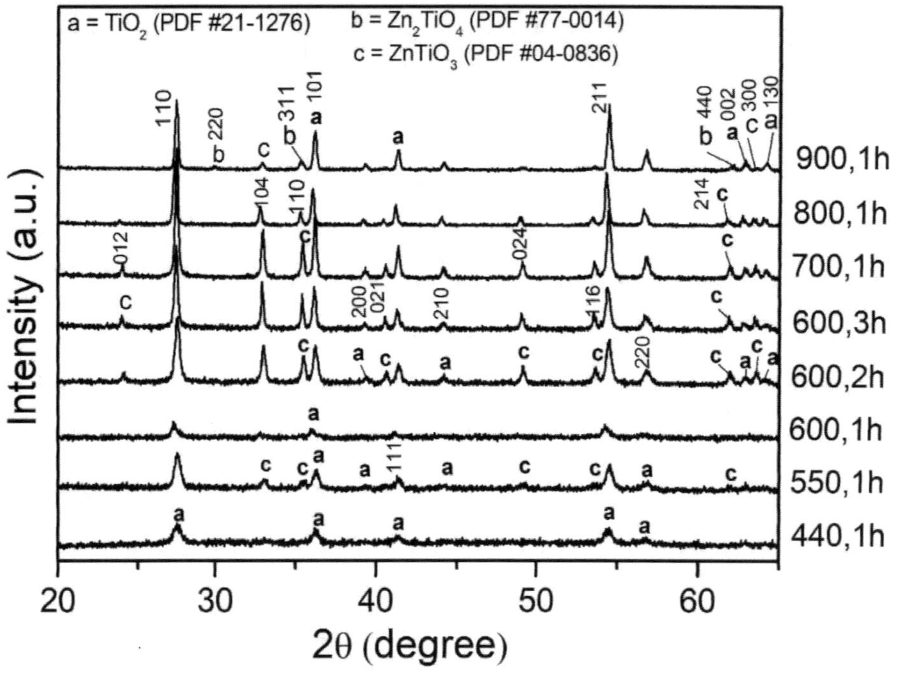

Figure 3. The associated XRD spectra of the calcined powders in Figure 2.

As shown in Figure 3, at 440 °C, TiO_2 rutile (Powder Diffraction File - PDF #21-1276) is the only phase detected. Further heating to 550 °C, some peaks belong to $ZnTiO_3$ phase (PDF #04-0836) were also recorded along the TiO_2 peaks. This suggests that using the liquid mixing method, $ZnTiO_3$ phase was initially formed at 550 °C. Kong et al. [1] reported that some XRD peaks of $ZnTiO_3$ was originally recorded at 600 °C for 2 h after being prepared by a modified alcoholysis method using $Zn(Ac)_2 \cdot 2H_2O$ and $Ti(OC_4H_9)_4$ as the raw materials, ethylene glycol as reagent/solvent and acetylacetone as stabilizer. Chai et al. [2] occupied sol-gel method and zinc acetate and ethylene glycol as the starting materials to prepare $ZnTiO_3$ powder. He found that the $ZnTiO_3$ peaks were identified at first at 700 °C for 6 h. Chang

IOP Conf. Series: Materials Science and Engineering **599** (2019) 012010 doi:10.1088/1757-899X/599/1/012010

et al. [3] prepared $ZnTO_3$ powder from the conventional solid-state reaction using 99.99 % pure ZnO and TiO_2 powders as the raw materials; he recorded the $ZnTiO_3$ peaks at 700 °C for 24 h in air. By following the sol-gel method, using the raw materials of zinc acetate dihydrate, titanium tetraisopropoxide and oxalic acid, and by experimenting various Zn:Ti precursor molar ratio to prepare zinc titanates, Nolan et al. [4] recorded the $ZnTiO_3$ peaks at temperatures of 600 – 900 °C. Based on these literatures, the liquid mixing method adopted in this work being a prospective method to synthesise $ZnTiO_3$ at a lower calcination temperature.

In Figure 3, however, the intensities of the XRD spectrum labeled 600, 1h diminished due to unknown reasons. By increasing the calcination holding time to 2 and 3 h, the numbers of $ZnTiO_3$ peaks and the intensities increased considerably. The increase of TiO_2 peak intensities was also noted. Further heating to 700 °C for 1 h caused the intensities of $ZnTiO_3$ and TiO_2 peaks continued to increase. The increase in the peak intensities of $ZnTiO_3$ and TiO_2 phases, however, slowed down when the sample was calcined at 800 for 1 h. At that temperature, the peak intensities of the two phases declined. Heating up to 900 °C, the intensities of TiO_2 went up again, but that of $ZnTiO_3$ weakened; and even some $ZnTiO_3$ peaks with the Miller indices of (012) at 2θ position of 23.92° and (214) at 62.80° are missing at 900 °C. At the same time, three new peaks at 2θ = 29.93°, 35.25° and 63.18° appeared. These three new peaks were identified as belonging to Zn_2TiO_4 phase (PDF #77-0014), each with Miller indices of (220), (311) and (440). These findings suggested that within 800 – 900 °C, some parts of $ZnTiO_3$ phase were decomposed into Zn_2TiO_4 via the reaction of $2ZnTiO_3 \rightarrow Zn_2TiO_4 + TiO_2$. That is the reason why at 900 °C: the intensities of TiO_2 increased as compared to that at 800 °C, the intensities of $ZnTiO_3$ diminished and simultaneously, some Zn_2TiO_4 peaks started to appear (see Figure 4). Equations 1-5 recapitulate the chemical reactions explained above.

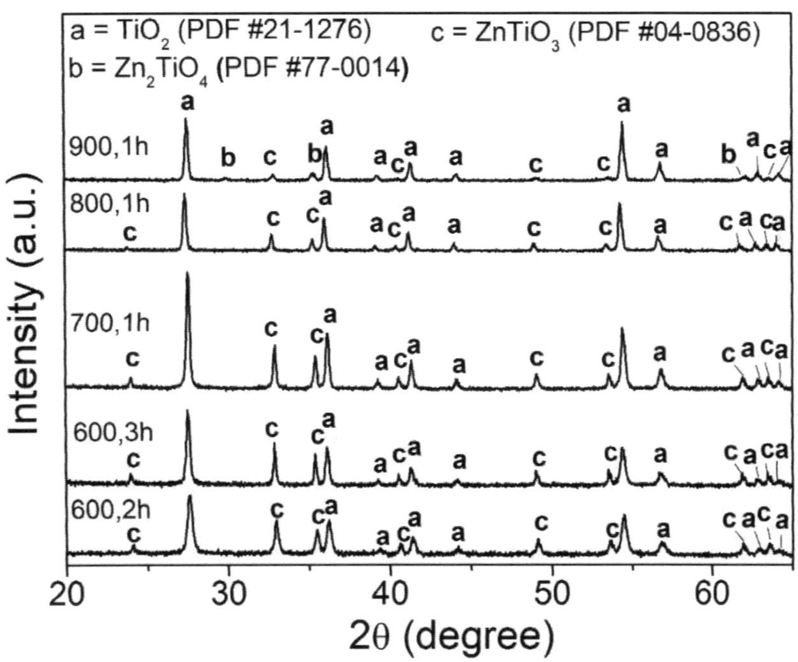

Figure 4. The similar XRD spectra in Figure 3, but only the spectra labeled 600, 2h up to 900, 1h.

$$\text{RT} - 243\ ^{\circ}\text{C}:\quad ZnCl_2 \cdot 5H_2O + TiCl_4 \rightarrow ZnCl_2 \cdot H_2O + 2H_2O^{\uparrow} + TiO_2 + 4HCl^{\uparrow} \quad (1)$$

$$243 - 440\ ^{\circ}\text{C}:\quad ZnCl_2 \cdot H_2O + TiO_2 \rightarrow Zn(OH)Cl + TiO_2 + HCl^{\uparrow} \quad (2)$$

$$440 - 600\ ^{\circ}\text{C}:\quad Zn(OH)Cl + 2TiO_2 \rightarrow TiO_2 + ZnTiO_3 + HCl^{\uparrow} \quad (3)$$

$$600 - 700\ ^{\circ}\text{C}:\quad TiO_2 + ZnTiO_3 \rightarrow TiO_2 + ZnTiO_3 \quad (4)$$

$$800 - 900\ ^{\circ}\text{C}:\quad TiO_2 + 2ZnTiO_3 \rightarrow Zn_2TiO_4 + 2TiO_2 \quad (5)$$

3.2 Crystallinity and particle size distribution

Figure 5 depicts the crystallinity of $ZnTiO_3$ phase as a function of calcination temperature from 600, 2h to 900 °C, 1h calculated from the ratio of peak intensity and of peak area of $ZnTiO_3$ phase with respect to that of the other phases present in the spectra in Figure 4. Since $ZnTiO_3$ is not the major phase, the crystallinity of the phase is limited as compared to that of the main phase of TiO_2 rutile. At 600, 2h in Figure 5, the crystallinity of $ZnTiO_3$ is 36 and 41 %. By prolonging the calcination holding time to 3 h, the crystallinity increased to 38 and 46 % before reaching to the maximum, 45 and 50 %, at 700 °C. Further, owing to the decomposition process of $ZnTiO_3$ into Zn_2TiO_4 and TiO_2 at higher temperatures, the crystallinity of $ZnTiO_3$ phase was affected. The values went down significantly to 27 and 35 % at 800 °C and 11 and 20 % at 900 °C.

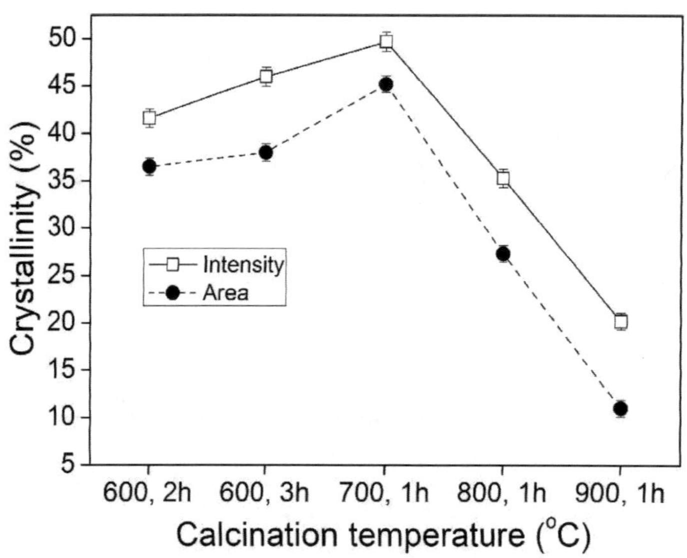

Figure 5. Crystallinity of $ZnTiO_3$ phase in the powders in Figure 4 after heat-treated at various temperatures from 600 to 900 °C.

Chai et al. [2] also reported the crystallinity of $ZnTiO_3$ phase calculated using the same approach, *i.e.* by taking the ratio of XRD peak intensities and of peak area of $ZnTiO_3$ phase relating to that of the other phases (TiO_2 rutile and α-Zn_2TiO_4). Within the same temperature range, from 600 to 900 °C, Chai et al. [2] obtained that the crystallinity of $ZnTiO_3$ increased from 77 and 83 % at 600 °C to 92 and 93 % at 800 °C. These values slightly declined to 90 and 91 % at 900 °C. In the Chai et al. case, $ZnTiO_3$ was the main phase. In the liquid mixing method, as seen in Figures 3 and 4, TiO_2 rutile phase has been formed since the beginning of the reaction. The presence of TiO_2 phase dominates even when the calcination temperature increases and this inhibits the formation of $ZnTiO_3$ phase. To overcome this, an attempt was initiated, i.e. by adding the stoichiometrically excess of Zn during the synthesis process.

Figure 6 shows the effect of the addition of stoichiometrically excess of Zn, in this case Zn: Ti = 1.05: 1, to the reduction of TiO_2 rutile phase dominance, as well as to the formation of $ZnTiO_3$ as the main phase in the system. The powder sample in Figure 6 was calcined at 600 °C for 3 h. As seen, by testing the ratio of Zn: Ti = 1.05: 1, the intensity of the highest peak belonging to the TiO_2 phase at $2\theta = 27.5°$ (*hkl* = 110) in Figure 6 becomes lower than the intensity of the highest peak belonging to $ZnTiO_3$ phase at $2\theta = 32.8°$ (*hkl* = 104). In other words, the (110) peak of TiO_2 phase in Figure 6 is not dominant anymore. This fact is very contrary to the result shown in Figure 4 labeled 600, 3h, where the TiO_2 peak intensity at $2\theta = 27.5°$ is much higher than the $ZnTiO_3$ peak intensity at $2\theta = 32.8°$. Reducing the dominance of TiO_2 phase was also detected at the peak at $2\theta = 36.1°$ with *hkl* = (101). At the same time, $ZnTiO_3$ peak at $2\theta = 35.3°$ with *hkl* = (110) becoming dominant. The addition of stoichiometrically excess of Zn has driven the formation of $ZnTiO_3$ phase; further addition of stoichiometrically excess of Zn potentially has further diminished TiO_2 phase and increased the formation of $ZnTiO_3$ phase. Figures 7a-c show the distribution of particle size in the powders in Figure 4 labeled 700, 800 and 900, 1 h.

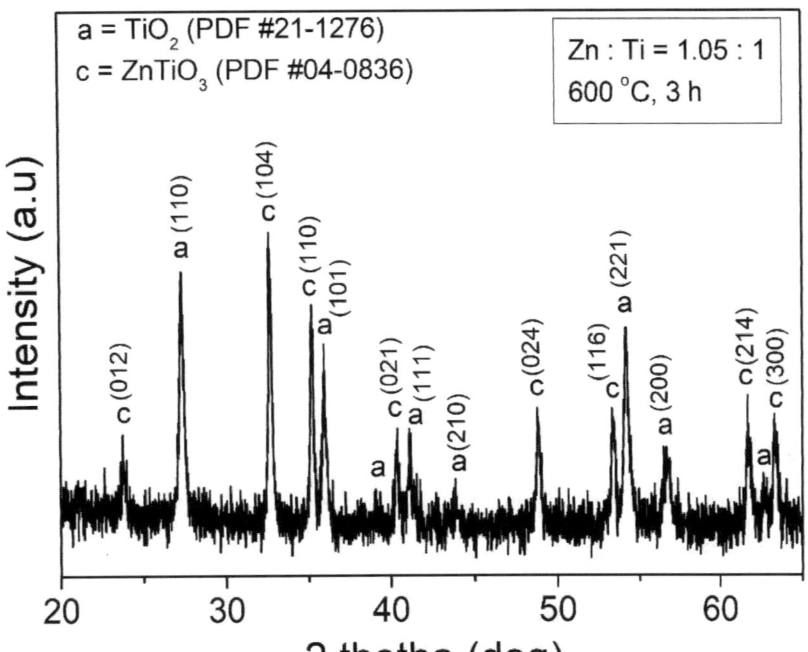

Figure 6. The XRD pattern of $ZnTiO_3$ powder with the addition of stoichiometrically excess of Zn, i.e. Zn: Ti = 1.05: 1. The powder was calcined at 600 °C, 3 h.

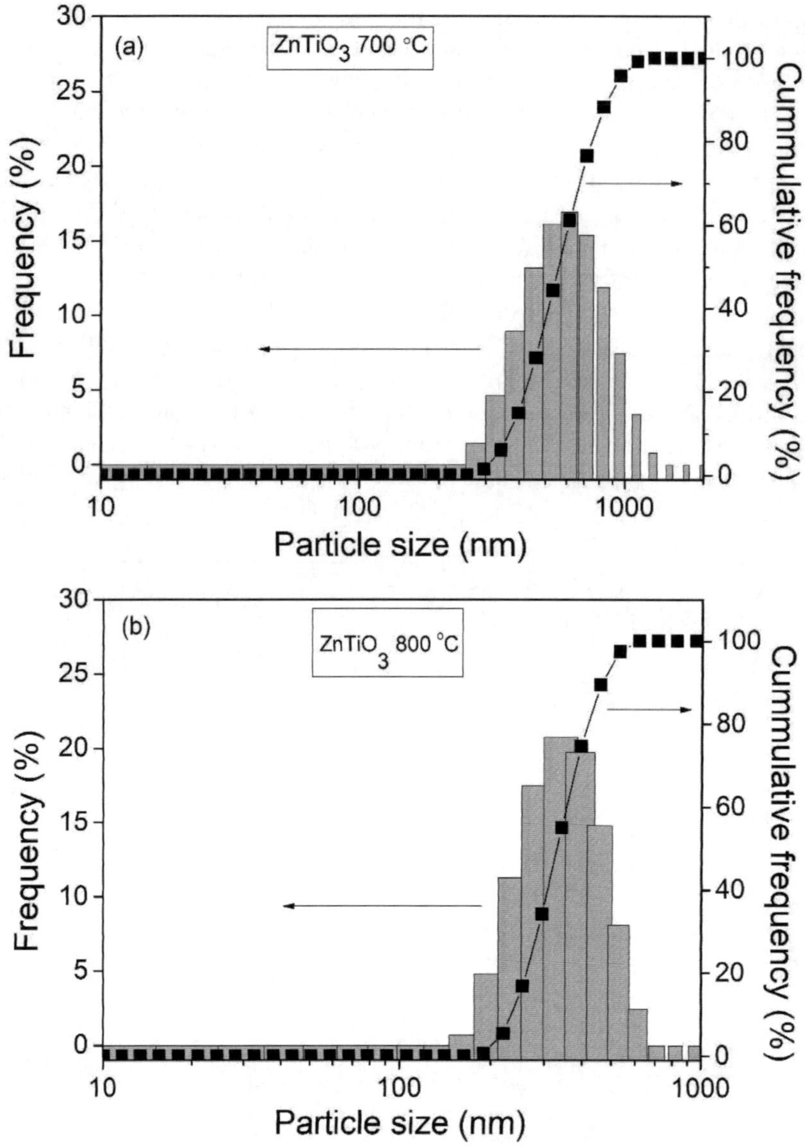

Figures 7a-b. Particle size distribution in $ZnTiO_3$ powder in Figure 4 labeled: (a) 700, 1h and (b) 800, 1h.

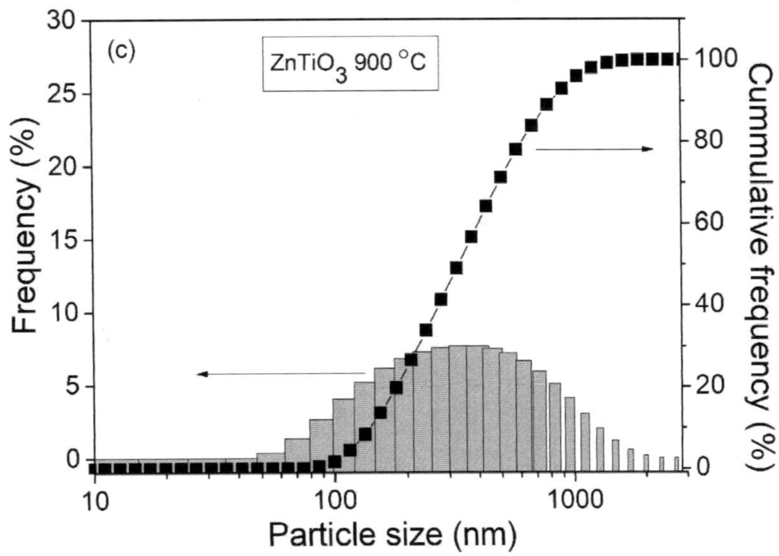

Figure 7c. Particle size distribution in $ZnTiO_3$ powders in Figure 4 labeled 900, 1h.

The particle size distribution in Figures 7a-c are all normal-single mode ranging from 295 – 1280 nm at 700 °C, 190 – 615 nm at 800 °C and 91 – 2304 nm at 900 °C. The average particle size at 700 °C is 683 nm, at 800 °C is 370 nm and at 900 °C is 709 nm, each with deviation of around 4 %. Using the fact that the distribution curves are single modes and the deviation of the average size is diminutive, it can be concluded that the particles in the above samples are all homogeneously distributed. Chai et al. [5] measured the particle size of the $ZnTiO_3$ powders directly from TEM image and obtained that the average size were about 30 nm after heating at 800 °C for 6 h.

4. Conclusion

The liquid mixing method has been demonstrated as a promising synthesis route to produce $ZnTiO_3$ phase because the desired phase can be formed at a temperature as low as 550 °C. Besides, particles in the samples are all homogeneously distributed with diminutive deviation for the average particle size. The problem faced by this method is when using the 'normal' stoichiometric ratio of Zn: Ti = 1: 1; TiO_2 rutile phase was formed as the main phase while the $ZnTiO_3$ phase is only a secondary phase. This situation causes limited crystallinity of $ZnTiO_3$ phase; the value is no more than 50 % and that was achieved when the powder was calcined at 700 °C. Above 700 °C, the crystallinity of $ZnTiO_3$ phase dropped significantly because some of the $ZnTiO_3$ phase decomposed into Zn_2TiO_4 and TiO_2 phases. The practical best solution to encourage the formation of $ZnTiO_3$ phase as the main phase is by adding the stoichiometrically excess of Zn during the synthesis process. The experiment using the ratio of Zn: Ti = 1.05: 1 has proved it.

5. References

[1] Kong J-Z, Li A-D, Zhai H-F, Li H, Yan Q-Y, Ma J and Wu D. 2009 *J. Haz. Mat.* **171** 918-23
[2] Chai Y-L, Chang Y-S, Chen G-J and Hsiao Y-J. 2007 *Mater. Res. Bull.* **43** 1066-73
[3] Chang Y-S, Chang Y-H, Chen I-G, Chen G-J, Chai Y-L, Fang, T-H and Wu, S. 2004 *Ceram. Int.* **30** 2183-89
[4] Nolan N T, Seery M K and Pillai S C. 2011 *Chem. Mater.* **23** 1496-1504
[5] Ermawati F U, Suasmoro S and Pratapa S. 2015 *Adv. Mater. Res.* **1112** 47-52
[6] Ermawati F U, Pratapa S, Suasmoro S, Hübert T and Banach U. 2016 *J. Mater. Sci. Mater. El.* **27** 6637-45
[7] Ermawati F U, Suasmoro S and Pratapa S. 2017 *Omega: Jurnal Fisika dan Pendidikan Fisika.* **3** 15-22

[8] Ermawati F U and Suasmoro S. 2018 *J. Phys: Conf. Ser.* **997** 012005
[9] Ermawati F U, Supardi Z A I, Suasmoro S, Pratapa S and Hübert T. 2018 *IOP Conf. Ser.: Mater. Sci. Eng.* **367** 012003
[10] Hennings E, Schmidt, H and Voigt, W. 2014 *Acta Crystallogr. Sect. E.* **E**70 515–18
[11] Haghighi, H, Chapoy, A and Tohidi, B. 2008. *Ind. Eng. Chem. Res.* **47**, 3983–89.
[12] Sutton, L E and Truter, M R 2007. *Molecular Structure by Diffraction Methods*, Volume 5 (Great Britain: Royal Society of Chemistry) p 173
[13] Brauer, G and Herrmann, W A. 2011 *Synthetic Methods of Organometallic and Inorganic Chemistry: Catalysis*, Volume 10 (The University of Carolina: Thieme) p 123
[14] Zhou, S, Patty, A and Chen S. 2015 *Advances in Energy Science and Equipment Engineering: Proceedings of the International Conference on Energy Equipment Science and Engineering*, (ICEESE 2015), May 30-31, 2015, Guangzhou, China. (Great Britain: CRC Press) p 31-34
[15] George S. 2001 *Infrared and Raman Characteristic Group Frequencies: Tables and Charts*. 3rd Ed. (West Sussex, England: John Wiley and Sons) p 322-26

Acknowledgement
The authors wish to thank Mr. Sholih of Material Laboratory, Physics Dept. Institute Technology Sepuluh Nopember Surabaya for the assistance in running the particle size analyzer experiment.

Synthesis and Characterization of Hydroxyapatite-Zinc Oxide (HAp-ZnO) as Antibacterial Biomaterial

Charlena[1*], Irma Herawati Suparto[1*], Eldora Kurniawan[1]
Department of Chemistry, Bogor Agricultural University, INDONESIA

Email: charlena@apps.ipb.ac.id, irma.suparto@gmail.com

Abstract. Hydroxyapatite (HAp) is an important biomaterial widely used as implants for bone and teeth. Biomaterial with antibacterial activities beneficial to prevent infection post implantation. Therefore, the present study aims to synthesized and characterized HAp composite with zinc oxide (ZnO) that has antibacterial properties. The HAp was synthesized through wet precipitation method using shell of *Bellamya javanica* as source of calcium. The zinc oxide was added with HAp using *in situ* method. The crystallinity of the material was decreased after added with ZnO which indicate Zn ion suppressed the crystal growth. Antibacterial abilities were tested against *Eschericia coli* and *Staphylococcus aureus.* Inhibition against bacteria increased when the HAp was composited with ZnO. The HAp-ZnO composite showed better inhibition toward *Escherichia coli*.

1. Introduction

Implantation technique is one of the solution to repair damaged bone and teeth. Material selection for the implant is an important factor to assure the success of the implantation. Research and studies have been done to find material that have an active interaction with the target tissue without causing undesirable reaction on the tissue. Calcium phosphate based material is one of the material which is widely studied. When implanted to the bone, it will form physicochemical bond with bone tissue [1].

Hydroxyapatite (HAp) is the most stable phase of calcium phosphate and the most studied biomaterial as bone and teeth implant. Synthetic HAp is osteoconductive and thermodinamically stable in physiological pH [2]. Natural resources with the high calcium content such as egg shell, fish bone, and the shell of tutut snail (*Bellamya javanica*) could be used as the precursor for HAp synthesis. Independently, HAp does not have a good antibacterial properties. It can be improve by combining inorganic ions with antibacterial properties. This can be solved by creating a composite of HAp and added with the antibacterial material. Existence of metal ions such as Ag nanoparticles, CuO and ZnO in HAp structure has been proved to increase its antibacterial material [3-5].

Tutut (*Bellamya javanica*) shell was chosen as starting material for this study because of its high calcium content and not yet produced commercially. Wet precipitation method was used to synthesize HAp. Zinc oxide (ZnO) was synthesized through low hydrothermal method using zinc acetate dihydrate and sodium hydroxide as the starting material. The composite was made using *in situ* method.

2. Methods

Zinc oxide was synthesized using $Zn(CH_3COO)_2 \cdot 2H_2O$ and NaOH as starting materials. $Zn(CH_3COO)_2 \cdot 2H_2O$ solution was made by dissolving 1.0957 g $Zn(CH_3COO)_2 \cdot 2H_2O$ in 50 mL flask with deionized water. Meanwhile, 0.2500 g NaOH was dissolved in another 50 mL flask with deionized water. $Zn(CH_3COO)_2 \cdot 2H_2O$ solution was added dropwise into NaOH solution while stirred at room temperature. The mixture then heated in oven at 90 °C for 2 hours and centrifuged at 3000 rpm for 30 minutes. Supernatant that contain ZnO was seperated. [6]

The mixture of $Ca(OH)_2$ 0.5 M and $(NH_4)_2HPO_4$ 0.3 M was decantated for 120 hours. The solution of ZnO in water was added into the suspension then reheated to 80-90 °C, stirred for 1 hour, and vaccum filtered after 12 hour. Filtrate was dried at 120 °C for 4 hours [5]. The obtained powder were annealed at 900 °C, 1000 °C, 1090°C for 2 hours . Synthesized powder was characterized using X-ray diffraction, Fourier Transform Infra Red, and antibacterial activities assay.

3. Results and Discussion

Calcium source that was used in this study was from the shell of tutut (*Bellamya javanica*). Calcium that obtained from tutut shell was in form of calcium carbonate. After calcination process, $CaCO_3$ was converted to CaO according to this reaction:

$$CaCO_{3(s)} \rightarrow CaO_{(s)} + CO_{2(g)}$$

CaO powder was left in open air for one week and coverted to $Ca(OH)_2$. Calcium content of the $Ca(OH)_2$ powder from tutut shell was 32.97%. Hydroxyapatite was synthesized through precipitation method [7] using $Ca(OH)_2$ and $(NH_4)_2.HPO_4$ as precursor.

$$10Ca(OH)_2 + 6(NH_4)_2HPO_4 \rightarrow Ca_{10}(PO_4)_6(OH)_2 + 6H_2O + 12\ NH_4OH$$

Hydroxyapatite was characterized using XRD. HAp has characteristic peaks at 2θ 31.80°- 34.00° [8]. XRD spectrum of synthesized HAp powder showed strong peaks at 2θ 31.80°, 32.95°, 32.21° (Fig 1).

Zinc oxide was obtained in the form of white powder. The powder was characterized using XRD and showed strong peaks at 2θ 36.22°, 33.01°, 31.72°. These peaks are characteristic peaks of ZnO. XRD spectrum also showed a peak at 2θ 13.28°. This peak indicated that there was zinc acetate dihydrate which was left after synthesis. (Fig.2)

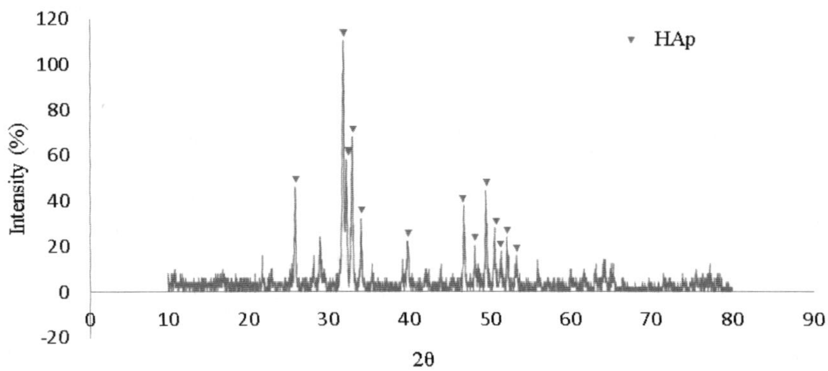

Figure 1. XRD pattern of Hap.

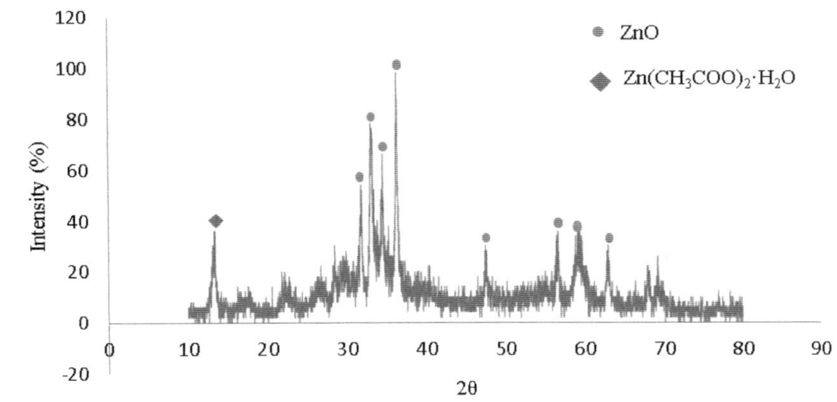

Figure 2. XRD pattern of ZnO.

HAp-ZnO composite was characterized using XRD. The crystallinity of the material was decreasing after the addition of ZnO. Synthesized HAp had crystallinity of 75.44% meanwhile HAp-ZnO composite had crystallinity of 60.02 %. The decrease of crystallinity could be caused by the substitution of calcium ions by zinc ions. The substitution would suppress the crystal growth [9] .

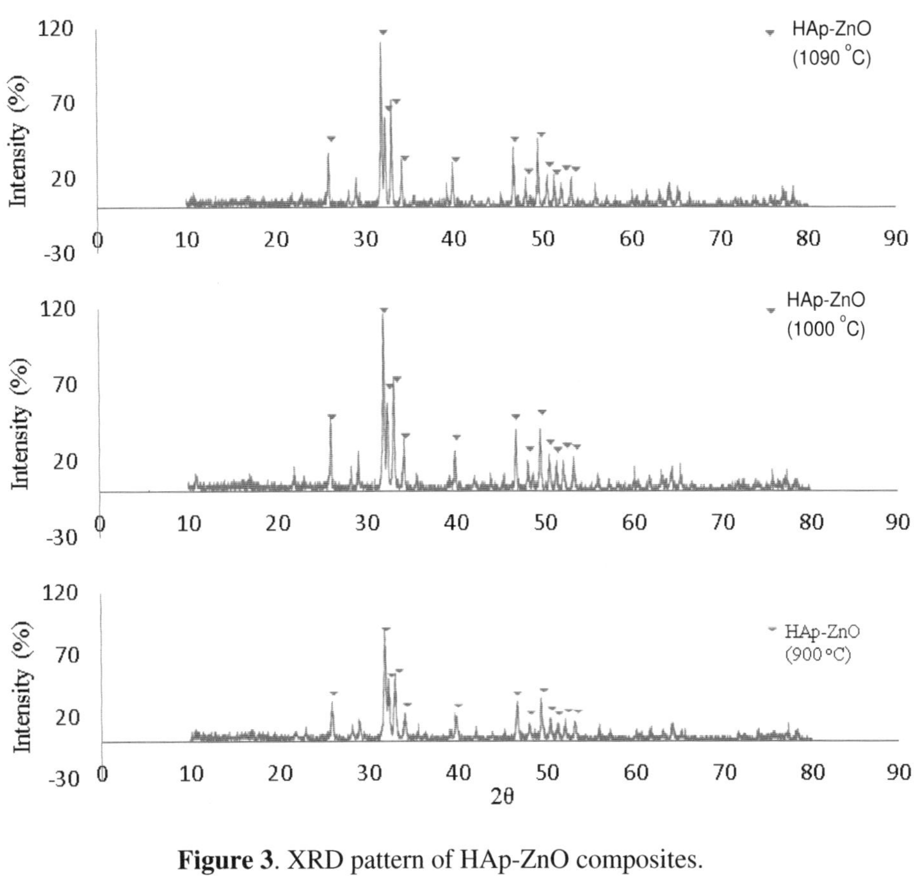

Figure 3. XRD pattern of HAp-ZnO composites.

In this study, composite materials were calcinated at 3 different temperature. The degree of crystalinity was increased when the material was calcinated at higher temperature. The degree of crystallinity of HAp-ZnO composite that was calcinated at 1000 °C was 76. 5849%, while the one that calcinated at 1090 °C had crystallinity of 86.6479%. XRD pattern showed that material that was calcinated at higher temperature had narrow peak width (Fig. 3).

Hydroxyapatite and its composite was characterized using FTIR. The spectrum showed absorption bands at 1400-1500 cm^{-1} indicated the existence of functional group $-CO_3$. Absorption bands that showed the presence of –OH was observed at wavenumber 630, 1600, 1900, and 3500 cm^{-1}. Wide absorption bands at wavenumber 3440 cm^{-1} showed the presence of adsorped water molecule [10]. The functional group of $-PO_4$ showed by the absorption band at 470.63, 567.07, 960.55, 1050 cm^{-1}.

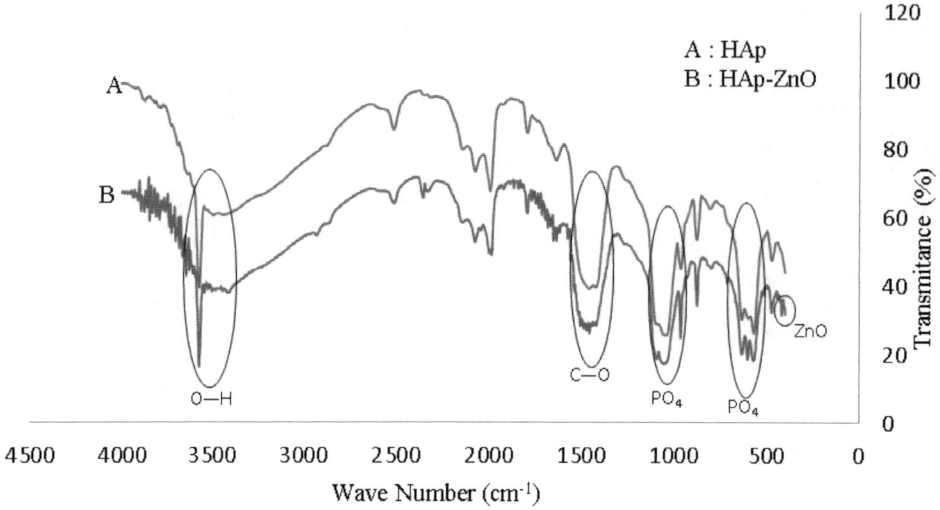

Figure 4. FTIR spectrum of HAp and HAp-ZnO composite that was calcinated at 900 °C.

Zinc oxide showed vibration at the 400-600 cm^{-1}[11]. FTIR spectrum of HAp-ZnO composite showed absoption band at 401-416 cm^{-1} (Fig.4). This absorption band did not appear at HAp FTIR spectrum so it could be concluded that ZnO was successfully composited with HAp. FTIR spectrum of the composite that was calcinated at 3 different temperatures showed the difference at the absorption bands at 3500 cm^{-1}. The intensity of the bands was decreasing which was caused by the interaction of ZnO with OH group from HAp[5]. High calcination temperature also affected interaction of ZnO with PO_4 group (Fig. 5). The absorption peaks of PO_4 at 560 cm^{-1} decreased when material was calcinated at 1000 and 1090 °C. This was caused by interaction of ZnO with PO_4 from hydroxyapatite with bidentate coordination of Zn[5].

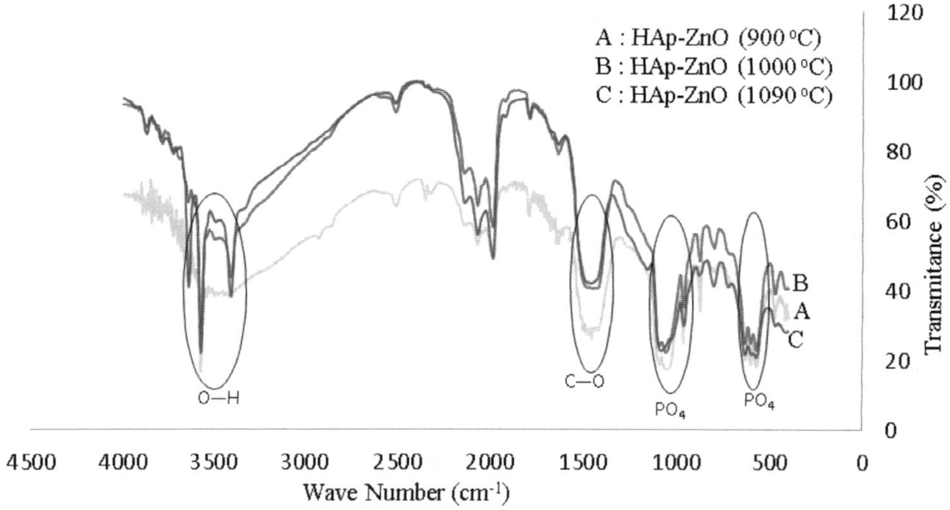

Figure 5. FTIR spectrum of HAp-ZnO composite which was calcinated at different temperature.

Synthesized ZnO, HAp, and HAp-ZnO were tested to observe the antibacterial capability. The result showed the increase of antibacterial activity after ZnO addition.

Table 1. Inhibition Index of HAp and HAp-ZnO for *E.coli* and *S.aureus*

	E.coli		*S.aureus*	
	Inhibition Diameters (mm)	Inhibition Index	Inhibition Diameters (mm)	Inhibition Index
ZnO	14	0.56	15	0.67
HAp	11	0.22	10	0.11
HAp-ZnO 900	13	0.44	-	-
HAp-ZnO 1000	16	0.78	12	0.33
HAp-ZnO 1090	14	0.56	12	0.33

The increase of antibacterial ability was proved by the widening of its inhibition zone (Fig.6).

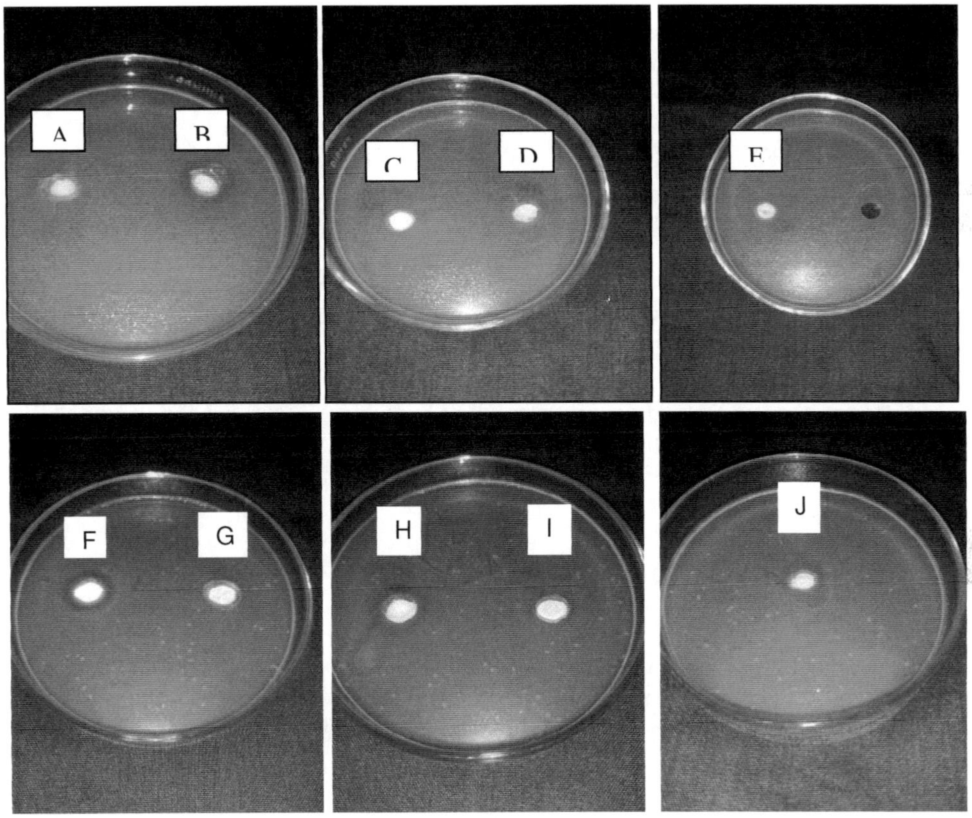

Figure 6. Antibacterial Test on *Staphylococcus aureus* on (A)HAp (B)ZnO (C)HAp-ZnO 1000 (D)HAp-ZnO 900 (E)HAp-ZnO 1090 and Antibacterial Test on *Eschericia coli* on (F)ZnO (G)HAp (H)HAp-ZnO 1000 (I)HAp-ZnO 900 (J)HAp-ZnO 1090.

Inhibition index of the HAp-ZnO composite towards *E.coli* was increased until 0.56 higher than HAp without ZnO addition. Antibacterial test towards *S.aureus* was also increased. The test showed that HAp-ZnO composite inhibited *E.coli* more effectively that *S.aureus* . But HAp-ZnO that synthesized in this study had lower inhibition index that HAp-CuO composite which was done by Yuliandari (2016).

4. Conclusion
HAp-ZnO composite was synthesized through low hydrothermal method. Zinc oxide addition caused the decrease of the sample crystallinity. Higher calcination temperature caused the increase of crystallinity and the narrowing of the XRD peak width. Antibacterial test of HAp-ZnO composite towards *E.coli* and *S.aureus* showed the increase of inhibition index. Composite material with the best inhibition index was the composite that was calcinated at 1000 °C, which was 0.78.

5. References
[1] Bagambisa FB, Joos U 1990 Preliminary studies on the phenomenological behaviour of osteoblast cultured on hydroxyapatite ceramics *Biomaterials* **11** 50-56
[2] Nayak AK 2010 Hydroxyapatite synthesis methodologies: An overview. *Int J. ChemTech Res* **2**(2)903-907
[3] Charlena, Nuzulia NA, Handika 2017 Synthesis and characterization of composite hydroxyapatite-silver nanoparticles *IOP Conf. Series: Earth and Environmental Science* **58** 1-9
[4] Yuliandari E. 2016. Sintesis dan pencirian komposit hidroksiapatit-tembaga(II) oksida (HAp-CuO) sebagai biomaterial antibakteri [skripsi]. Bogor (ID): Institut Pertanian Bogor.
[5] Molodovan M, Prodan D, Popescu V, Prejemerean C, Sarosi C, Saplontai M, Talu S, Vasile E. 2015 Structural and morphological properties of HA-ZnO powders prepared for biomaterials *Open Chemistry* **13** 725-733
[6] Osman DAM, Mustafa MA 2015 Synthesis and characterization of Zinc Oxide Nanoparticles using zinc acetate dihydrate and sodium hydroxide *Journal of Nanoscience and Nanoengineering* **1** 248-251
[7] Santos MH, de Oliveira M, Souzal PF, Mansur HS, Vasconcelos WL 2004 Synthesis control and characterization of hydroxyapatite prepared by wet precipitation process. *Materials Research* **7**(4) 625-630
[8] Elhadad AA, Barranco V, Jiménez-Morales A. Peon E, Galvan JC. 2007. Multifunctional sol-gel derived thin film based on nanocrystaline hydroxyapatite powders. *Journal of Physics: Conference Series* **252**(1) 1-8
[9] Galindo TGP, Kataoka T, Tagaya M 2015 Morphosynthesis of Zn-substituted stoichiometric and carbonate hydroxyapatite nanoparticles and their cytotoxicity in fibroblasts. *Journal of Nanomaterials* **2015** 1-8
[10] Esfahani H, Salahi E, Tayebifard A, Rahimipour MR, Keyanpour-Rad M 2016 Structural and morphological analysis of zinc incorporated nonstoichiometric hydroxyapatite nano powders *Revista Materia* **21** 569576
[11] Azam A, Ahmed AS, Oves M, Khan MS, Habib SS, Memic A 2012 Antimicrobial activity of metal oxide nanoparticles against gram-positive and gram-negative bacteria:a comparative study. *International Journal of Nanomedicine* **7** 6003-6009

The Impact Resistance, Flammability, and Hardness of Polypropylene/ Unidirectional Sumberejo Kenaf Fiber Composites

M I Maulana and A L Juwono*

Department of Physics, Faculty of Mathematics and Natural Sciences, Universitas Indonesia, Depok 16424, West Java, Indonesia

*E-mail: ariadne.laksmidevi@ui.ac.id, muhamad.ichsan41@sci.ui.ac.id

Abstract. The use of natural fibers to substitute synthetic fibers on composites has been an interest to be developed. Kenaf is a plant that abundant in Indonesia because it's adaptive and its fiber has good strength. The purpose of this study was to determine the value of impact resistance, flammability, and hardness of polypropylene/ unidirectional Sumberejo kenaf fiber composite boards in accordance with Standar Nasional Indonesia (SNI). Composites were fabricated by the variation of fiber, which were 30 wt%, 40 wt%, and 50 wt%. Pure polypropylene samples were also fabricated for comparison. The fabrication used a compression molding method. The best results were obtained from polypropylene / 50 wt% kenaf fiber composites with impact energy, flammability, and hardness values were (47.54 ± 10.7) J/ cm^2, (7.1±5.1) mm/ minute, and (66 ± 0.8) HD respectively. Optical microscope observation on the impacted surface was a destruction of the matrix-fiber bonding and fiber breakage, in which showed the energy of impact was absorbed by kenaf fiber. The observation on the burned surface was burnt kenaf fibers and rest of PP that bonding with kenaf fiber, this indicated the kenaf fiber acted as an oxygen barrier.

Keywords: unidirectional kenaf fiber, polypropylene, composite, impact resistance, flammability, hardness, fiber breakage

1. Introduction

The development of the use of natural fibers to substitute synthetic fibers on fiber reinforced composites has been an interest to the researchers. As a tropical country, Indonesia is a fertile land for various types of plants such as pineapple, bamboo, kenaf where those plant fibers can be used as a substitute for glass fiber [1]. The advantages of using natural fibers were eco-friendly, abundant of its availability, and relatively cheap [2].

The hydrophilic on natural fibers and the hydrophobic on polymers that act as the matrix decreased the bond strength between matrices and fibers [3]. Beside that, the bonds are influenced by several basic factors, including fiber roughness topology, type, and model of fiber fracture and adhesion between matrix surfaces and fibers [4]. To increase this bond, lignin and hemicellulose on natural fibers should be reduced by giving the alkaline treatment to the fiber which the best concentration was 5% for kenaf fiber (KF) [5, 6]. The fiber tensile strength was related to the mechanical strength of composites. Asim (2016) reported that the alkaline treatment increased the fiber tensile strength 161% from 282.60 MPa to 455.74 MPa after the treatment [3].

The impact resistance of recycled polypropylene (PP)/ chopped kenaf fiber composites reported that the 20 % weight fraction of KF increase 15 % the impact resistance than the recycled PP itself [7]. Pornwannachai (2018) studied flax/ PP composites and it showed that the flammability test result was (19.3 ± 1.3) mm/ minute with the random fiber oriented [8]. The effect of natural fiber reinforced composite to the hardness test reported by Lusiani (2015), that the longer fiber length increase the hardness of composites [9]. The fiber length and its oriented were the main factors of composites mechanical strength [10].

Previously study showed the weight fraction of 40 wt% Sumberejo kenaf fiber on PP/ SKF composites obtained the best tensile strength and temperature deflection [11]. This study was about to study the continuous unidirectional fiber oriented on the composites. Impact resistance, flammability, and hardness properties of PP/Sumberejo Kenaf fiber (SKF) composites have not been studied. Therefore, in this study the impact resistance, flammability, and hardness of PP/ SKF composites with fiber fractions of 30 wt%, 40 wt%, and 50 wt% of unidirectional orientation fiber were reported.

2. Materials and Methods

2.1 Materials and Fabrication

Kenaf fiber was obtained from Sumberejo, Central Java, Indonesia, and the matrix was BI32AN PP pellet from PT Chandra Asri Petrochemical. For alkaline treatment, NaOH was used from PT Merck. Kenaf fiber was cut into a length of the composite mold, then it was soaked in 5 % NaOH solution at room temperature for 24 hours. After 24 hours, the fiber was washed with aqua bidest for 5-7 times until the washed-water was colorless. Then, it was dried at room temperature for 40 hours and followed at 60 °C for 24 hours. The single fiber test was carried out for raw SKF and treated SKF. SNI 08-1112-1989 was applied for the standard procedure in which the fiber as a bunch length of 5 cm and width of 0,5 cm for five specimens at room temperature.

PP pellet was fabricated into sheets in a mold (20 cm x 30 cm) by compression molding method with a thickness of 1 mm. The temperature of compression molding was 190 °C for 10 minutes. The fiber fractions of 30 wt%, 40 wt%, and 50 wt% were prepared for unidirectional fiber oriented and arranged in between PP sheets. The fabrication of composites used compression molding method then fabricated with a pressure of 0 MPa for 10 minutes and 5 MPa for the next 10 minutes at the temperature of 190 °C, followed by cold press method for 15 minutes on the room temperature. The samples were identified as PP, PP/30SKF, PP/40SKF, and PP/50SKF.

2.2 Sample Testings and Microscope Observation

The samples of impact resistance test were prepared according to ASTM D6110-10 standard for Charpy method with the dimension of 60 mm x 10 mm and a thickness of 4 mm. A notch of 2 mm thickness with an angle of 45° was created as required by the standard. This test was carried out on a GT-7045 Gotech Testing Machine. Before and after impact tested samples were observed by optical microscope using a Zeiss Axiocam microscope camera.

The standard procedure for flammability test was UL94 Horizontal Burning from The UL Thermoplastics Testing Center. The dimension of the specimen was 125 mm x 13 mm with a thickness of 4 mm. The measurement was counted on the time whe the flame burning on the specimen length of 75 mm or until the flame stopped before it reached the length. The test was carried out on a flammability tester. For the comparison, the flamed samples were observed by the microscope.

ASTM D2240 standard was used for hardness test on a Krisbow Durometer shore D from Krisbow. Ten specimens were prepared for this test. The minimum thickness according to the standard is 60 mm and to reach it can pile several specimens.

3. Result and Discussion

3.1 The Single-Fiber Test

The tensile strengths of kenaf fiber before and after alkaline treatment were 83.57 MPa and 186.18 MPa respectively. The treatment made the fiber stronger by 123 % after the lignin and hemicellulose were removed from the fiber. This result agreed well with the previous result of Asim (2016) which the value KF fiber tensile strength increased 123 % after alkaline treatment [3].

3.2 The Impact Resistance Test, Flammability test, and Hardness test

Table 1 shows the result of impact resistance, flammability, and hardness tests. The PP/ 50SKF composites had the best composites. The continuous unidirectional SKF affected the impact resistance because the continuous fiber acted as an energy impact absorber. In flammability test, the SKF acted as an oxygen barrier which made the flame inconvenienced to reach the deepest part of composites. In hardness test, the fiber withstood the load of Durometer sensor. The result of impact resistance test similar to Suharty (2016) with the addition KF to PP made the composites 15% more resist the energy impact [7]. Values of flammability test similar to Pornwannachai (2018) research of natural fiber as flame resistant [8]. While hardness values agreed well to Lusiani (2015) which more longer length of natural fiber made higher hardness values of composites [9].

Table 1. The results of the impact resistance, the flammability, and the hardness tests

	Impact resistance (J/cm^2)	Flammability (mm/minute)	Hardness (HD)
PP	7.17	25	40.2
PP/30SKF	24.86	13.6	58.1
PP/40SKF	32.6	12.2	61.1
PP/50SKF	47.54	7.1	66

3.3 Optical Microscope Observation specimen

Figure 1 shows the opical microscope observation for PP/ 50SKF before and after flammability (a, b) and impact resistance (c, d) tests. The after flammability test sample shows the burned polypropylene and stacked on fiber's surface. The fiber acted as an oxygen barrier in which the oxygen was needed for the burning. The after impact resistance test sample shows fiber breakage occured as a proof that the fiber absorbed the impact energy. From this observation, it showed that the continuous unidirectional fiber played important roles as energy absorber and oxygen barrier.

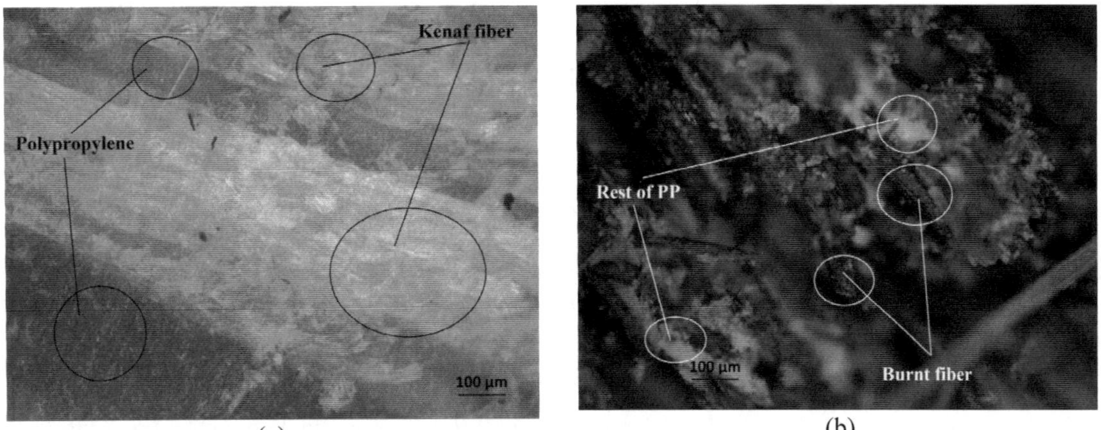

(a) (b)

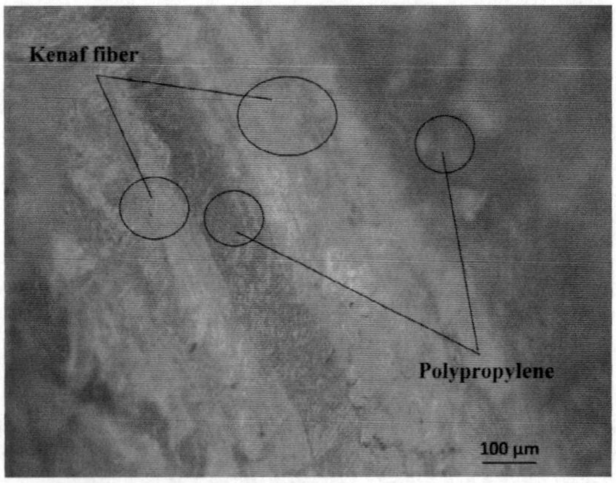

(c)

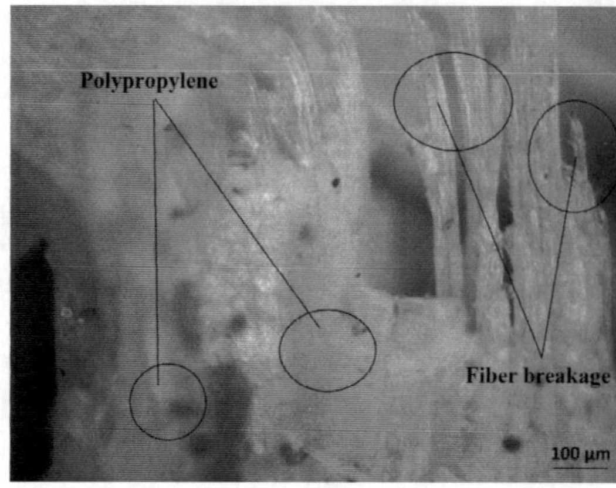

(d)

Figure. 1 Optical microscope observation of PP/50SKF surface samples. (a) before flammability test (b) after flammability test (c) before impact resistance test (d) after impact resistance test.

4. Conclusion

Alkaline treatment was proven to increase the tensile strength of SKF by 123% higher than kenaf fiber before alkaline treatment. The addition of SKF made the ability to impact resistance, flammability, and hardness increased. The PP/50wt% SKF composites had the highest energy impact of (47.54±10.7) J/cm^2, the lowest flame rate of (7.1±5.1) mm/minute and highest hardness value of (66±0.8) HD. The continuous unidirectional SKF acted as an oxygen barrier and impact energy absorber.

5. References

[1] Sudjindro 2011 Prospek serat alam untuk bahan baku kertas uang *Perspektif* **10** 92-104

[2] Lestari F P 2008 Pengaruh Temperatur Sinter dan Fraksi Volume Penguat AL2O3 terhadap Karakteristik Komposit Laminat Hibrid AlSiC-Al/Al2O3 Produk Metalurgi Serbuk *Thesis* Universitas Indonesia, Depok

[3] Asim M, Jawaid M, Abdan K, Ishak M R 2016 Effect of Alkali and Silane Treatments on Mechanical and Fibre-matrix Bond Strength of Kenaf and Pineapple Leaf Fibres *J. Bionic Eng.* **13** 426-435

[4] Eichhorn S J, Baillie C A, Zafeiropoulos N M, Waikambo L Y, Ansell M P, Dufrense A, Entwistle K M, Herrera F P J, Escamilla G C, Groom L M, Hughes M, Hill C, Rials T G, Wild P M 2001 Review Current International Research into Cellulosic Fibers and Composites *Journal of Material Science* **36** 2107-2113

[5] Setyanyo H R, Diharjo K, Miasa M I, Setyono P 2013 A Preliminary Study: The Influence of Alkali Treatment on Physical and Mechanical Properties of Coir Fiber *Journal of Material Science Research* **2** 80-88

[6] Ramesh M 2016 Kenaf (Hibiscus cannabinus L.) fiber based bio-materials: A review on processing and properties *Prog. Mat. Sci.* **78** 1-92

[7] Suharty N S, Ismail H, Diharjo K, Handayani D S, Firdaus M 2016 Effect of Kenaf Fiber as a Reinforcement on the Tensile, Flexural Strength and Toughness Properties of Recycled Polypropylene/Halloysite Composites *Proc. Chem.* **19** 252-258

[8] Pornwannachai W, Ebdon J R, Kandola B K 2018 Fire-resistant natural fibre-reinforced composites from flame retarded textiles *Poly.Deg. and Stab.* **154** 115-123

[9] Lusiani R, Sunardi S, Ardiansah Y 2015 Pemanfaatan Limbah Tandan Kosong Kelapa Sawit Sebagai Papan Komposit dengan Variasi Panjang Serat *Jurnal Teknik Mesin Untirta* **2** 1

[10] Matthews F L, Rawlings R D 1999 Composite materials: engineering and science. *CRC press*

[11] Ollivia S L, Juwono A L, and Roseno S 2017 Tensile Properties and Deflection Temperature of Polypropylene / Subang Pineapple Leaf Fiber Composites with Fiber Content Variation *IOP Conf. Ser.: Mater. Sci. Eng.* **196** 012031

Acknowledgment
This research was supported by Hibah PITTA 2018 funded by DRPM Universitas Indonesia No.5000/UN2.R3.1/HKP.05.00/2018

The Effect of NaOH Immersion Time to Physical Properties of *Purun Tikus* (*Eleocharis dulcis*) Cement Board Composite

Ninis Hadi Haryanti[1, a], Henry Wardhana[2], Suryajaya[1, b]

[1]Program Studi Fisika, Fakultas Matematika dan Ilmu Pengetahuan Alam, Universitas Lambung Mangkurat Banjarbaru, Jl. A.Yani Km. 36, Indonesia
[2]Program Studi Teknik Sipil, Fakultas Teknik, Universitas Lambung Mangkurat Banjarbaru, Jl. A.Yani Km.36, Indonesia

E-mail: [a]ninishadiharyanti@gmail.com or [b]suryajaya@gmail.com

Abstract. A research in cement board composite made from *purun tikus* (Eleocharis dulcis) has been conducted. The aims are to determine its physical properties as variations of *purun tikus* fibers mass and NaOH immersion time. The results were yielded 5.00 - 15.15%, 0.52 - 0.76 g / cm3 and 0.05 - 0.18%, respectively for water content, density and thickness development. For moisture content and density values, the results were not met the standard of SNI 03-2104-1991, while for thickness development of all samples met the standards. The results showed fluctuating changes in water content, density and thickness development, which were influenced by the immersion time and mass of *purun tikus* fibers. From the results, it can be concluded that shorter immersion time and less *purun tikus* fibers yielded a smaller value of water content and thickness development, while density was become higher.

Keywords: *purun tikus*, NaOH, cement board composite, water content, thickness development

1. Introduction

At present, the use and utilization of natural fibers as material reinforcement were growth continuously and very much in demand by the industrial world. the natural fibers have low density, biodegradable, recycled easily, low energy production, have good mechanical properties and could be renewed as they come from nature [1]. besides that, the basic advantages of natural fibers are their abundant quantities and low costs. composite materials with natural fiber reinforcement such as bamboo, hemp, and banana have been applied to the automotive world as panel reinforcement for doors, rear seats, dashboards, and other interior devices [2].

Indonesia has a wide range of biodiversity so it has a great opportunity to explore the use of natural fiber materials. In terms of the availability of raw materials for natural fiber, in South Kalimantan the raw material of *purun tikus* plants (*Eleocharis dulcis*) is quite abundant. *Purun tikus* are wild plants that lead to weeds in open locations. These *purun tikus* plants can be said to be acid sulphate-specific, because they are resistant to high acidity (pH 2.5 -3.5). Therefore these plants can be used as indicator vegetation for acid soils [3, 4]. The morphological characteristics of *purun tikus* plants are straight stems, no branched, gray to shiny green with a length of 50-200 cm and thickness of 2-8 mm. While the leaves shrink to the basal, thin midrib like a membrane, the ends are asymmetrical, reddish brown [5]. Based on data from Industry and Investment Office of Barito Kuala in 2006, the distribution of purun plants has an areas about 713 Ha, covering purun danau ± 641 Ha and *purun tikus* ± 72 Ha [6].

To get high tensile strength properties, natural fibers are usually given a variety of treatments. The treatment given was alkaline immersion which aims to improve adhesive properties. Adhesiveness is the attachment of the interface surface of the elements that are joined together. The interface referred to in the composite is a surface formed by a joint bond between the reinforcement (fiber) and the matrix that forms the intermediate bond needed for load transfer. *Purun tikus* fiber is a cellulose material, so it has a complex structure. For this reason additional special treatment is needed to remove lignin. Alkaline ingredients, such as NaOH, are expected to eliminate the content that binds to cellulose in *purun tikus* fibers.

Cement board composite is a panel made from particles of ligni-cellulose and Portland cement as a binder. The comparison between particles and cement in board fabrication is 1: 2.75. The board has several advantages, which are, resistance to moisture, fire, mold and destructive insects [7]. Research in the fabrication of cement board composites made from *purun tikus* fiber with no immersion treatment was already carried out. The chemical analysis of *purun tikus* such as, water content (9.50% on shoots, extractive content (4.45% at base), low holo-cellulose content (52.62% on shoots) and moderate levels of lignin (25.80% on the stem) were already obtained [8]. From the test of the mechanical properties of *purun tikus*, the samples from the bottom of *purun tikus* were yielded diameter of 2.158 mm and tensile strength of 3.63 Mpa. While samples of the top of *purun tikus* were yielded 2.064 mm in diameter and tensile strength of 4.21 Mpa. The *purun tikus* is possible to be used as natural fiber material in the manufacture of construction materials. It is recommended to use the top of *purun tikus* because of its tensile strength is 116.15% more compare to the bottom of the *purun tikus* [9]. Wardhana, H and Ninis H Haryanti (2017) did a research on the characteristics of cement boards made from *purun tikus* particles without immersion treatment. From the results of testing of *purun tikus* particles, they have met the requirements of SNI 03-2104-1991 on moisture content, density and thickness development. The composition of *purun tikus* fiber 100 g in the form of matrix reinforcement is the right composition to be used as cement board composite material [10].

Purun tikus fiber as a strengthening material would greatly determine the properties of the cement board composite because it would continue the load that distributed by the matrix. *Purun tikus* fiber with alkali immersion treatment (NaOH) is expected to produce maximum physical properties of cement board composites, therefore research were needed. In this research, the physical properties of cement board composite as the effect of immersion time and *purun tikus* composition would be investigated. The expected benefit is *purun tikus* can be used as natural fiber in making cement boards. It is useful for building planners to reduce the construction burden.

2. Research Method

This research is a continuation of previous research. The results from the previous study which are *purun tikus* characteristic test in terms of chemical, physical and mechanical properties were used as the basis for this study [8, 9]. Besides that the composition of the materials used is also based on the results of previous studies [10, 11].

Purun tikus in the form of fibers was used as material for manufacturing the Cement Board composites. *Purun tikus* were immersed in 50 ml of NaOH 5% solution mixed with 1 liter of distilled water. The immersed time were varied for 1 hour, 2 hours, 4 hours and 6 hours. Then *purun tikus* were mixed with PPC Cement (Pozolan Portland Cement). The variation of *purun tikus* fibers mass used were 50 gr and 100 gr while PPC Cement (Pozolan Portland Cement) were 575 gr and 525 gr. Besides that, 110 mL Aquades and 10 mL Adhesives (Addition H.E, ASTM (494-81, Type A) were also added. Then the mixture was poured in composite board mold of 25 cm x 25 cm x 1 cm in size. The investigation parameters were water content, density and thick development. In measuring the physical properties of the cement board, the sample used was 5 cm x 5 cm x 1 cm. The result would be compared to Indonesian National Standards 03-2104-1991.

3. Results and discussions

The most problem of natural fiber composites is the lack of good bonding between the matrix and the fiber resulting in poor composite properties. This deficiency is caused by the nature of natural fibers which could still absorb water. The water would enter the bond between the matrix and fiber. In turn, it would be affecting the physical properties of the composite. The right chemical treatment could increase the bond between fiber and matrix, so that the composite properties are getting better.

Kuncoro conducted a study to increase the bond between fiber and matrix (adhesive) where the clean fibers were soaked in alkaline solution (NaOH 5%). The immersion time variations were 0, 2, 4, and 6 hours. As the results, the optimal tensile strength was yielded by composite materials which are reinforced by hemp fiber with a 2-hours alkali treatment [12].

The water content of the cement board composite as a function of the variations of immersion time and *purun tikus* mass was shown in Figure 1. It can be seen that the immersion time and composition of *purun tikus* and cement could affect the water content. The water content of the cement board composite has increased as the immersion time increased. As can be seen in the graph, for samples with 100 grams of *purun tikus* and 525 grams of cement have higher water content as compare to samples with 50 grams of *purun tikus*. This was caused by the composite would absorb water when it was immersed in distilled water and increased the thickness. Therefore the composition which has more fiber would have more water content.

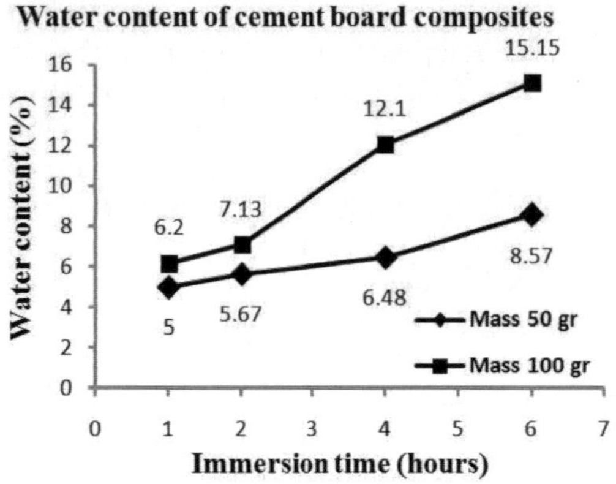

Figure 1. The Water content of composite to immersion time and purun mass variation.

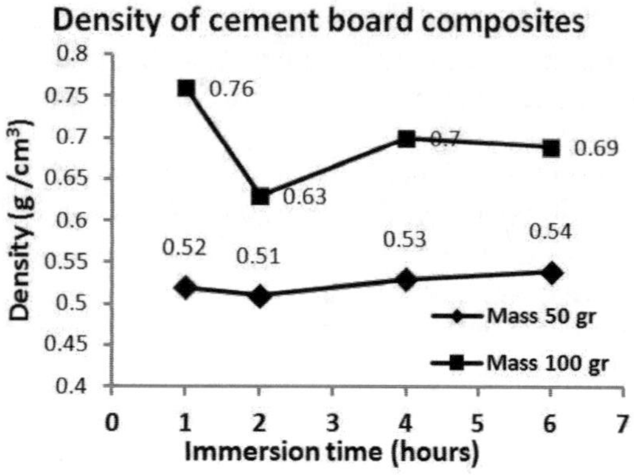

Figure 2. The composite density to immersion time and purun mass variation.

The water content of the cement board composite obtained was in the range of 5.00 to 15.15%. The highest value of water content was obtained for 100 grams of *purun tikus* and 525 grams of cement with 6-hours immersion time, which was 15.15%. While the lowest water content value was obtained for 50 grams of *purun tikus* and 575 grams cement with immersion time 1 hour, which was 5.00%. According to Purwanto [13], the water content value of cement board composite is influenced by several factors. Firstly, the condition of the storage because the cement board has hygroscopic properties towards the surrounding water vapor, the others are influenced by the composition of cement used and particle size. Research conducted by Violet in 2012 said that the use of more cement can cause particle cell cavity filled with less water because the cavity has been filled with cement that has hardened. It is presumed that more cement would cause the water evaporated easily thus produced lower water content. From the results, it can be seen that not all samples were met the requirements of SNI 03-2104-1991 which has the maximum value for water content of 14% [14]. A good cement composite board should have low water content. The results of the water content of samples show that for short immersion time and light *purun tikus* fibers mass would has a low water content value. Therefore the composite board with 50 grams of *purun tikus* fibers and 575 grams of cement was a good composite board. In this study, the best composite board was made from 50 grams of *purun tikus* and 575 grams of cement with 1 hour immersion time which has the smallest water content, about 5.00%.

The density of cement board composite as variations of immersion time and *purun tikus* mass is shown in Figure 2. From Figure 2, the density of the cement board composite is obtained in the range of 0.52–0.76 g / cm^3. Based on SNI 03-2104-1991 the minimum value for density is 0.57 g / cm^3, thus not all samples meet SNI requirements. The cement board composite was thicker than expected. It is assumed that when the composite released from molds and clamps, the particles that were not perfectly bound with the cement were detached. As a result, the arrangement of material components in the board is not tight and allows the existence of pores. This causes the board's volume becomes larger and the density becomes smaller.

The highest density value was the sample of 50 grams of *purun tikus* and 575 grams of cement at 1 hour immersion time, which was 0.76 g / cm^3. While the lowest density value was obtained for 100 grams of *purun tikus* and 525 grams of cement with 2 hours immersion time, about 0.51 g / cm^3. If the cement composition is getting higher, the density would also become higher. It was caused by the composite has a lot of binding material, which is cement. It allows the formation of good bonds. This is in line with Simbolon et al. higher cement composition would give higher density, because the weight of the cement board would be increased as factor of the amount of cement used [15]. Purwanto said that a high density of cement board can be attributed to the properties of the raw materials of particles used which could increase bonding with cement [16].

It can be seen from Figure 2, all cement board composite samples with 50 grams of *purun tikus* and 575 grams of cement at immersion time of 1 to 6 hours obtained high density, which were 0.69-0.76 g / cm^3. While the minimum value of SNI standard for density is 0.57 g / cm^3. Whereas for 100 grams of *purun tikus* and 525 grams of cement with immersion time of 1 to 6 hours obtained average density of 0.5 g / cm^3. From the results of the density test, it was found that shorter of immersion time and smaller the mass of the purun fiber (higher cement mass) gave higher density. The highest density value was yielded by cement board composites with 50 grams of *purun tikus* and 575 grams of cement. So, the board composites, which met the minimum standard of SNI in density, were the mixture of 50 grams of *purun tikus* and 575 grams of cement at immersion time of 1 to 6 hours.

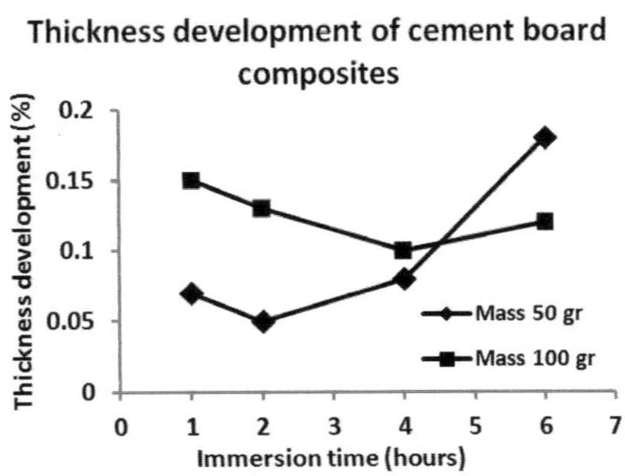

Figure 3. The results of thickness measurement to immersion time and purun mass variation.

The measurement of cement board composite thickness as variations of immersion time and *purun tikus* mass is shown in Figure 3. The values of the development of the cement board composite thickness were 0.05–0.18%. Based on SNI 03-2104-1991 the maximum thickness development (swelling) is 12%. Thus, all cement board composite samples meet SNI standards. The highest thick development value was obtained for 50 grams of *purun tikus* and 575 grams of cement with immersion time of 6 hours, which was 0.18%. While the lowest thick development value was obtained for 50 grams of *purun tikus* and 575 grams of cement with 2 hours immersion time, which was 0.05%. The thickness development value after 2 hours of immersion is smaller than the 6 hours immersion. This is due to the immersion time affecting the amount of water that could be penetrated through the pores of composite board.

Based on the results, there was a tendency that the thickness development was decreased as the cement composition increased. This is due to the increase of ratio cement used makes the particles able to be tied better and makes the board structure denser so the physical properties of the board increase. As a result, when the board was immersed, water is hard to penetrate into the pores of the board, causing low thick development values. This is in line with Simbolon et al. (2015) which states that the more cement composition is in used, the more particles could be bound by the cement [15].

Olufemi et al (2012), said that the development of cement board thickness was also influenced by the comparison of the mixture of raw materials with cement [17]. The greater cement content on the board will reduce thick development, also the use of a mixture of raw materials [17].

As can be seen from Figure 3, there was a decrease in the thickness of the *purun tikus* mass 50 and 100 grams at immersion times of 2 and 4 hours, but increased at 6 hours immersion time. From the results, the thickness development would become higher when samples have longer immersion time and smaller purun mass. Theoretically, smaller value of thickness development would give better cement board composite. From the results, cement board composites of 50 grams of *purun tikus* and 575 grams of cement at 2 hours immersion time was the best composite obtained. It has the smallest thickness development which was equal to 0.05%.

The nature of *purun tikus* as natural fibers is hydrophilic. The effect of alkali treatment on the nature of the surface of natural fibers, the optimum content of water can be reduced so that the nature of hydrophilic fibers can provide interfacial bonds with the matrix optimally. NaOH is an alkaline solution that is classified as easily soluble in water and is a strong base that can be ionized perfectly. NaOH is also moist liquid and spontaneously absorbs carbon dioxide from free air.

Alkalinization of fiber is a process of modifying the fiber surface by immersing fiber into alkaline bases. The following reaction describes the process that occurs at the alkali treatment of fibers:

$$\text{Fibers-OH} + \text{NaOH} \text{ ----- } \text{Fibers-O-Na}_{+} + \text{H2O}$$

The purpose of the alkalization process is to reduce the fiber constituent components which are less effective in determining interface strength, which are hemicelluloses and lignin or pectin. As the components of lignin and hemicelluloses were reduced, it would produce a better fiber surface structure and easier to be moistened by resin, resulting in better mechanical interlocking. When there is a chemical reaction between fiber and alkali, a reaction time is needed, if the reaction time is less than is required the reaction is not optimal while if the reaction time is excessive, the fiber would be damaged. The process of immersion treatment has an influence on the surface of the fiber, the length of time will make the fiber surface cleaner and the surface of the fiber becomes more rough so that the fiber bond with the matrix will be better (more adhesive), increasing the physical properties of the composite formed.

4. Conclusions

Cement boards composite made from *purun tikus* fibers were successfully fabricated. The physical properties of water content, density and thickness development were yielded 5.00 - 15.15%, 0.52 - 0.76 g / cm^3 and 0.05 - 0.18%, respectively. Comparing to SNI, only thickness development was meet the standard. The effects of NaOH immersion time on cement board composites have been observed. Shorter immersion time would cause lower the water content, higher the density and also increase the thickness development. The effects of material composition on the cement board composites were also observed. The composition of each material would be affecting homogeneity on the surface of the *purun tikus* cement board composite. When the composition of *purun tikus* fibers is smaller than the cement, the value of the water content is low, the density is high and the thickness development value is small. A good composite is a composite with a low water content and thick development, while its density is high. As for suggestions, further research needs to be done using other alkaline materials, for example $KMnO_4$, with varying immersion time variation.

5. References

[1] Wang B, Panigrahi S, Tabil L, Crerar W J, and Sokhansanj S 2003 Modification of flax fibers by chemical treatment *Journal The Society for Eng. In Agricultural, Food, and Biological Systems* **3** 337
[2] Boeman R G and Johnson N L 2002 Development of a cost competitive, composite intensive, body-in-white *Journal SAE* (2002-01-1905)
[3] Steenis C G G J Van 2006 *Flora* (Jakarta: PT Pradnya Paramita)

[4] Brecht J K 1998 *Water chestnut* Horticultural Sciences Department University of Florida http://www.hortisci.org
[5] Noor M 2004 *Lahan Rawa Sifat dan Pengelolaan Tanah Bermasalah Sulfat Masam* (Jakarta: PT Raja Grafindo Persada)
[6] Rahadi 2007 Penelitian *Penyebaran jenis tumbuhan purun* (Barito Kuala Kalsel)
[7] Suhasman 2012 Perbandingan karakteristik papan semen dari batang dan cabang kayu asal hutan rakyat *Jurnal Perennial* **8** (1) 30-5
[8] Wardhana H et all 2015 Chemical, physical, and mechanical features of *purun tikus* (*Eloecharis dulcis*) fiber *Asian Academic Research Journal of Multidisciplinary (AARJMD)* **2** (3) 127-34
[9] Haryanti N H and Wardhana H 2016 *Prosiding Simposium Fisika Nasional SFN XXIX Physics For Future Development of Science and Technology* 19-21 September 2016 *Sifat mekanik purun tikus sebagai bahan komposit* p 50-54, Universitas Hasanuddin Makassar Indonesia
[10] Haryanti N H dan Henry W 2017 *Purun tikus* (*Eleocharis dulcis*) fiber composition as cement board composite material, *Journal of Biodiversity and Environmental Sciences (JBES)* **11** (3) 137-42
[11] Wardhana H and Haryanti N H 2017 The characteristics of *purun tikus* particle board cement board *IOSR Journal of Applied Chemistry (IOSR-JAC)* **10** (1) 01-4
[12] Kuncoro D 2008 Pengaruh perlakuan alkali terhadap sifat tarik bahan komposit serat rami-polyester *Jurnal Teknik Mesin* **8** (1) 8-13
[13] Purwanto D 2016. Sifat fisis dan mekanis papan semen dari serat tandan kosong kelapa sawit. *Jurnal Riset Industri Hasil Hutan* **8** 43-52
[14] Violet 2012 Sifat Fisika dan keteguhan patah papan semen partikel dari pelepah kelapa sawit (*Elaeis guineensis* jack) *Jurnal Hutan Tropis* **13**. 38-44
[15] Simbolon IL, Sucipto T, Hartono R, 2015 Pengaruh ukuran partikel dan komposisi semen partikel terhadap kualitas papan semen dari cangkang kemiri (*Aleurites Moluccana* Wild) *Peronema Forestry Science Journal* **4** (1) 41-8
[16] Purwanto D 2014 Sifat fisik mekanik pada papan semen dari limbah kulit kayu galam *Jurnal Riset Industri* **8** (3) 197–204
[17] Olufemi AS, Abiodu O, Omajor, Paul FA, 2012 Evaluation of cement-bonded particle board Produced from Afzelia Africana wood residues *Journal of Engineering Science and Technology* **7** (6) 732-43

Acknowledgement
Thank you to all those who have helped in this research, especially to the Faculty of Engineering, Lambung Mangkurat University, which has provided funding for this research.

Water Absorption and Flexural Property of Unidirectional Polypropylene/Sumberejo Kenaf Fiber Composites

A P Firdausi, A L Juwono*
Department of Physics, Faculty of Mathematics and Natural Sciences, Universitas Indonesia, Depok 16424, Indonesia

*Corresponding Author: ariadne.laksmidevi@ui.ac.id

Abstract. Kenaf fiber can be used as an alternative to synthetic fiber reinforcement in composites. This research was conducted in order to determine the flexural and diffusion properties of the composite by varying the fiber fractions. Before being used as a reinforcement, kenaf fiber was firstly treated with alkali treatment in NaOH solution for 24 hours. Polypropylene and kenaf fibers were fabricated with a hot press machine. The fiber fractions were 30 wt%, 40 wt%, 50 wt%, and pure polypropylene samples were also fabricated as a comparison. The highest flexural strength, the minimum water absorption and water content were found in the polypropylene/30 wt% kenaf fibre composites with the value of (4.77 ± 0.799) MPa, (3.61 ± 0.823)% and (0.192 ± 0.154)% respectively.

Keywords: kenaf fiber, polypropylene, flexural strength, water content, water absorption

1. Introduction

Nowadays, composites become one of materials that are widely used in daily necessities, from household appliances to automotive. Synthetic fibers have been used in composites as reinforcements but it have non-biodegradable properties so they are not environmentally friendly materials [1]. Composites with natural fiber reinforcement can be an alternative to replace synthetic fibers that can cause damage to nature. The use of natural fibers as reinforcement in composites has been widely researched and developed at this time. Increased utilization of natural fibers was due to low costs, high specific modulus, lightness, and lower energy requirement [2].

Table 1. Chemical composition of kenaf fiber [4].

Cellulose (%)	45.0 – 70.6
Hemicellulose (%)	12.3 – 21.5
Lignin (%)	8.0 – 13.0
Pectin (%)	3.0 – 5.0

Kenaf fiber is one of the natural materials that can be used as an alternative to synthetic fibers. Kenaf is a non-wood lignocellulose material because its main constituent materials are cellulose, hemicellulose, and lignin [3]. Table 1 shows the chemical composition of KF [4].

Before being used as a reinforcement in composites, natural fibers need alkali treatment. Mechanical properties such as strength, flexibility, and stiffness of natural fibers can be improved by this treatment [5]. To obtain optimal strength of kenaf fiber composite, NaOH 5% aqueous solution was used [6]. Fiber fractions were varied to determine which fiber fractions is the best. Based on research conducted by Ollivia [7], the highest tensile strength and deflection temperature were obtained from PP/40wt% Sumberejo kenaf fiber composites. There is a lack of study in water content and flexural properties of PP/Sumberejo kenaf fiber composites. The aim of this research was to determine the value of water content, water absorption, and flexural properties of PP/Sumberejo kenaf fiber composites to meet the requirements of SNI 01-4449-2006.

2. Experimental Method

2.1. Preparation of Kenaf Fiber

Kenaf fibers (KF) and polypropylene (PP) were materials used in this research. KF was initially treated with alkali treatment to remove lignin by soaking it in NaOH 5% solution for 24 hours. Treated KF was rinsed with distilled water until the distilled water was colorless. After that, KF was dried at room temperature for 48 hours and in the oven with the temperature of 60°C for 24 hours. Both untreated and treated KF were analyzed using Fourier Transform Infrared Spectroscopy (FTIR) to identify the chemical change of the fibers in a range of 400 – 4000 cm^{-1}. For this measurement, pelletized mixtures of KF samples and potassium bromide (KBr) powder were used.

2.2. Fabrication of Composite

PP pellets were extruded into sheet forms using a hot press machine at 190 °C for 6-7 minutes then were cooled using a cold press machine for 2-3 minutes. The purpose of turning PP into sheet forms was to make PP easily dispersed and wetted the KF. PP and KF respectively weighed and arranged unidirectionally into molds in accordance with certain weight fraction and it was fabricated using a hot press machine with a pressure of 5 MN/m^2, temperature of 190°C for 6-7 minutes. Fiber fraction used in this experiment was 30 wt% (PP/KF30), 40 wt% (PP/KF40), and 50 wt% (PP/KF50).

Referring to ASTM D7264, the flexural test was carried out by cutting the specimens with size ratio thickness: length equal to 1:20. Then, the specimens was given a dot right in the center. The flexural test conducted with pressure speed of 10mm/minute.

SNI 01-4449-2006 was used as a reference in this research. Specimens with a dimension of 50 mm x 50 mm were tested by soaking the specimens in water at room temperature for 24 hours. To calculate the value of water absorption in the specimens, Equation (1) was used.

$$Water\ Absorption\ (\%) = \frac{W_{before} - W_{after}}{W_{after}} x100 \quad (1)$$

According to SNI 01-4449-2006, water contents of composites were obtained based on the weight of specimens with a dimension of 50 mm x 50 mm before and after being heated in an oven at a temperature of 103 °C until the weight was constant. Equation (2) was applied to calculate the value of water content.

$$Water\ Content\ (\%) = \frac{W_{wet} - W_{dry}}{W_{dry}} x100 \quad (2)$$

Specimen morphology before and after flexural test was observed with an optical microscope (OM) to see the type of damage of the composite surfaces.

3. Result and Discussion

Figure 1 showed the Fourier Transform Infrared spectra of untreated and treated KFs by alkali treatment. The broad peak around 3300 cm^{-1} can be associated with the O-H group of raw kenaf fiber and was identified as cellulose. The peak shown at 2978 cm^{-1} suitable to the C-H group. The area between 2000 and 1000 cm^{-1} were stretched carbonyl group (C=O, 1745 cm^{-1}) and were related to the presence of lignin and/or hemicelluloses. The peak at 1732 cm^{-1} belongs to hemicelluloses and it was showed that hemicelluloses were removed from the fiber surface after the alkali treatment. C-O stretching of the lignin was found at peak 1243 cm^{-1} and the loss of this peak in treated KF confirmed the removal of lignin by alkali treatment. These results are similar to a research by Zarina *et al*, who conducted research on different treatments to kenaf fibers [8].

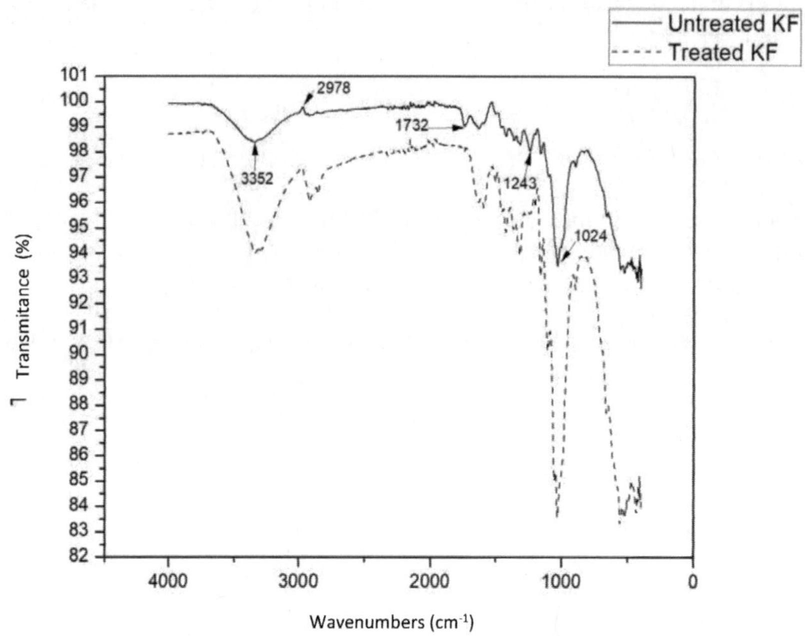

Figure 1. FTIR spectra of untreated and treated Sumberejo kenaf fiber.

Figure 2 represents the flexural strength and flexural modulus of PP/KF composites. Based on this research the highest flexural strength and flexural modulus were found in PP/KF30 with the value of (4.77 ± 0.799) MPa and (1.13 ± 0.574) MPa respectively. These results have similar trends with a research about epoxy/KF composites, conducted by Sapiai N *et al* [9]. In Sapiai research, it was indicated that the presence of kenaf fiber increased the flexural strength and flexural modulus.

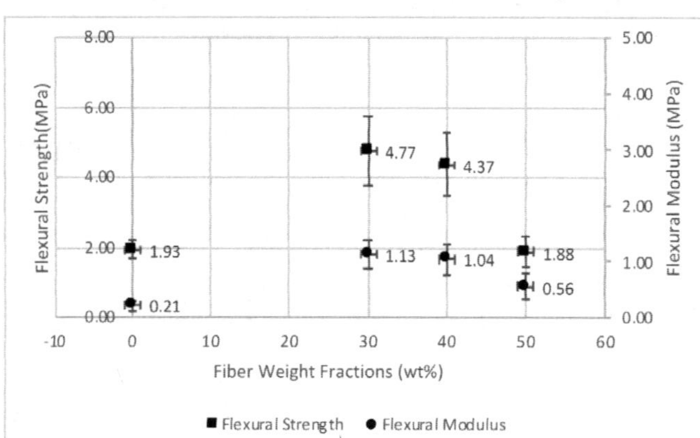

Figure 2. Flexural strength and flexural modulus of PP/Sumberejo kenaf fiber composites versus fiber fraction.

IOP Conf. Series: Materials Science and Engineering **599** (2019) 012014 doi:10.1088/1757-899X/599/1/012014

Figure 3 presents the water absorption of PP/KF composites at temperature of 25°C for 24 hours. PP/KF50 composites water up-take has a greater value because these composites contain the highest KF that absorbed more water. According to previous research on kenaf fiber/unsaturated polyester composites conducted by E. Osman [10], the higher the amount of fiber fraction, the higher the water absorption. PP/KF30 has the smallest value of (0.192 ± 0,154) %. All of the Sumberejo kenaf fiber composites fulfilled SNI 01-4449-2006.

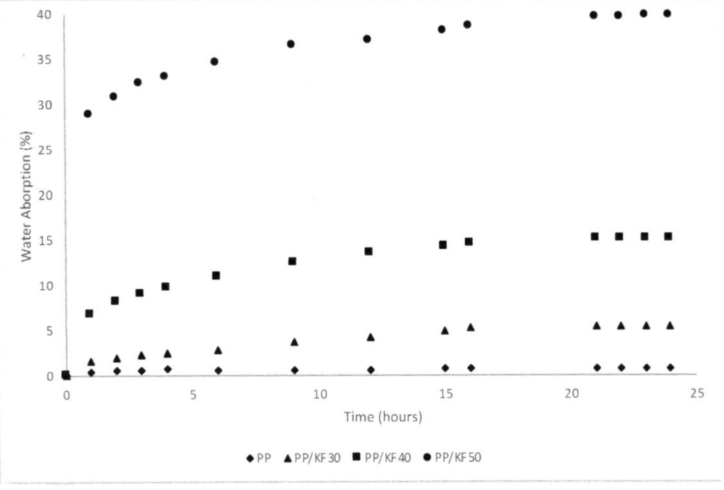

Figure 3. Water absorption test of PP/Sumberejo KF composites.

The result of water content test is illustrated by Figure 4. The water content of all specimens compatible with SNI 01-4449-2006. However, the PP/KF30 composites had the lowest value of water content because these composites had the lowest fiber loading. PP as a polymer does not contain water and KF was the one that released water from the composites.

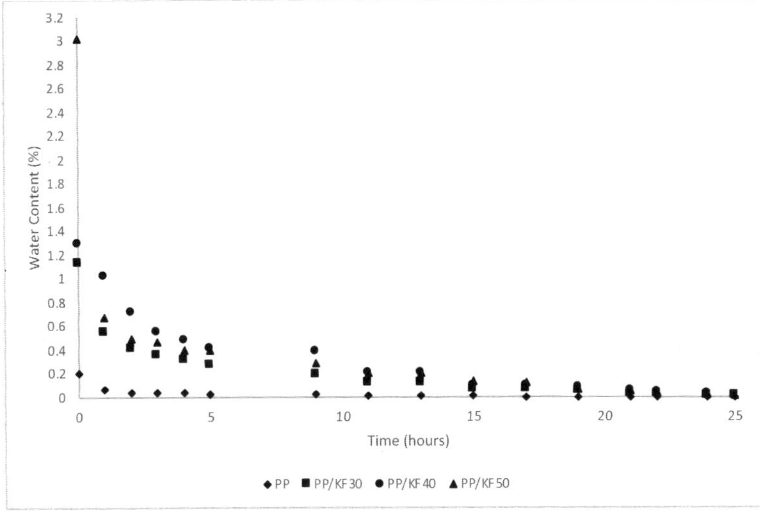

Figure 4. Water content test for PP/Sumberejo KF composites.

PP/KF composites were observed with OM before and after flexural test. Figures 3 (a), (c), (e) show the OM images of PP/KF30, PP/KF40, PP/KF50 respectively before flexural test. These images show a typical homogeneous PP/KF surfaces and this indicates that the matrix and the fiber has a good interface. While figures 3 (b), (d),(f) show the OM images of PP/KF30, PP/KF40, PP/KF50 after flexural test. These images exhibit that fiber pull-out occurred on all composites. This mode of failure indicates that the kenaf fiber was able to hold the flexural load.

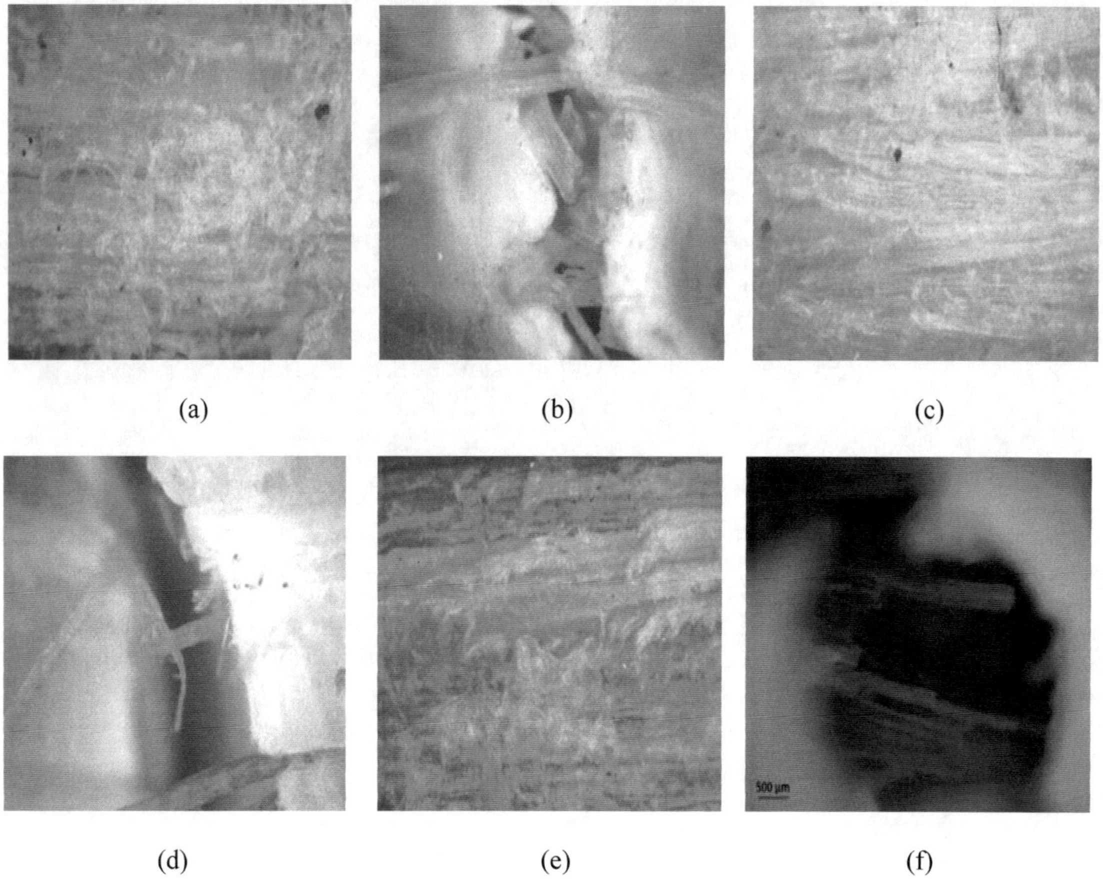

Figure 5. OM image of PP/KF30, PP/KF40, PP/KF50 respectively, (a), (c), (e) before flexural test, (b), (d), (f) after flexural test.

4. Conclusion

It can be concluded that the highest flexural strength, the minimum water content and water absorption were found in PP/Sumberejo KF30 composites with the values of (4.77 ± 0.799) MPa, (3.61 ± 0.823)% and (0.192 ± 0.154)%, respectively. The optical microscope observation indicated fiber pulled-out occurred after the flexural test.

5. References

[1] Haameem M J A, Abdul Majid M S, Afendi M, Marzuki H F A, Fahmi I. and Gibson A G 2016 *Compos. Struct.* **136** 1

[2] Zahari W Z W, Badri R N R L, Ardyananta H, Kurniawan D, and Nor F M 2015 *Procedia Manufacturing* **2** 573

[3] Izran K, Mohd Zharif T, Beyer G, Mohamad Jani S, Noor A A R, and Yanti A K 2014 *Journal of Science and Technology* **6** 41

[4] Faruk O, Sain M 2015 *Compos. Mater.* **7** 72

[5] Suharty N S, Ismail H, Diharjo K, Handayani D S, and Firdaus M 2015 *Proced. Chem.* **19** 253
[6] Asumni O M L, Reid R G, and Paskaramoorthy R 2012 *Composites: Part A* **43** 1431
[7] Ollivia S L, Juwono A L, and Roseno S 2016 *Conf. Ser: Mater. Sci. Eng* **196** 012023
[8] Zarina S and Ahmad I 2014 *Bioresources* **10** 256
[9] Sapiai N, Jumahat A, Mahmud J 2015 *Tech. Journal* **76** 115
[10] Osman E, Vakhguelt A, Sbarski I, S Mutasher 2011 *Compos. Mater*

Acknowledgments
This research was funded by Universitas Indonesia for PITTA grant in 2018.

IOP Conf. Series: Materials Science and Engineering **599** (2019) 012015 doi:10.1088/1757-899X/599/1/012015

Effect of Subang Pineapple Leaf Fiber Loading on Flammability and Mechanical Properties of Pineapple Leaf Fiber Reinforced Polypropylene Composites

Toho Dustin Sutomo, Ariadne L Juwono

Department of Physics, Faculty of Mathematics and Natural Sciences, Universitas Indonesia, Kampus UI Depok 16424, Indonesia

E-mail: toho.dustin@ui.ac.id
Corresponding author: ariadne.laksmidevi@ui.ac.id

Abstract. Environmentally friendly composites are composites with one or all of the constituent elements being natural materials. Indonesia with high biodiversity resources produces a lot of various types of natural fibers as reinforcements, one of them is pineapple leaf fiber (PALF). This study aimed to determine the effect of Subang PALF contents on flammability, impact resistance, and hardness of polypropylene (PP) / Subang PALF composites. Three compositions of PALF in PP were fabricated by hot press method. Pristine PP samples were also prepared for comparison. Flammability, impact, and hardness test results showed that composites with 40 wt.% PALF were the best with their respective values of (15.7 ± 0.4) mm/minute, (31.6 ± 4.5) J/cm^{2,} and (60.5 ± 2.01) HD.

1. Introduction

Synthetic fibers that have been used for reinforcement in composites cannot be degraded technically. The use of synthetic fibers is not environmentally friendly. This problem attracts attention and environmental issues such as pollution and pollution from fuel. Therefore it is expected that natural fibers can be an alternative material for synthetic fibers and provide better impact on the environment [1]. Indonesia is one of the countries that has a potential to develop natural fibers such as jute, flax, pineapple, cotton, abaca, agave, linum, rosella, and kenaf [2]. Pineapple or Ananas comosus is an alternative fiber-producing plant which has only been used as a textile fiber-producing material [3]. Utilization of pineapple leaf as reinforcing fibers in composites makes pineapple leaf waste has added value, one example is as composite boards [4].

Pineapple leaf fibers (PALFs) have good mechanical properties associated with high cellulose levels [5]. PALFs are used effectively in polymer matrix to develop composites with better mechanical properties [6]. The most common treatment to increase fiber and matrix adhesion interaction is using chemical treatment. The chemical treatment modifies the fiber surface is called alkali treatment using NaOH solution. Alkali treatment on fiber can increase roughness on the fiber surface and remove lignin, wax, and oil from fiber cell walls. Increasing the effective contact area of the fiber creates higher adhesion between the fiber and the polymer matrix and increase the mechanical properties [7,8].

Several previous studies have been conducted to determine the flammability and mechanical properties of natural fiber reinforced polymer composites [9-11]. The impact strength of PALF/PP composites were studied on various fiber loading and showed that the impact strength of the composites increased with fiber loading [9]. Rockwell hardness test was tested on PALF reinforced epoxy resin composites and showed that the hardness of composites increased with the fiber loading [10]. Flammability tests were carried out on PP/sisal composites using horizontal and vertical burning test [11]. There is a lack of study in PP/PALF composites. The aim of this study is to investigate the effect of weight fractions of Subang PALF on the flammability and mechanical properties of PALF/PP composites.

2. Materials and Method

2.1. Materials and Fiber Pre-treatment

Polypropylene resin was supplied by PT Chandra Asri Petrochemical Tbk, Indonesia. The polypropylene used has type B132AN. Pineapple leaf fibers were harvested from Subang, Indonesia. The PALFs were soaked in the 5% of the NaOH solution in water reservoir for 5 hours at room temperature. After that, fibers were rinsed several times with distilled water until the distilled water was colourless and then the fibers were dried at room temperature for 24 hours. PALFs were heated in an oven for 24 hours.

2.2. Preparation of Composites

PP as matrix in the form of pellets were melted into sheet form using a hot press method at 190°C. PP which was already in sheet forms were combined with PALFs based on the variations in fiber weight, namely 20 wt.%, 30 wt.%, and 40 wt.%. PP/PALF composites were fabricated by using a 220 mm x 130 mm x 4.5 mm stainless metal plate. The plate was placed into the hydraulic hot press at 190°C temperature. The stainless steel plate was removed from the hot press machine after 20 minutes and kept in room temperature for cooling.

3. Testing

3.1. Fourier Transform Infrared Spectroscopy

Fourier Transform Infrared Spectroscopy (FTIR) measurement was conducted on PALFs using Parkin Elmer tool. The test was conducted to analyze the organic compounds present in PALFs before and after PALFs alkali treatment. The measurements was carried out in the wavelength region of 500 until 4000 cm^{-1}.

3.2. Flammability Test

Flammability tests were conducted using flammability tester according to UL-94 HB. The dimension of the specimens used was 125 mm x 13 mm x 4.2 mm. The specimen was burnt horizontally from one end of the specimen.

3.3. Impact Test

Impact tests were conducted according to ASTM D 6110-10. The impact strength of the composites was measured using GOTECH GT-7045 impact testing machine. The dimension of the specimen used was 125 mm x 12.7 mm x 3 mm.

3.4. Hardness Test

The hardness of the composites was measured using D shore durometer according to ASTM D2240. Durometer used to measure hardness was attached to the surface of the sample to be measured at room temperature. Hardness test was carried out on composite samples which had a thickness of 4.2 mm.

4. Results and Discussion

4.1. Fourier Transform Infrared Spectroscopy

FTIR spectrum of PALFs before and after alkali treatment were shown in Figure 1. The FTIR spectra showed an interesting comparison about several peaks. The presence of hydroxyl (-OH), C-H stretching of methyl groups and C-H bending vibrations bands can be observed at 3339 cm^{-1}, 2892 cm^{-1}, and 1316 cm^{-1} respectively. These peaks belong to cellulose [12]. Lignin was indicated by the presence of peaks in the range of 1200-1300 cm^{-1} [13]. Figure 1 exhibits that the peaks in the range of 1200-1300 cm^{-1} appeared on the before alkali treatment PALFs spectrum, while these peaks disappeared on the after alkali treatment PALFs spectrum. Comparison of the two FTIR spectra for both PALFs showed that alkali treatment caused a decrease in lignin content in PALFs.

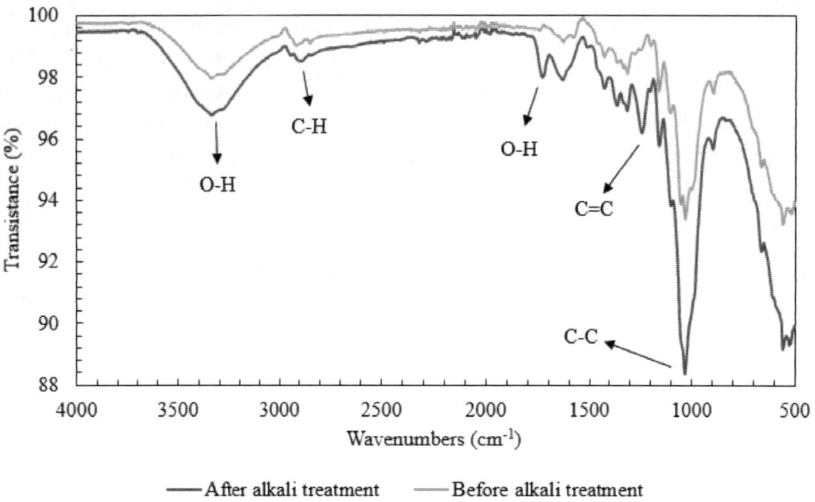

Figure 1. FTIR spectra of before and after alkali treatment Subang PALFs.

4.2. Flammability Properties

The horizontal UL-94 tests were conducted to quantify the flame rate of the composites. The results of the UL-94 tests for the polypropylene and PALF/PP composites are presented in the Figure 2. The horizontal test shows that the flame rate of PP is the highest. It is a fact that the PALF/PP 40% composites show the lowest combustion rate. It could be observed that, the flame retardancy of composites increase with the fiber loading. The flame retardancy increased because fibers in the composites inhibit oxygen in the composite combustion process. From this test, it was revealed that the more fiber loading, the less oxygen played a role in composites combustion process. This process agreed well with previous research by Barath *et al* [14].

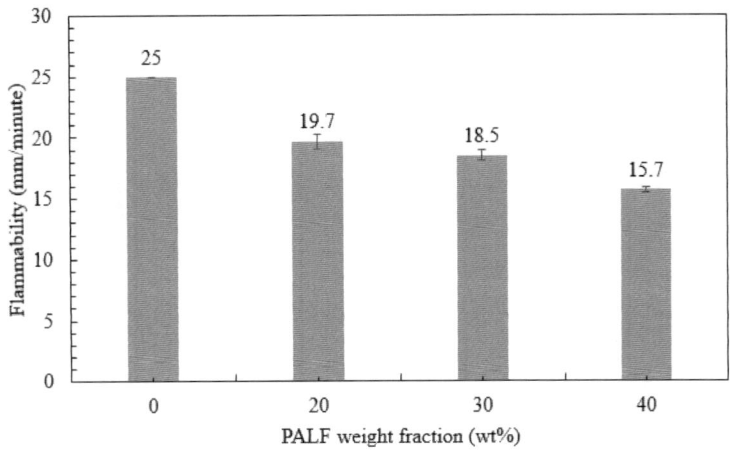

Figure 2. Flammability of Subang PALF/PP at different fiber loading.

4.3. Impact Strength Results

Impact strength of PP and PALF/PP composites at different fiber loading are shown in Figure 3. The highest impact strength is found in PALF/PP 40% composites. According to Figure 2, impact strength of PALF/PP composites increases as the fiber loading increases. The increase in impact strength occurred because with increased fiber loading, more energy is needed to hit the sample. From this test, it can be seen that the function of the fiber as the recipient of the load from the matrix takes place well. These results agree well with the previous study about PALF/PP composites [9].

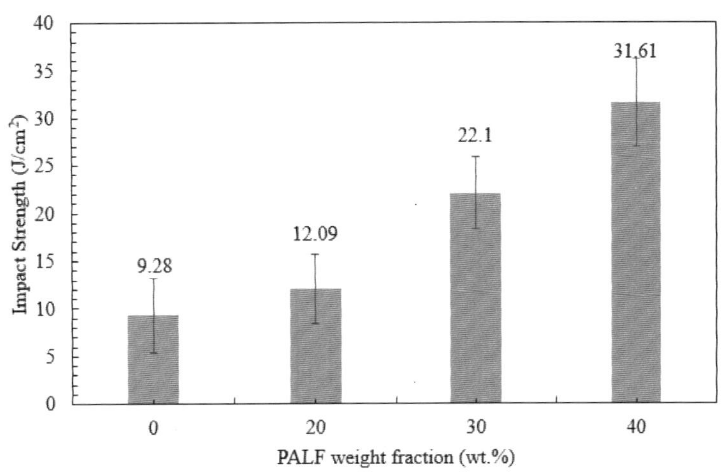

Figure 3. Variation of impact strength at different fiber loading.

4.4. Hardness Results

Hardness test of the PP and PALF/PP composites were determined and shown in Figure 4. The more the fiber loading, the higher the hardness value. PALF/PP 40% composites have the highest hardness number. The increased in hardness of the composites because the higher content of PALF which hold the load. The increment of composites hardness due to fiber loading were confirmed on PALF/epoxy composites as the PALFs had similar function [10].

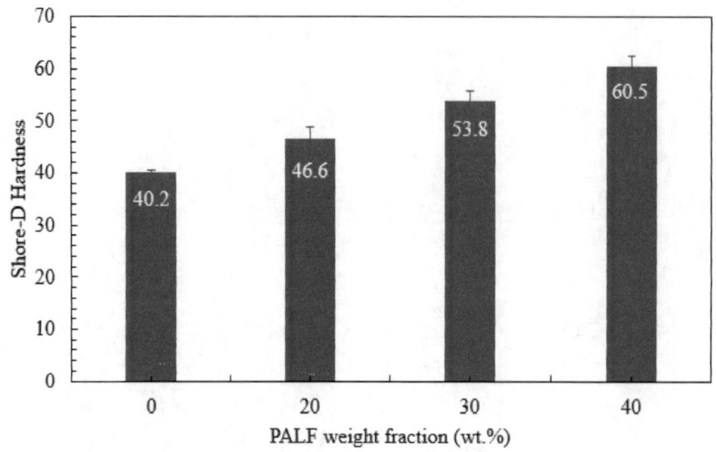

Figure 4. Variation of hardness at different fiber loading.

5. Conclusion

The effects of pineapple leaf fiber loading on flammability and mechanical properties on Subang PALF/PP composites were studied. It was concluded that PALF loading increased the mechanical properties and flame retardancy of Subang PALF/PP composites. Flammability, impact, and hardness test results showed that 40 wt.% Subang PALF/PP composites were the best composites with their respective values of (15.7 ± 0.4) mm/minute, (31.6 ± 4.5) J/cm^{2,} and (60.5 ± 2.01) HD.

6. References

[1] M Haameem J A, Majid M S A, Afendi M, Marzuki H F A, Fahmi I, Gibson A G 2015 Mechanical properties of napier grass fibre/polyester composites *Comp. Struct* **136** 1-10

[2] Sudjindro 2011 Prospek serat alam untuk bahan baku kertas uang. *Perspektif.* **10** 92-104

[3] Hidayat P 2008 Teknologi Pemanfaatan Serat Daun Nanas Sebagai Alternatif Bahan Baku Tekstil *Teknoin* **13** 31-35

[4] Hafizhah R, Juwono A L, Roseno S 2017 Study of Tensile Properties and Deflection Temperature of Polypropylene/Subang Pineapple Leaf Fiber Composites *IOP Conf. Series:*

[5] Odusote J K, Oyewo A T 2016 Mechanical Properties of Pineapple Leaf Fiber Reinforced Polymer Composites for Application as A Prosthetic Socket *Journals of Engineering and Technology* **7** 125-139

[6] Panyasart K, Chaiyut N, Amornsakchai T 2014 Effect of Surface Treatment on The Properties of Pineapple Leaf Fibers Reinforced Polyamide 6 Composites *Energi Procedia* **56** 406-413

[7] Li Y, Sreekala M S, Jacob M 2009 Textile composite based on natural fibres, in *Natural fibrereinforced polymer composites from macro to nanoscales* S Thomas, L A Pothan, ed Old City Publishing: Philadelphia chapter 8 pp 202-227

[8] Li X, Canada A, Tabil L 2007 Chemical Treatments of Natural Fiber for Use in Natural Fiber-Reinfored Composites: A Review *Journal of Polymers and The Environment* **15** 25-33

[9] Mochtar M, Rahmat A R, Hassan A 2007 *Characterization and Treatments of Pineapple Leaf Fiber Thermoplastic Composite for Construction Application Research VOT* **75147** Universitas Teknologi Malaysia chapter 4 pp 51-52

[10] Kumar S, Praveen B A, Aithal K, Keempaiah U N 2015 Development of Pineapple Leaf Fiber Reinforced Epoxy Resin Composites *International Research Journal of Engineering and Technology* **02** 2190-2193

[11] Jeencham R, Suppakarn N, Jarukumjorn K 2014 Effect of flame retardants on flameretardant,mechanical, and thermal properties of sisal fiber / polypropylene composites *Compos. Part B* **56** 249-253

[12] Lojewska J, Miskowiec P, Lojewski T, Proniewicz L M 2005 Cellulose oxidative and hydrolytic degradation: In situ FTIR approach *Polymer Degradation and Stability* **88** 512-520

[13] Reddy N, Yang Y 2005 Structure and properties of high quality natural cellulose fiber fromcorntalks *Polymer* **46** 5494-5500

[14] Bharath K N, Basavarajappa S 2014 Flammability Characteristics of Chemical Treated Woven Natural Fabric Reinforced Phenol Formaldehyde Composites *Procedia Materials Science* **5** 1880-1886

Acknowledgement

This research was funded by PITTA 2018 Universitas Indonesia.

IOP Conf. Series: Materials Science and Engineering **599** (2019) 012016 doi:10.1088/1757-899X/599/1/012016

Flexural Property and Water Content of Unidirectional Polypropylene/ Subang Pineapple Leaf Fiber Composites

A Nugraha, A L Juwono*

Department of Physics, Faculty of Mathematics and Natural Sciences, Universitas Indonesia, Depok 16424, Indonesia

*E-mail: ariadne.laksmidevi @ui.ac.id, azizi.nugraha96@gmail.com

Abstract. The development of natural fiber reinforced composite research has shown significant results. Pineapple leaf fiber reinforced composites for non-structural application provide added value for Subang pineapple leaf fiber (PALF) waste. This study aimed to observe flexural strength, water absorption and desorption of composites which referred to the Indonesian National Standard (SNI) for composite board. Alkali treatment on Subang PALF was performed to improve the interface bonding between PALF with polypropylene (PP), in which the PALF was soaked in 5% NaOH solution for 24 hours. Treated PALF and PP sheets were fabricated using compression molding (hot press) technique with the variation of fiber fraction to polypropylene were 20 wt%, 30 wt% and 40 wt%. Flexural property, desorption and absorption water tests were conducted on PP and PP/PALF composites, and Optical Microscope (OM) observation was performed on the surfaces after the flexural test was conducted. The highest flexural strength and flexural modulus were (4.89 ± 0.37) MPa and (1.13 ± 0.43) Mpa respectively, these values shown an increased respectively by 153,37 % compared to the pure PP. Then the desorption and the absorption water tests-were (1.08 ± 0,12) % and (8,83 ± 3,35) % respectively. The morphology of flexural tested composites indicated that matrix failure; and it showed that there was a strong interface bonding between the fiber and matrix.

1. Introduction

Technological developments and awareness of environmental conditions have resulted in some types of material considered to be environmentally unacceptable. Indonesia has many types of natural fiber and these fiber have the potential to be developed as raw materials in several industries such as pineapple, hemp, kenaf, bamboo because of its advantages of having low density, biodegradation, environmental friendliness and lower production costs than synthetic fibers [1], [2]. Cellulose from fibers act as a composite reinforcement replacing synthetic fibers[2]. Recycle materials that are considered waste into materials that have higher economic value and as an alternative way to reduce environmental damage. Pineapple leaf fiber (PALF) has a cellulose content of about (70-8%), lignin (5-12%) and ash (1.1%) [3].

PALF has the potential for better mechanical properties compared to other types of fibers which have lower cellulose content such as jute fiber [4]. Problems occurred when combining the polymer matrix and the natural fiber got a poor adhesion bonding between matrix and fiber, thereby reducing the load transfer capability from the matrix phase to the fiber phase [4], [5]. It is necessary to make chemical modifications to the fibers with the purpose to increasing the interface bond between the polymer-fibers, namely by alkaline treatment. Alkaline treatment aimed to remove lignin, wax, oil, and other impurities that interfere with cellulose from natural fibers by soaking the fibers into a chemical solution. The effect of alkaline treatment on fiber is to reduce the elasticity of the fiber so that the tensile strength will increase [6]. The dried fiber was purpose to minimize air bubbles during composite fabrication.

Previously, PP/PALF20, PP/PALF30 and PP/PALF40 composites that were fabricated by compression molding method were prepared Rana Hafizhah *et al* [7]. The PP/PALF30 composites had the optimum properties in tensile strength and deflection temperature. Research on flexural and diffusion properties on PP/Subang PALF composites ha not conducted yet. The desorption water test was the novelty of research from no one who analyzed the desorption water in polymer reinforced pineapple fibers or other natural fibers composite. So that, the aimed of this research was to obtain flexural strength, desorption and absorption rate of the PP/Subang PALF composites.

2. Experimental method

2.1Alkaline treatment

PALF was soaked in 5% NaOH solution for 5 hours at room temperature 25°C. The fiber was washed in distilled water until the fiber was clear from NaOH and the fiber color was changed to white. Then, the fiber was stored at room temperature 25°C for 24 hours, and then dried at 75°C for 24 hours. Single fiber test on untreated and treated PALFs were carried out in bundles with a length of 5 cm and a width of 0.5 cm specimen size, according to SNI 08-1112-1989 standard.

2.2Composite fabrication

PP pellets were processed into sheets by compression molding method with a temperature of 190°C. Composites were fabricated liked sandwiches; the fibers were unidirectional arrange in between the PP sheets. The fiber fraction to the weight of PP were 20 wt%, 30 wt% and 40 wt%. The composites were fabricated by compression molding method with a temperature of 190°C with a pressure of 0 MN/m^2 for 10 minutes and 4.9 MN/m^2 for 10 minutes. The results of composites fabrication were identified PP, PP/PALF20, PP/PALF30 and PP/PALF40.

2.3Flexural test

Three-point bending test was conducted according to ASTM D7264 on a Universal Testing Machine (UTM). The specimens were pressed transversally in the middle at a speed of 10 mm/minute until the specimens reached the maximum curve point. Then the fractured specimens were observed at its curved point by the Optical Microscope.

2.4Water Content test

Water content test was performed using SNI 01-4449-2006 reference to the method of drying ovens and soaking in water for 5 specimens. The desorption composite specimens were heated with a temperature of 103°C for 24 hours. The absorption composite specimens were soaked in water at 25°C water for 24 hours, with each specimen was weighed until the absolute weight of the specimens was obtained. Equations (1) and (2) were applied to calculate the water desorption and absorption respectively.

$$\text{Water desorption } (\%) = \frac{W_0 - W_t}{W_t} \times 100 \quad (1)$$

$$\text{Water absorption } (\%) = \frac{W_t - W_0}{W_0} \times 100 \quad (2)$$

3. Results and Discussion

3.1Alkali Treatment.

The single fiber test results showed that the tensile strength of untreated and treated PALFs were (109.92 ± 39.08) MPa and (378.94 ± 101.7) MPa respectively. The surface of the fiber changed from smooth to become rough surface. The tensile strength of treated PALF increases. This result agrees well with the study conducted by Asim *et al* [6]; in which PALF tensile strength increased after alkali treatment from 290.61 MPa to 432.01 MPa.

3.2Flexural test

Figure 1 shows the flexural strength and flexural modulus of PP/Subang PALF composites versus fiber fraction. It shows that the PP/PALF40 composite has the maximum value of the flexural strength which is 3.22 MPa and flexural modulus which is 1.13 MPa than PP/PALF20 and PP/PALF30. Meanwhile, S. Dhakal *et al* [8] obtained the flexural strength value of composite with 20wt% fiber fraction was 124.61 MPa. This difference was due in the quality of PP, fiber, and composite fabrication used. Composites that used in Dhakal's study had better fiber-matrix bonds than that use in this study, which is the most important thing in acquiring flexural strength of a composite, resulting in greater result of flexural strength. The bond between fiber and matrix is discussed in the next section. However, both Dhakal's study and this study have a same trend; the greater the fiber content, the higher the flexural strength. This study also shows that fiber was able to enhance the composite ability to load bearing and hold the matrix to not failure. Failure matrix in composite indicated the limit to bearing the maximum load.

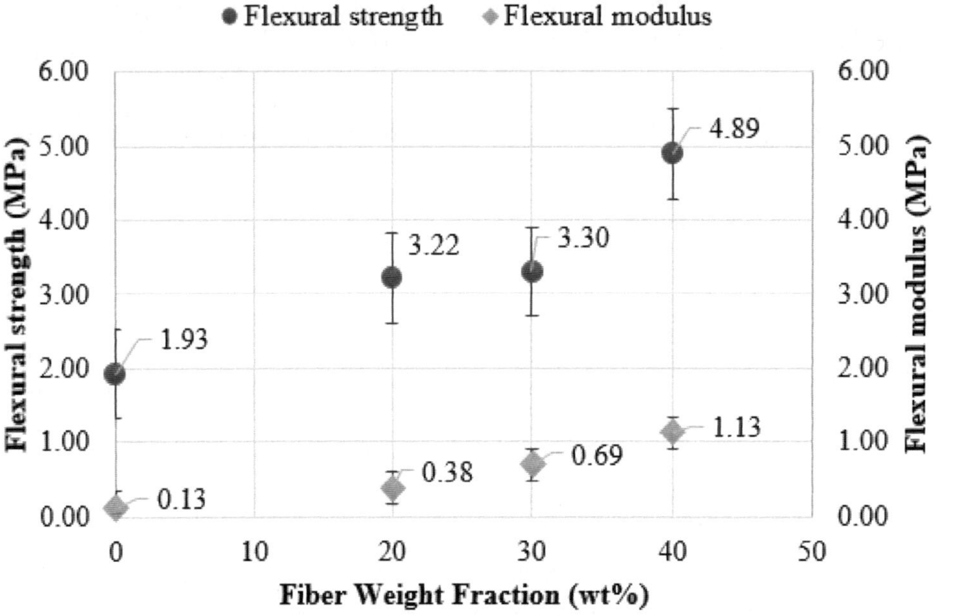

Figure 1. Flexural strength and flexural modulus of PP/PALF composite versus fiber weight fraction.

3.3Water Content Test

PP has hydrophobic properties that is difficult to absorption and desorption in water. Absorption is an ability to hold water and desorption is the ability to release water from a composite. The rate of slope in water desorption is based on SNI procedure standards. Figures 2 (a) and 2 (b) prove that the fiber used for composite fabrication still has a water content. After the water desorption test was carried out to reach the absolute weight of the composite, all composites had the same desorption rate as pristine PP. The composites reached saturation state of (1.08 ± 0.52) % after 15 hours drying. Figure 2 (b) shows the composites reached saturation state after 16 hours soaking in the value of (8.83 ± 5.71) %. The graph plot in Figure 2 (b) agree with the results of a study conducted by Osman *et al* [9] on polyester/kenaf composites with the same temperature. In Osman research, composites with the highest fiber content had the highest water absorption rate. In this study, the desorption and water absorption values meet the criteria of SNI fiberboard.

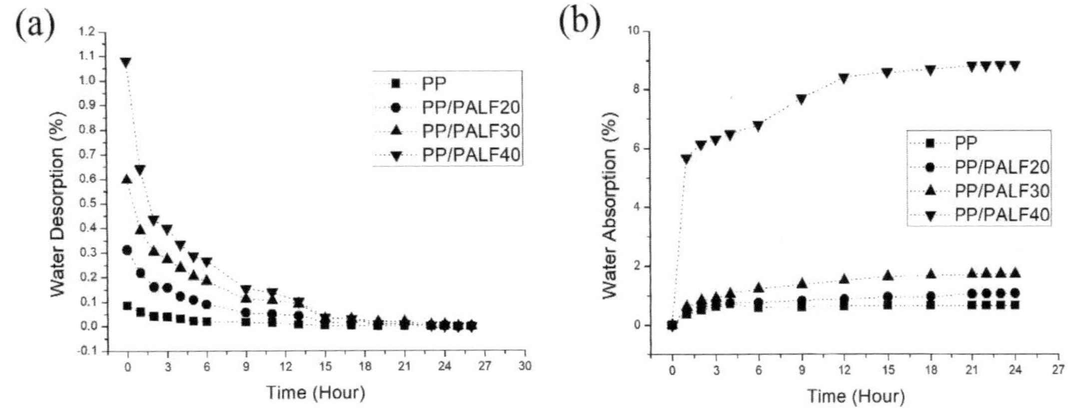

Figure 2. Results of (a) water desorption test, (b) water absorption test.

3.4Optical Microscope (OM)

Figures 3 (a) – 3 (d) shows the observations of optical microscopes for flexural tested composites for all fiber fractions. Figure 3 (a) shows that the pristine PP was failed. Figures 3 (b), 3 (c), and 3 (d) show matrix failure for all composites while fiber-pull-out did not occurred. From OM observation, it was confirmed that there was a strong interface bonding between matrix-fiber. It was clear that the function of fiber in composite is as the load barrier. The results of current study were opposite to the results of Jaramillo *et al* [10] on PP/PALF composites that were observed by SEM. In Jaramillo study, it was showed that there was no strong bond between matrix and fiber. Fiber-pull-out was occurred because fiber was not wetted by PP. The efficiency of stress-strain on the matrix and fiber was important in holding the load.

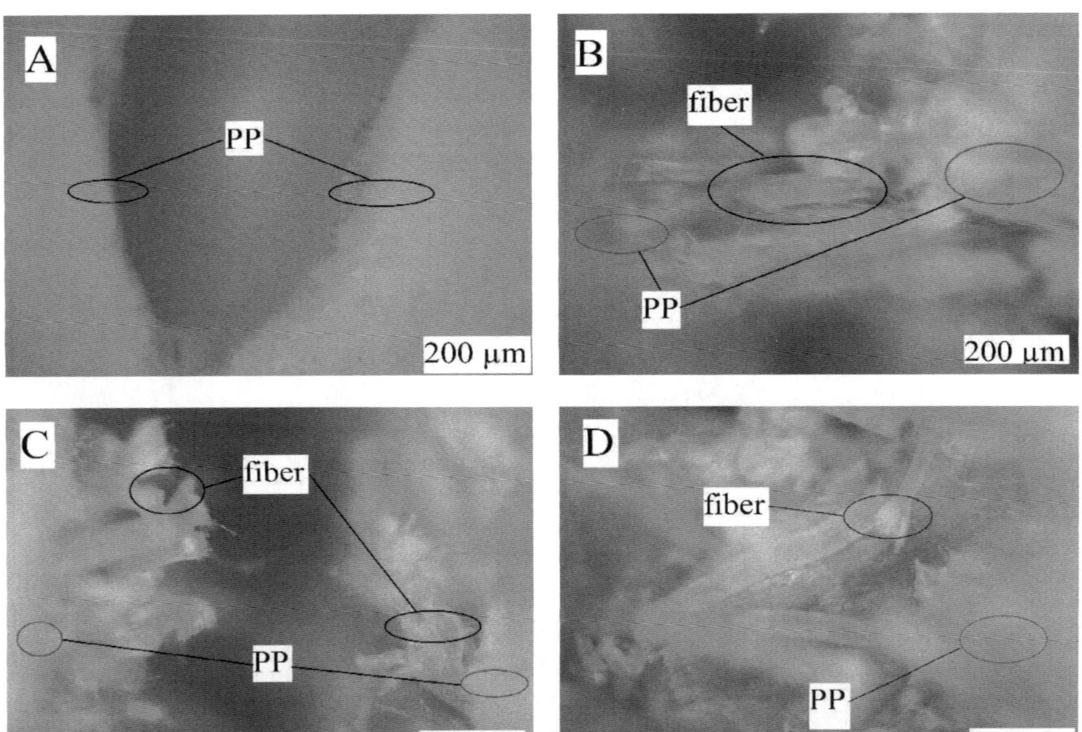

Figure 3. Optical microscope observation at break point of (a) pristine PP, (b) PP/PALF20, (c) PP/PALF30, (d) PP/PALF40.

4. Conclusion

The tensile strength of treated PALF was increased by 244,74% compared to untreated PALF, from (109.92 ± 39.08) MPa to (378.94 ± 101.7) MPa. The flexural strength and flexural modulus values of (4.89 ± 0.37) MPa and (1.13 ± 0.42) MPa were possessed by the PP/PALF40 composites. The desorption water value was (1.08 ± 0.52) % and the composites reached saturation state after 15 hours drying. The absorption water value was (8.83 ± 5.71) % and the composites reached saturation state after 16 hours soaking. The flexural strength, desorption water and absorption water values met the Indonesian National Standard (SNI) for composite board. Optical Microscope observation showed that matrix failure occurred in all composites and confirmed that there was strong interface bonding between matrix-fiber.

5. References

[1] Venkateshwaran N, Elayaperumal A, Alavudeen A, and Thiruchitrambalam M 2011 *Mater. Des.* **32** 7
[2] Elanchezhian C, Ramnath B. V, Ramakrishnan G, Rajendrakumar M, Naveenkumar V, and Saravanakumar M. K 2018 *Material Today Proc.* **5** 1
[3] Mishra S 2001 *Journal of Reinforced Plastics and Composites* **20** 04
[4] Asim M *et al.* 2015 *Int. Journal Polymer Science* **2015** 6
[5] Arib R. M. N, Sapuan S. M, and Ahmad M. M. H. M 2006 *Material & Design* **27** 391–396.
[6] Asim M, Jawaid M., Abdan K., and Ishak M. R 2016 *Journal Bionic Engineering* **13** 3
[7] Hafizhah R., Juwono A. L, and Roseno S 2017 *IOP Conf. Series: Material Science and Engineering* **196** 1-4
[8] Dhakal S and Gowda B. S. K 2017 *Material Today Proc.* **4** 8
[9] E. Osman, A. Vakhguelt, I. Sbarski, S. Mutasher 2011 *18th Int. Conf. Composite Mateials* **2** 7
[10] N. Jaramillo, D. Hoyos, and J. F. Santa 2016 *Ingeniería y Competitividad* **162** 2

Acknowledgement
This work was funded by PITTA 2018 Universitas Indonesia.

Flexural Strength of Hybrid Mendong – C-glass/Epoxy Thermoseting Composites

L Banowati*, T K Iskandar

Aeronotic Department, Nurtanio University, Bandung, Indonesia

*E-mail: liesbano@gmail.com

Abstract. Hybrid composite materials are those *composites* which have a combination of two or more reinforcement fibre in order to better mechanical properties. This paper concerns with development of hybrid composite materials using mendong and C-glass as the reinforcement and epoxy bakellite EPR 174 resin as the matrix, using hand lamination process. The mendong fibre was treated with alkali (NaOH). Hybrid mendong – C-glass/epoxy composite specimens were arranged to make a unidirectional (0°), bidirectional (0°/90°) and bidirectional (± 45°) fibre orientations with a 50% v/v fibre fraction content. The specimens were then tested using flexural tests, and the results were analyzed using two parameters Weibull distribution. The result shows that the flexural strength of hybrid mendong – C-glass/epoxy composites with a unidirectional (0°) is higher than other fiber orientations.

Keywords: hybrid, mendong – C-glass, Weibull, flexural

1. Introduction

Indonesia is a country with the potential for abundant natural fibers, the types of natural fibres such as pineapple, banana, ramie, coconut, bamboo, water hyacinth, and mendong. The advantages of natural fiber composites is a material that can be recycled, more environmentally friendly, potentially abundant, less costly than synthetic fibers and green composites [1].

The usage of natural fibres can be found widely in a variety of industries, such as in automotive, aerospace, building, plastic and textile. One of the noticeable applications of natural fibres is to replace synthetic fibres or man-made fibres as reinforcement in the composites production. There has also been an increasing interest in the use of polymer fibre composites reinforced natural fibres. Polymer composites have widely been less environmental impact compare to synthetic fibres [2].

Mendong fiber is a cultivated plant that have good economic potential, but during the use of the fibre itself only as a floor mat, strap, as well as handicraft products. Due to the potential of the mendong fibre is so large in Indonesia, the researchers attempt to improve the role of mendong not only as a traditional product, but can be used as raw material of natural fibre composites.

The main disadvantages is the poor compatibility between hydrophobic polymer matrix and the hydrophilic fibres. This leads to the formation of weak interfaces, which result in poor mechanical properties of the composites [3]. Alkali treatment of natural fibres is one of most common treatments of the natural fibres [4] that increased the interfacial bonding between the fibre and the matrix.

In recent years, the natural fibre composites have received a full attention in many applications, natural fibers are satisfying both economic and ecological interest [5]. The use of natural fibre mendong as reinforcement in composites is to increase the use of natural fibres which is abundant in Indonesia become a hybrid composite that is by incoparate glass fibres. Hybridization with glass fibre provides a method to improve the mechanical properties of natural fibre composites and its effect in different modes of stress depends on the design and construction of the composites [6].

The development of composite materials improving their performance limits based on the reinforcement of two or more fibres (synthetic fibre with another synthetic fibre or synthetic fibre with natural fibre or synthetic fibre with metallic fibres) in a single polymeric matrix, which leads to the advanced material system called hybrid composites with a great diversity of material properties, is still in its infancy [7].

The purpose of the present research is to develop a flexural strength of hybrid mendong − C-glass/epoxy thermosetting composites which consist of mendong and C-glass fibres as a reinforcement and epoxy bakellite EPR 174 as a matrix. Some previous studies have been conducted to determine the flexural strength of glass fibre and graphite fibre composites for ($0^o/90^o$) and ($0^o/45^o$) fibre orientation [8] and glass/epoxy composites for ($0^o/90^o$) and (± 45°) fibre orientation [9]. In this research, the fibre orientations, such as unidirectional (0°), bidirectional $(0^o/90^o)_s$ and bidirectional $(\pm 45°)_s$ were selected to make the hybrid composite laminates.

The fibre volume fraction used was 50%. Such considerations are based on [10] [11], where the failure mode occurring in ramie/epoxy composites is brittle failure for 10-30% fiber volume fraction, debonding and delamination at 40-50% fiber volume fraction. In the present study, two parameters Weibull distribution was used to analyze the test results [12] [13 [14] which is in the form of Eq.1:

$$F(x;b,c) = 1 - \exp\left(-\left(\frac{x}{b}\right)^c\right), b > 0, c > 0 \qquad (1)$$

F(x;b,c) represents the probability that the fracture strength is equal to or less than *x*. Using the equality *F(x;b,c)* + *R(x;b,c)* = 1, the reability *R(x;b,c)* is given by Eq.2:

$$R(x;b,c) = \exp\left(-\left(\frac{x}{b}\right)^C\right), b > 0, c > 0 \qquad (2)$$

Several methods have been used to estimate the two parameters *b* and *c*. In this study, a linear regression method was chosen for practical purposes. Hence, a linear regression model in the form of $Y = m X + r$ is obtained as in Eq.3:

$$\ln\left[\ln\left(\frac{1}{1-F(x;b.c)}\right)\right] = c\ln(x) - c\ln(b) \qquad (3)$$

Since *F(x;b,c)* is unknown in Eq.3, therefore, it is estimated from observed values: order *n* observations from smallest to largest, and let $x_{(i)}$ denote the smallest observation (*i*=1 correspond to the smal lest and $i = n$ corresponds to the largest). Then a good estimator of *F(x;b,c)* is the median rank of $x_{(i)}$ as in Eq.4:

$$F(x_{(i)};b,c) = \frac{i-0,3}{(n+0.4)} \qquad (4)$$

By using Eq.3 and Eq.4 and based on least square minimization of linear regression, both parameters *b* and *c* can be estimated. Then, by using Eq.2, the reliability curve can be drawn.

2. Materials and Methods

The fibers are used in this research is C-glass and mendong and the type of matrix is epoxy bakellite EPR 174. Mendong fiber was soaked in alkali solution containing NaOH concentrations of 5% for 2 hours at ambient temperature [15], then dried under the sun for 6 hours. The composite specimen consists of number of layers in which mendong fibre layers and C-glass fibre layers are placed alternatively by adding the required amount of epoxy bakellite EPR 174 resin until we obtain required thickness.

The processing method was using a hand lay up with the optimum fiber volume fraction is 50%. The fibers was arranged as such that in the end it will produce a unidirectional 0°, bidirectional $(0°/90°)_s$ and ±45°. The flexural tests were carried out according to ASTM D-7264 using universal testing machine Tensilon with a maximum capacity of 10,000 N.

Figure. 1 and 2 shows a mendong fibre that has been arranged and the hybrid mendong–C-glass/epoxy thermosetting composites lamination results respectively then the composite was formed into a flexural test specimen.

Figure 1. Mendong fiber.

Figure 2. Hybrid mendong–C-glass/epoxy thermosetting composites.

3. Results and Discussion

The final specimen test results and typical stress-strain curves for unidirectional (0°), bidirectional $(0°/90°)_s$ and ±45° fibre orientation specimens for flexural tests respectively are given in Figure 3, 4, 5 and Figure 6,7,8.

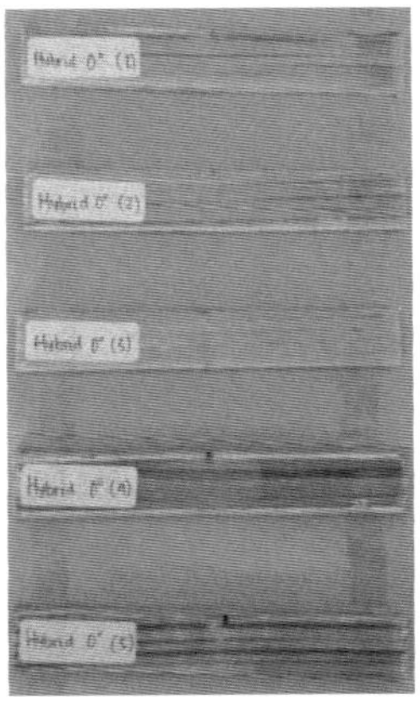

Figure 3. Specimen unidirectional (0°).

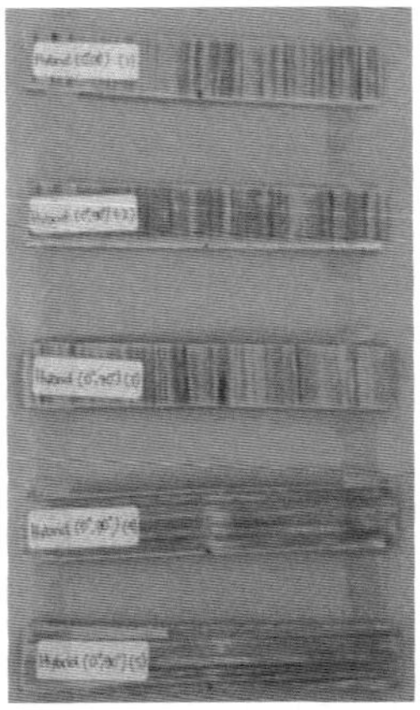

Figure 4. Specimen bidirectional (0°/90°).

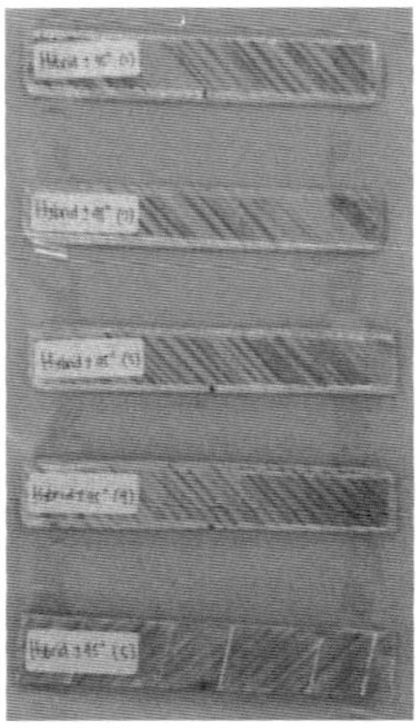

Figure 5. Specimen bidirectional (±45°).

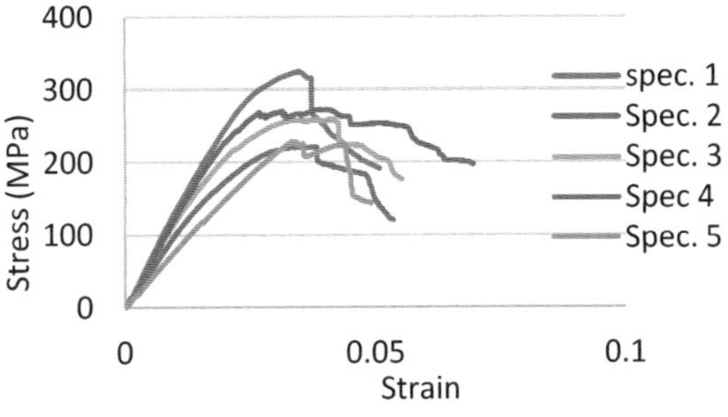

Figure 6. Stress-strain unidirectional (0°).

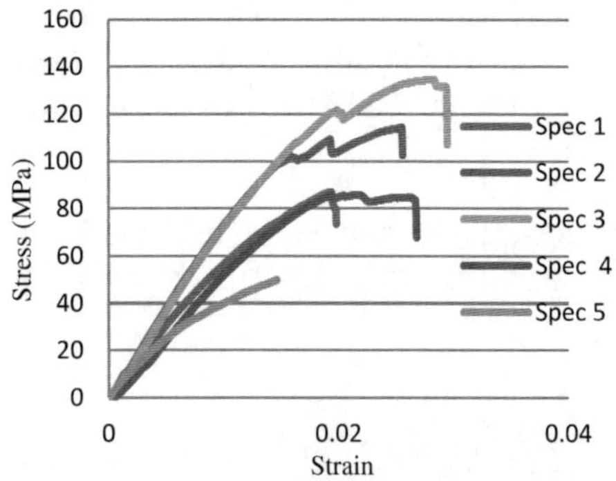

Figure 7. Stress-strain bidirectional (0°/90°).

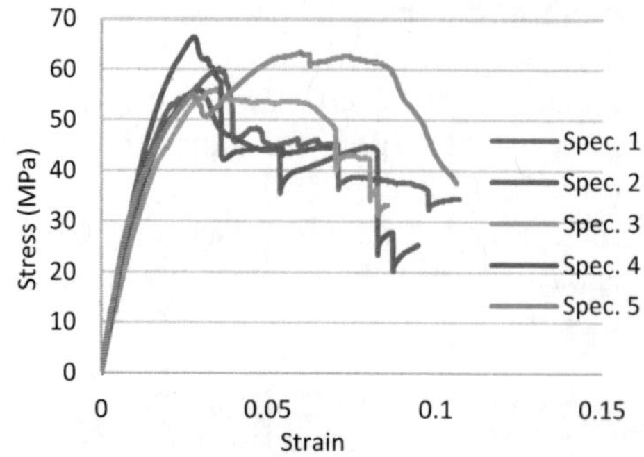

Figure 8. Stress-strain bidirectional (±45°).

The example of final reliability distribution shows in Figure.9 for linear regression of unidirectional 0° hybrid mendong– C-glass/epoxy thermosetting composite specimens. Whereas Figure.10, 11 and 12 shows the reliability curve for unidirectional (0°), bidirectional (0°/90°), and bidirectional (±45°) composite specimens.

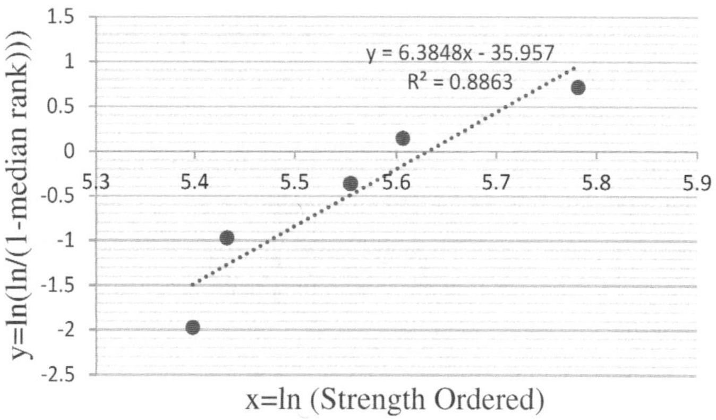

Figure 9. Linear regression of unidirectional (0°).

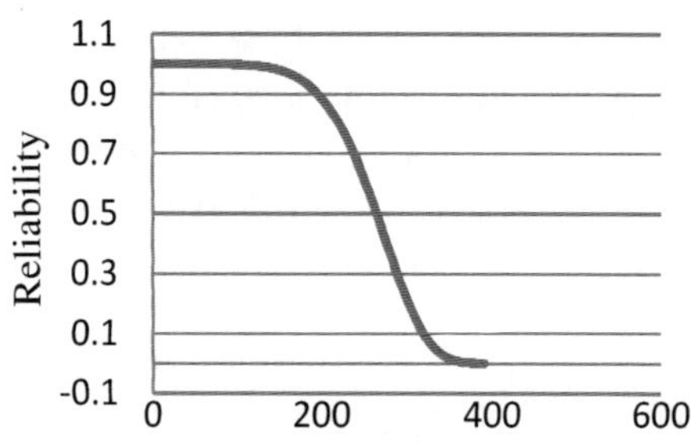

Figure 10. Reliability curve unidirectional (0°).

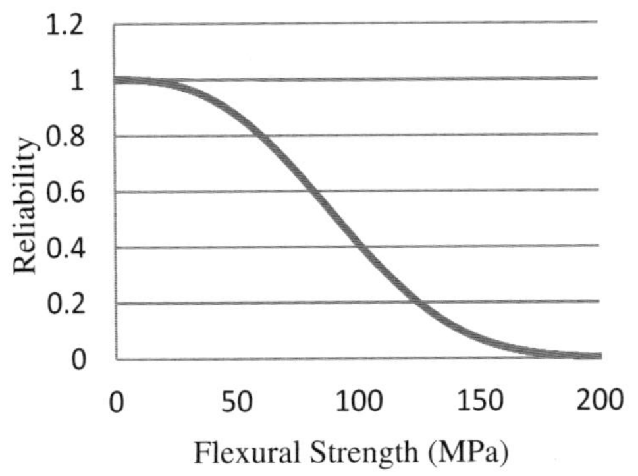

Figure 11. Reliability curve bidirectional (0°/90°).

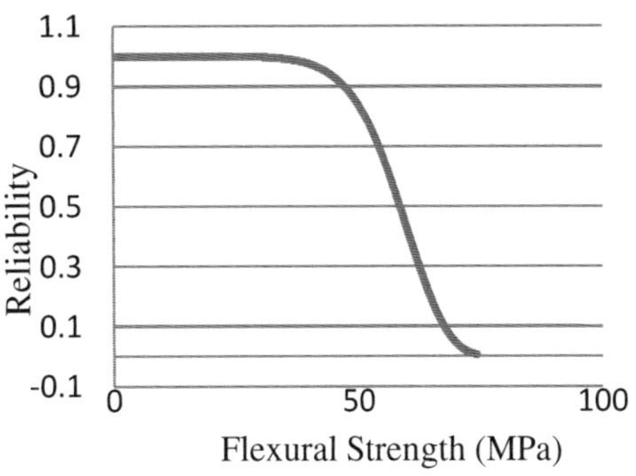

Figure 12. Reliability curve bidirectional (±45°).

Figure. 10, 11 and 12 shows the reliability of fracture of the specimens, it means that the probability of the specimens to fail is 50% in a particular stress level. In this research, the reliability chosen is 50%. Table 1. shows the comparison the flexural strength of hybrid mendong– C-glass/epoxy and mendong/ epoxy thermosetting composites [16] for 50% reliability level.

Table 1. Comparison the flexural strength of hybrid mendong – C-glass/epoxy and mendong/ epoxy thermosetting composites.

Fibre orientation	Flexural strength hybrid mendong-- C-glass/epoxy (MPa)	Flexural strength mendong/ epoxy (MPa)
Unidirectional (0°)	263	123 [16]
Bidirectional (0°/90°)	91	-
Bidirectional (±45°)	62	-

Based on the table shows of both fibre orientations and flexural strength and its improvement of bidirectional (±45°), bidirectional (0°/90°), unidirectional (0°) and 62, 91, 263 MPa. The increase of flexural strength was caused by fibre orientations. It shows that the hybrid mendong – C-glass/epoxy thermosetting composites with unidirectional (0°) having higher flexural strengths when compared to bidirectional (0°/90°) and bidirectional ±45° fibre orientations. These result related to the ability of fiber to withstand againts affective loads where reinforcement fibre is working to rise of most those loads, this ability decreases with increasing of fibre orientation angle at (0°/90°) and (±45°).

Whereas based on the results of the literature study shows that mendong/epoxy thermosetting composites having low flexural strengths when compared to hybrid mendong – C-glass/epoxy thermosetting composites with unidirectional 0° fibre orientations. Therefore, it can be concluded that with the addition a C- glass fiber and thus the flexural strength of the composite was increased.

It is important also to compare the results of the current study with similar flexural strengths of natural hybrid natural-glass fiber composites available in the literature. Table 2. shows such a comparison.

Table 2. Comparison the flexural strength with other hybrid natural-glass fiber composite systems.

Composites	Flexural strength (MPa)
Glass EC 2400 P207-Ramie/polyester UCEFLEX UC 5530 -M	105 [17]
E-Glass - Jute/epoxy LY556	11 [18]
Glass-Palmyra/epoxy	59 [19]

It shows that the mechanical properties found in the present study are able to compete with other hybrid natural and glass fiber composites. Therefore, it can be concluded that hybrid mendong – C-glass/epoxy has a good potential to be developed further for wide application, whether in the furniture industries as well as in the building structure equipment.

4. Conclusion

The present study concludes that hybrid mendong – C-glass/epoxy have a better mechanical properties compared with mendong/epoxy and hybrid natural-glass fiber composite systems. So by adding a C-glass fiber and thus the flexural strength of the composite was increased. Furthermore, the flexural test of composites gives best results at unidirectional 0° fibre orientations. Despite the simple manufacturing processes, the specimens produce good qualities with moderate flexural strengths. Therefore, mendong – C-glass/epoxy has a good potential to be developed further, for wider application in the furniture industries and building structure equipment.

5. References

[1] Rashdi A A A, Sapuan S M, Ahmad M M H M, Khalina, A 2010 Combined effects of water absorption due to water immersion, soil buried and natural weather on mechanical properties of kenaf fibre unsaturated polyester composites (KFUPC) *International Journal of Mechanical and Materials Engineering* **5**(1) p 11 -17
[2] Hoo Tien Nicholas Kuan, Meng Chuen Lee 2014 Tensile Properties of Pandanus Astrocarpus based Composites *Journal of Applied Science & Process Engineering* **1**(1)
[3] Thi Thu Loan D 2006 Investigation on jute fibres and their composites based on polypropylene and epoxy matrices *Dissertation* (Dresden. Der Fakultat Machinenwesen Der Technischen Universitat)
[4] Mishar S, Misra M, Tripathy S S, Nayak S K and Mohanty A K 2001 Graft copolymerization of acrylonitrile on chemically modified sisal fibers *Macromolecular Material and Engineering* **286**(2) 107-133
[5] A Athijayamani, Mani Thiruchitrambalam, Manikandan Vairavan and B Pazhanivel 2010. Mechanical properties of natural fibers reinforced polyester hybrid composites *Int J Plast Technol* **13**(1) 1
[6] B Singh, M Gupta and A Verma 1995 Mechanical behaviour of particulate hybrid composite laminates as potential building materials *Const. Build. Mater* **9** 39
[7] R T Durai Prabhakaran, Tom L Andersen, C M Markussen, Bo Madsen and Hans Lilholt 2013 Tensile and Compression Properties of Hybrid Composites-A Comparative Study *The 19 th International Conference On Composites Materials* (Montreal. Canada)
[8] G Rathnakar 2013 Fiber Orientation and Its Influence on the Flexural Strength of Glass fibre and Graphite fiber reinforced polymer composites *International Journal of Innovative Research in Science, Engineering and Technology* **2**(3)
[9] Sandeep M B, D Choudhary, Md Nizamuddin Inamdar, Md Qalequr Rahaman 2014 Experimental Study of Effectof Fiber Orientation on The Flexural Strength of Glass/Epoxy Composite Material *International Journal of Research in Engineering and Technology* 3(9)

[10] Wambua P, Ivens J andVerpoest I 2001 Some mechanical properties of kenaf/polypropylene composites prepared using a film stacking technique *In: 13th international conference on composite materials conference proceedings* (Beijing, China. The Chinese Society for Composite Materials and China Universities Alumni Association)

[11] Folkes M 1985 *Short fibre reinforced thermoplastics* ed Belvis M (Herts: John Wiley & Sons Ltd.) p.151

[12] W Weibull 1999 *A Statistical Theory of the Strength of Materials* (Royal Swedish Institute)

[13] M H Dirikolu, A Aktas, B Birgoren and Turkish J 2002 *Eng Env Sci* **26** 45-48.

[14] Ghosh A 1999 A FORTRAN Program for Fitting Weibull Distribution an Generating samples *Computer and Geoscience* **25** 729-735

[15] Heru Suryanto, Sukarni Sukarni, Yanuar Rohmat Aji Pradana and Uun Yanuhar 2017 Effect of Mercerization on Properties of Mendong (Fimbristylis globulosa) Fiber *Songklanakarin Journal of Science and Technology* p14 (in press)

[16] Vicky Firdaus, Lies Banowati and Ruslan Abdul Gani 2016 Flexural Strength Analysis of Mendong/Bakelite EPR 174 Epoxy Composites *National Seminar Technology and Science II* (Jakarta. Tarumanagara Univesity) p 23

[17] Daiane Romanzini, Hitor Loiz Ornaghi Junior, Sandro Campos Amico and Ademir Jose Zatterra 2012 Preparation Characterization of Ramie-Glass Fiber Reinforced Polymer Matrix Hybrid Composites *Materials Research* **15** 418

[18] Sanjay M R, Yogesha B 2016 Studies on Mechanical Properties of Jute/E-Glass Fiber Reinforced Epoxy Hybrid Composites *Journal of Minerals and Materials Characterization and Engineering* **4** 22

[19] R Velmurugan, V Manikandan 2007 Mechanical Properties of Palmyra/Glass Fiber Hybrid Composites *Composites Part A: applied science and manufacturing* **38** 2217

Structural and Optical Studies of Zinc doped $BiFeO_3$ on the Bi-site

Hamdan Akbar Notonegoro[1,2], Bambang Soegijono[1*], Rifqi Almusawi Rafsanjani[1], Isom Mudzakir[1], Teguh Yoga Raksa[1]

[1]Dept. of Physic, Universitas Indonesia, Depok 16424, Indonesia
[2]Mechanical Engineering Dept., Universitas Sultan Ageng Tirtayasa, Cilegon 42435, Indonesia

*Corresponding author: naufal@ui.ac.id

Abstract. Zinc-doped bismuth ferrite ($Bi_{1-x}Zn_xFeO_3$, x = 0, 0.05, 0.1) powdered were synthesized through sol-gel auto combustion methods to investigate the optical properties while maintaining the whole structure of $BiFeO_3$ (BFO). The Zn doped induced the reveal of $Bi_{25}FeO_{40}$ as a secondary phase and decreased the bond angle of Bi–O–Fe in the BFO phase structure which affect the electron density formation inside of the lattice. The presence of Zn content has also conducted the shifting of optical absorption in the UV-Vis region and increasing the dielectric constant value. This property is potent in visible light photonic applications.

1. Introduction

Multiferroic materials are materials that exhibit several ferroic behaviors, usually (anti)ferromagnetic and ferroelectric [1]. One of them is bismuth ferrite ($BiFeO_3$, BFO) which have ABO_3 perovskite type that mostly studied in recent years. Some of the important features of bismuth ferrite materials are optical behavior with a large ferroelectric polarization ($P \approx 90 - 100\ \mu C/cm^2$) [2,3]. Strong coupling between those properties suggested in several studies makes it become a good candidate for advanced photonic application [3–5].

Structural engineering by doping BFO on Bi/Fe-site to improve its optical performance has been studied. Fatima *et al.* have reported that substitution of La and Mn on Bi/Fe-site in BFO were enhanced dye removal compared to pure BFO [8]. Meanwhile, Liu et al. were reported that Zn doped on Fe-site could improve the crystallinity and show a blue shift in the fundamental absorption edge [6]. Li–doped on Fe–site absorb in UV-Vis wavelength region with a red–shifted and blue – shifted were also being reported by Billah [7]. In the other hand, the presence of the oxygen vacancies in the lattice of the BFO doping sample could be determined by the frequency dependencies of the dielectric permittivity [9]. This vacancy could affect optical behavior, especially in the ability of light absorption [2].

In order to doped the BFO sample, we could use a sol-gel auto-combustion method. This method is a common technique to synthesized a perovskite – sample in many studies [1,10,11]. This method uses simple equipment, result in powder samples, and suitable for manufacturing an oxide material.

In this work, we have organized the study of structural and optical properties of Zn doped BFO in $Bi_{1-x}Zn_xFeO_3$ (x = 0, 0.05, 0.1). Whereare, doping Bi-sites with Zn is still rarely found in the previous studies. A destructive method was also used as an alternative technique to prepare Zn–metal as a precursor in the sol–gel process. By this work, the effect of doping Zn in Bi-site of the BFO to the optical behavior expected could be identified.

2. Experimental procedures

Zinc doped bismuth ferrite ($Bi_{1-x}Zn_xFeO_3$; x =0, 0.05, 0.1) powdered were synthesized by sol–gel auto–combustion methods. Bismuth nitrate ($Bi(NO_3)_3 \cdot 6H_2O$), iron (III) nitrate ($Fe(NO_3)_3 \cdot 9H_2O$) from Sigma–Aldrich, citric acid and aqua bidest were used as a precursor. A destructed zinc–metal in nitric acid were used as precursor to accompany a sol–gel methods. All the precursor in a proportional composition were converted to gel phase by stirring and heating at 80 °C till it became light brownish colour, then following with dried process at 110 °C for 24 h, afterwards ground and calcined at 600◦C for 5 h.

The structure of obtained BFO powder samples was carried out by X-ray diffraction on an XRD PANanalytical Diffractometer (Model: X'Pert Pro; Cu-Kα_1, λ = 1.5405 Å) at room temperature. This XRD data result was refined using the HighScore Plus. The 3D modeling base on Rietveld refinement result was developed by Vesta Ver. 3.3.9 software. The light absorption spectra of the sample were measured by Agilent Cary 100/300S UV–Vis spectrophotometer. The dielectric measurement was carried out by the four-point probe system on Agilent E4980A LCR meter over the frequency range from 50 kHz to 2 MHz at room temperatures.

3. Results and discussion

3.1. X-ray diffraction analysis

The crystalline structure and the phases present in the sample were investigated by XRD within the range of $2\theta = 10^o - 70^o$. Figure 1 shows the peak pattern of $Bi_{1-x}Zn_xFeO_3$ (x = 0, 0.05, 0.1) powdered samples. The peak pattern affirms the presence of high (112), (104), (110), (024), (214), (202), (116) peaks for the existence of BFO phase. This phase recognized as the rhombohedral structure with space groups (s.g) *R3c*. The reveal of $Bi_{25}FeO_{40}$ as a second phase has also occurred with the presence of high (013), (022), (222), (123) peaks (x = 0.05, 0.1). This phase has recognized with a cubic structure and s.g. *I23*.

Table 1 were obtained a parameter result of both phase present. There is a good agreement between calculated and observed XRD pattern with a low value of Goodness of Fit (χ^2). We can see from the table, a small amount of Zn in BFO structure had reduced the lattice volume which causes an increase the lattice density. This condition indicated that some Zn with smaller atomic radii and smaller valence than Bi had replaced on its site and let a formation of $Bi_{25}FeO_{40}$ phase [7].

The 3D–imaging of the BFO modeling base on refinement result using Vesta Ver. 3.3.9 are shown in figure 2. The shifting of the atomic position as a caused of Zn doped was observed in the gray circle area (figure 2i). A change of tilt angle between Bi/Zn–O–Fe (figure 2ii) and O–O–O (figure 2iii) which correlated with the Bi/Zn–Fe (figure 2iv) distance position could be determined this shifting.

Table 2 had shown that Zn doped were decreasing the tilt angle of Bi/Zn1–O1–Fe1 bond and O1–O1–O1 bond followed by decreases of Bi/Zn–Fe distance position. In between x = 0.05 and x = 0.1 of the doping sample, we can see that the tilt angle of O1–O1–O1 bond has the same value, but the tilt angle of Bi/Zn1–O1–Fe1 bond becomes shrink with $\Delta\theta = 0.0008°$. This condition was correlated with the increase of the BFO phase portion ration in the sample.

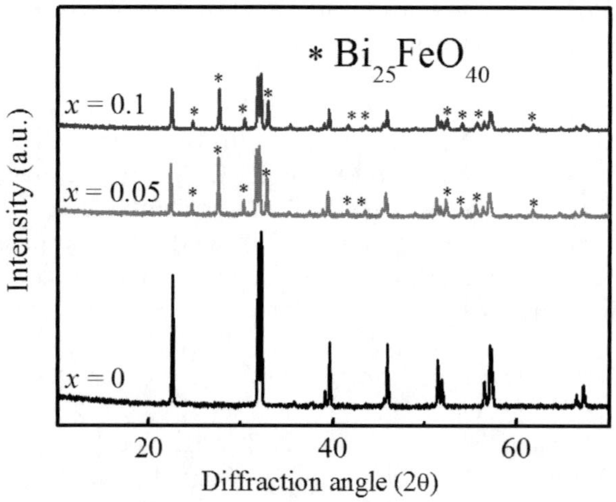

Figure 1. XRD pattern of $Bi_{1-z}Zn_zFeO_3$ (x = 0, 0.05, 0.1) respectively prepared by sol-gel auto-combustion method. The Peak pattern shows $BiFeO_3$ phase at x = 0. When x =0.05 and x = 0.1, the pattern of $Bi_{25}FeO_{40}$ phase clearly revealed.

Table 1. Rietveld refinement result of $Bi_{1-x}Zn_xFeO_3$ (x = 0, 0.05, 0.1)

x	Parameter of $BiFeO_3$ *s.g.* R3c				Phase		Micro strain (%)	R_{wp} (%)	GOF (χ^2)
	a=b (Å)	c (Å)	V (Å^3)	ρ (g cm^{-3})	$BiFeO_3$ (%)	$Bi_{25}FeO_{40}$ (%)			
0	5.580	13.874	374.157	8.33	100	0	0.012	4.88	1.60
0.05	5.576	13.863	373.318	8.35	63.5	36.5	0.038	6.53	1.38
0.1	5.578	13.868	373.805	8.34	63.55	36.45	0.055	6.08	1.04
Cell parameters of $Bi_{25}FeO_{40}$ *s.g.* I23									
0.05	10.183		1055.758	9.31					
0.1	10.192		1058.661	9.28					

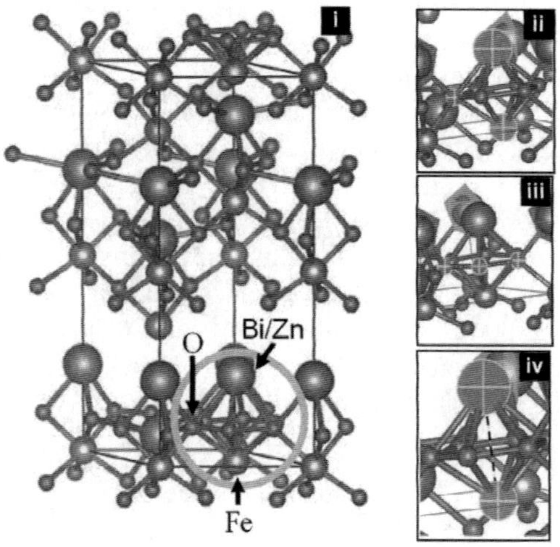

Figure 2. 3D modelling of BFO structure base on refinement result using Vesta Ver. 3.3.9 software. i) crystallite structure of BFO with a grey circle as an observed area, ii) Tilt angle of Bi/Zn–O–Fe, iii) Tilt angle of O–O–O, and iv) a distance of Bi/Zn–Fe displacement.

Table 2. Tilt angle (θ) of Bi/Zn–O–Fe bond, O–O–O bond and distance of Bi/Zn – Fe position

	Tilt Angle (°)		Distance (Å)
x	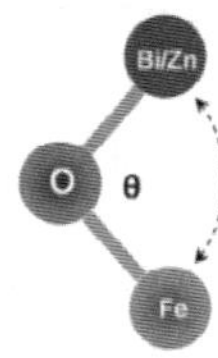	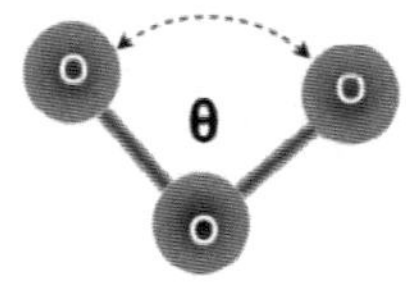	
	Bi1/Zn-O1-Fe1	O1-O1-O1	Bi1/Zn-Fe1
0	81.9961	132.8539	3.0588
0.05	82.7140	130.0406	3.0558
0.1	82.7132	130.0406	3.0573

Figure 3 shows electronic density of $Bi_{1-x}Zn_xFeO_3$ (x=0, 0.05, 0.1) on *hkl*–plane (0,0,0.2) at room temperature. When x = 0, we can see that the electron on the middle area (white arrow) was not too dense (figure 3a). The presence of Zn doped (x = 0.05) affected the increasing density on the middle area (white arrow) which followed by density alteration on the middle–right position (yellow arrow) (figure 3b). The increasing of Zn doped (x = 0.1) had caused a shifting density in the below corner position (grey arrow) (figure 3c). The change of electron density formation may govern its physical properties, especially in optical behavior.

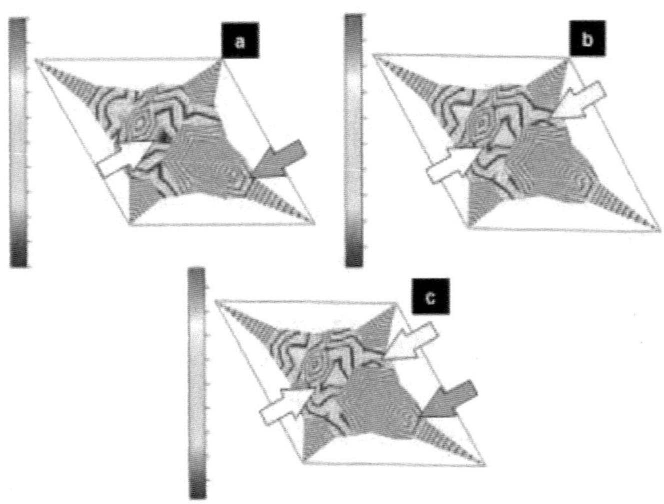

Figure 3. An electronic density of $Bi_{1-x}Zn_xFeO_3$ for hkl-plane (0,0,0.2) at room temperature. a) x= 0, b) x= 0.05, c) x= 0.1.

3.2. Optical characterizations

The optical property of Zn–doped BFO was studied by UV–Vis DRS absorption spectra. Figure 4 displays the UV–vis absorption results of $Bi_{1-x}Zn_xFeO_3$ (x=0, 0.05, 0.1). As seen in the figure, with the increasing concentration of Zn that the absorption slope is shifted toward the visible wavelength region as a green shift, in line with Liu et al. had reported [6]. This condition was also correlated with density change of electron inside of lattice structure. This matter indicated that the ability of BFO samples to absorb visible light increase with doping Zn. The result has shown that both samples have become a potential candidate for photonic applications.

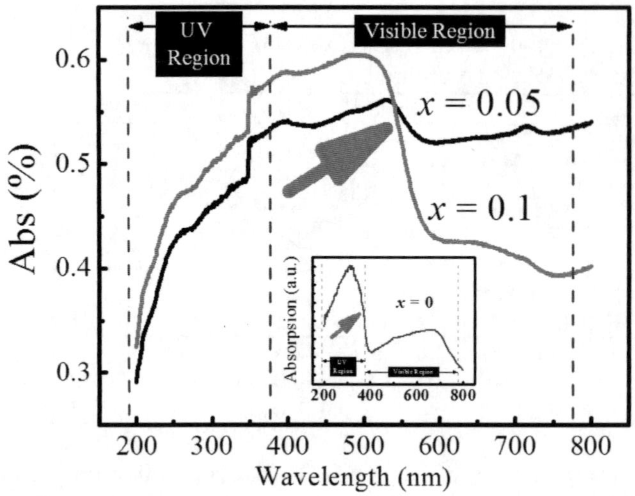

Figure 4. UV-Vis-DRS spectral of $Bi_{1-x}Zn_xFeO_3$ (x = 0, 0.05, 0.1) at room temperature.

3.3. Dielectrical studies

The dielectric measurement of Zn – doped BFO was performed over frequency range 50 kHz – 2 MHz at room temperature. Figure 5 shows the frequency dependence of the relative permitivity or dielectric constant (ε_r) of $Bi_{1-x}Zn_xFeO_3$ at x = 0, 0.05, 0.01, respectively. Dielectric behavior was known as low frequency space–charge relaxation. As shown in the picture, the value of dielectric constant of x = 0, 0.05 have an identical value in frequency above 100 kHz, but splits below 100 kHz. Its correlated with the presence of some low electron density region in x = 0 [9]. The increasing of doping to x = 0.1 were raised the dielectric constant value which inline with the shifted of electron densities.

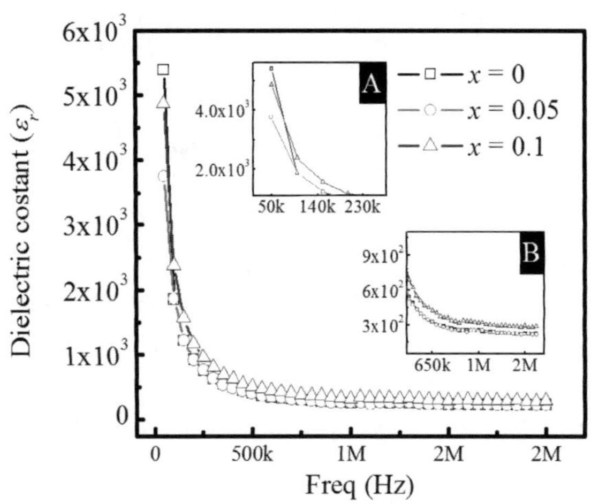

Figure 5. Dielectric constant as function of frequency for $Bi_{1-x}Zn_xFeO_3$ (x = 0, 0.05, 0.1) at room temperature. inset: A) Freq. below 140 kHz, B) Freq. above 140 kHz.

4. Conclusion

The effect of Zn–doped $BiFeO_3$ for the Bi–site on the crystal structure, optical properties, and dielectric behavior has been studied. It has been found that the increasing value of Zn–doped had to change the structure and affected the electron densities formation. The optical absorption was shifted in line with the rising amount of dopant, as well as the dielectric constant which also increased. The optical property of this material makes it becomes a potential candidate for visible light–photonic applications.

5. References

[1] Rafsanjani RA, Triyono D, Laysandra H. 2018 *AIP Conf Proc* 2018 **3** 020024.
[2] Haislmaier RC, Podraza NJ, Denev S, Melville A, Schlom DG, Gopalan V. 2013 *Appl Phys Lett* 2013 **103**.
[3] Allibe J, Jacquet E, Infante IC, Fusil S. 2012 2012 **182902** 2010–3.
[4] Choi T, Lee S, Choi YJ, Kiryukhin V, Cheong SW. 2009 *Science (80-)* 2009 **324** 63–6.
[5] Allibe J, Bougot-Robin K, Jacquet E, Infante IC, Fusil S, Carrétéro C, et al. 2010 *Appl Phys Lett* 2010 **96** 182902.
[6] Liu J, Deng H, Zhai X, Lin T, Meng X, Zhang Y, et al. 2014 *Mater Lett* 2014 **133** 49–52.
[7] Billah A. 2017 Investigation of Multiferroic and Photocatalytic Properties of Li Doped BiFeO3 Nanoparticles By. Bangladesh University of Engineering and Technology, 2017.
[8] Fatima S, Ali SI, Iqbal MZ, Rizwan S. 2017 *RSC Adv* 2017 **7** 35928–37.
[9] Khomchenko VA, Kiselev DA, Kopcewicz M, Maglione M, Shvartsman V V., Borisov P, et al. 2009 *J Magn Magn Mater* 2009 **321** 1692–8.
[10] Budiawanti S, Soegijono B, Mudzakir I. 2018 *J Phys Conf Ser* 2018 **985**.
[11] Notonegoro HA, Soegijono B, Mudzakir I. 2018 *MATEC Web Conf* 2018 **218** 1–6.

Acknowledgement

The author acknowledge with gratitude the financial support from the Ministry of Research, Technology and Higher Education of Indonesia (RISTEK-DIKTI), under grant Penelitian Dasar Unggulan Perguruan Tinggi (PDUPT), No. 366/UN2.R3.1/HKP05.00/2018).

Preliminary Study: Instrumentation Design for Studying Granular Segregation

Rahmawati Munir[1,a], Handika Dany Rahmayanti[1], Elfi Yuliza[1], Nadya Amalia[1], Fisca Dian Utami[1], Sparisoma Viridi[2] and Mikrajuddin Abdullah[1,b]

[1]Material and Electronic Physics Laboratory, Faculty of Mathematics and Natural Science, Bandung Institute of Technology, Jl. Ganesha 10 Bandung 40132, Indonesia
[2]Nuclear Physics and Biophysics Laboratory, Faculty of Mathematics and Natural Science, Bandung Institute of Technology, Jl. Ganesha 10 Bandung 40132, Indonesia

[a]rahma.mipaunmul@gmail.com
[b]din@fi.itb.ac.id

Abstrct. Rice winnowing is a process of separation of small and large rice grains by air flow practiced by farmes. This process shows interesting phenomenon but has never been documented by previous researchers. Based on the movements performed within rice winnowing, we tried to design a simulation instrument that resembled rice winnowing process. This research aims to investigate the process of granulars segregation with different size or density. We also observed carefully using manually released objects were captured for 60 seconds by digital camera (Nikon L840, 16 megapixel and 30 fps) in a video mode. Every video tracked using video tracker then analyzing the results. Furthermore, the results of study can be useful to develop a model of granulars segregation based on size or density.

1. Introduction

Granulars segregation is interesting physical phenomena to be investigated. Many researchers have developed granular segregation method such as kinetic sieving to produce a mechanism for granular segregation that is influenced by gravitational forces and the principle of granular flow [1]. This method is also used to observe granulars segregation phenomena in rivers [2] and studied particles separation with different sizes and densities [3]. Furthermore, the modulation method is used to control granular segregation [4]. In addition, another method developed are the experiment of granular flow as a mechanism for layer formation [5] and prediction of particle separation by sieving method in flour milling [6].

The purpose of this paper is to design an instrumentation for studying granulars segregation. This instrumentation resembles the traditional in rice winnowing by tray flapping. Rice winnowing process takes advantages of natural wind blowing that often used by farmers who are harvesting rice to separate the filled grains from empty grains [7] or unwanted matters like weeds, straws, sand, dust particles,etc to give cleaner outcome [8]. Lack of winnowing rice using natural wind is often less favorable for natural wind speeds such as direction and random wind intensity. Therefore, if it's done in a closed room where no natural wind blow, the common method for producing air flow is by flapping a tray above where the rice grains are placed.

Experimental method was also conducted to compare the results of granulars segregation by winnowing manually. The motion profile of the granulars is taken using a camera, and then processed using video processing by video tracker. The results of this study can be an alternative method for granulars segregation process with a very simple way and and can be useful to develop a model of granulars segregation based on different size and density.

2. Modeling

For simplicity, we assume the granulars segregation due to wind blowing is caused particle motion in vortex. Vortex motion is generated by winnowing motion using winnow flapping. The total force on particle motion in vortex are gravitational force (F_g), the force due to gradien leading to the center of vortex (F_p), the centrifugal force (F_c) and the air friction force (F_f).

$$\vec{F}_{tot} = \vec{F}_g + \vec{F}_p + \vec{F}_s + \vec{F}_v \tag{1}$$

With

$$\vec{F}_g = -mg\hat{j} \tag{2}$$

$$\vec{F}_p = -\frac{\pi}{6}\rho d_p^3 \frac{u^2}{r}\hat{r} \tag{3}$$

$$\vec{F}_s = \frac{\pi}{6}\rho_p d_p^3 \frac{v_t^2}{r}\hat{r} \tag{4}$$

$$\vec{F}_v = -C_D \frac{\pi\rho d_p^2}{8}|\vec{v} - \vec{u}|(\vec{v} - \vec{u}) \tag{5}$$

The equation must be solved based on the total force in particle motion in vortex are

$$a_x = -\frac{1}{\sqrt{2}}\frac{\rho}{\rho_p}\Omega^2 x + \frac{1}{2}\frac{(-yv_x + xv_y)^2 x}{(x^2+y^2)^{3/2}} - \frac{18\mu\rho}{d_p^2\rho_p}(v_x + \Omega y/\sqrt{2}) \tag{6}$$

$$a_y = -g - \frac{1}{\sqrt{2}}\frac{\rho}{\rho_p}\Omega^2 y + \frac{1}{2}\frac{(-yv_x + xv_y)^2 y}{(x^2+y^2)^{3/2}} - \frac{18\mu\rho}{d_p^2\rho_p}(v_y - \Omega x/\sqrt{2}) \tag{7}$$

3. Design Instrument

We designed the granulars segregation instrument using a motor gear box (1:25) which set on a series of devices connected to a power source. The gearbox pushes pully that connected to the tray due to it swings up and down to resemble the arms of human winnowing rice. At the bottom of tray was assembled the holder which connected to two springs (spring constant = 15 N/m). The springs were used to provide elastic force when tray moved up and down. Ilustration of procedures for experimental methods is shown in Figure 1.

Furthermore, we designed the experimental methods using hardware and software. Hardware include granular segregation instrument, one unit camera (Nikon L840, 16 Mpixel ; 30 fps) along with a tripod, a memory card, and a laptop (Intel Core I7 Spec.). The software were used OS Windows, video tracker and microsoft office excel 2016.

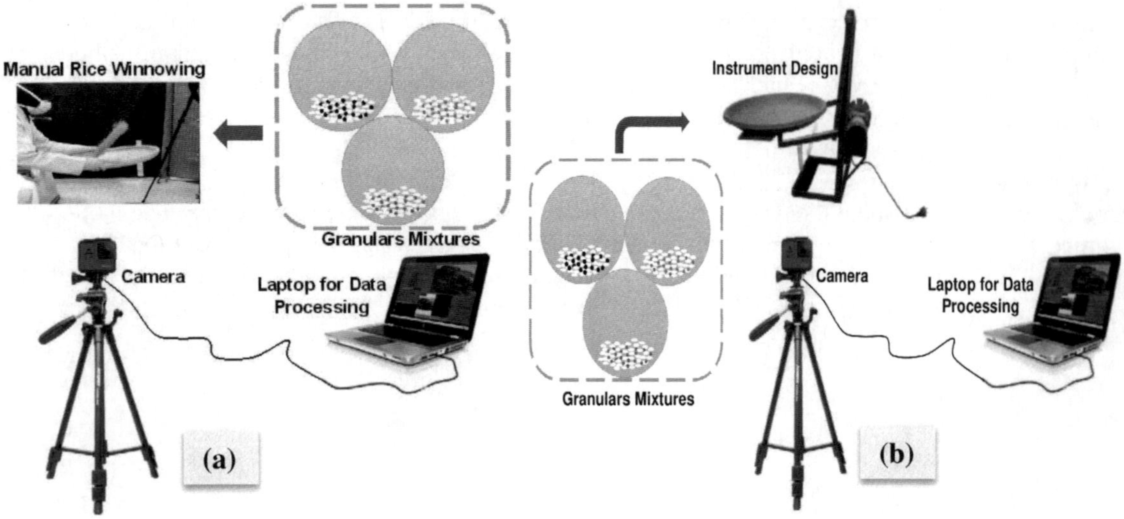

Figure 1. Ilustration of procedures for experimental methods. (a) Manually winnowing, (b) Instrument winnowing.

4. Experimental Results

Firstly, we demonstrated the effect of swings tray on the segregation of granulars mixtures from two different size and density (rice + basil seeds; soybean + basil seeds). The granulars mixtures were placed on a tray with diameter of around 0.6 m. Initially the granulars were mixed homogeneously on the tray surface. The granulars motion was recorded to obtain profile particle motion. We also observed the motion profile of each grain by tracking using a video tracker.

Figure 2 shows the profile of the granular mixture motion when manually winnowing from the initial position at maximum height then falls to the winnower surface. Large rice grain trajectories marked with blue circles and small basil seeds grain are marked with red triangle (see Figure 2 (a)).The Figure 2 (b) shows the condition of the granular mixture before winnowing process. The Figure show that both grains is mixed well. Figure 2 c shows the final position of the two types of granules after the swing movement execute within winnowing process. There was a change in each type of grains position, where smaller grains are pushed in front of the bigger grains. These results were achieved after 5 winnowing attempt.

IOP Conf. Series: Materials Science and Engineering **599** (2019) 012019 doi:10.1088/1757-899X/599/1/012019

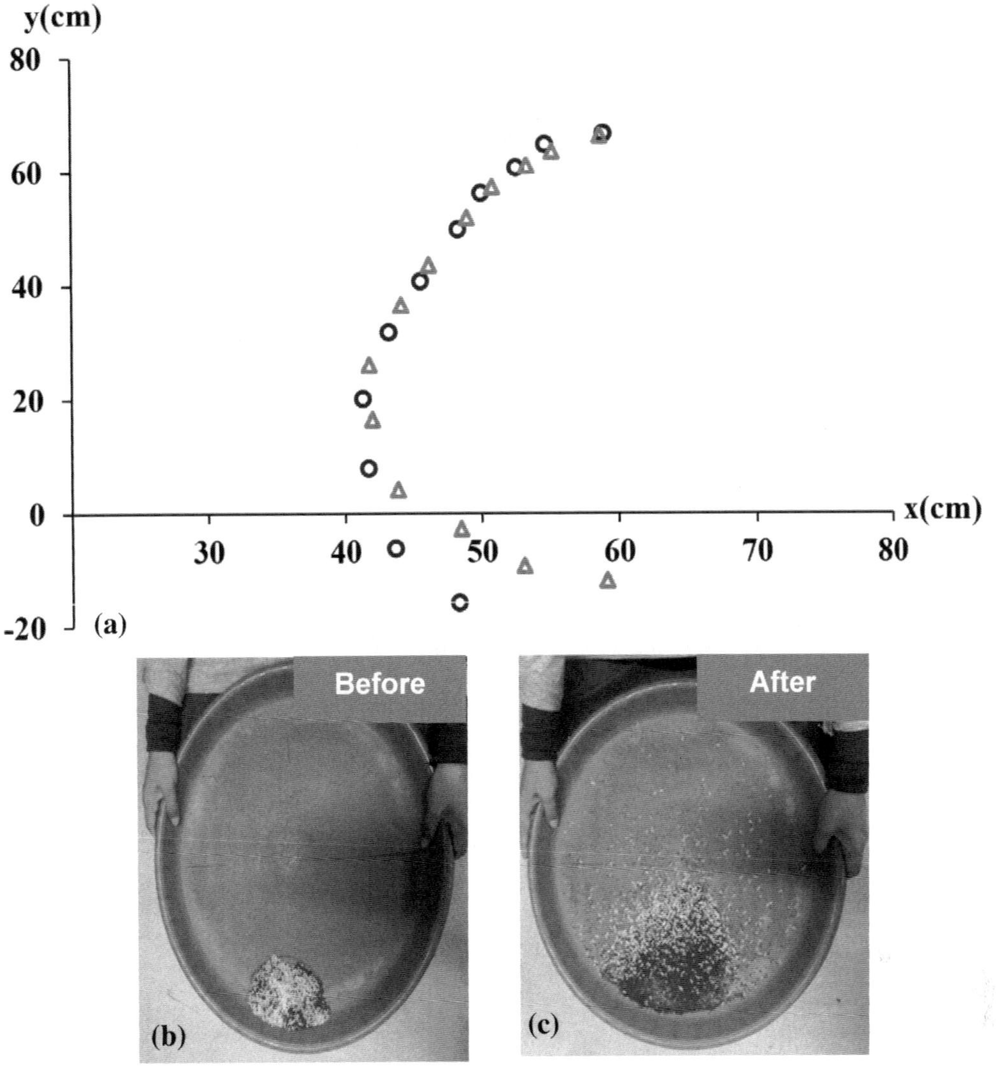

Figure 2. Profiles for falling motion of the mixture granulars by manually winnowing. (a) The trajectories of mixture granulars: rice (circle symbol, blue) and basil seeds (triangle symbols, read), (b) before winnowing and (c) after winnowing by 5 times.

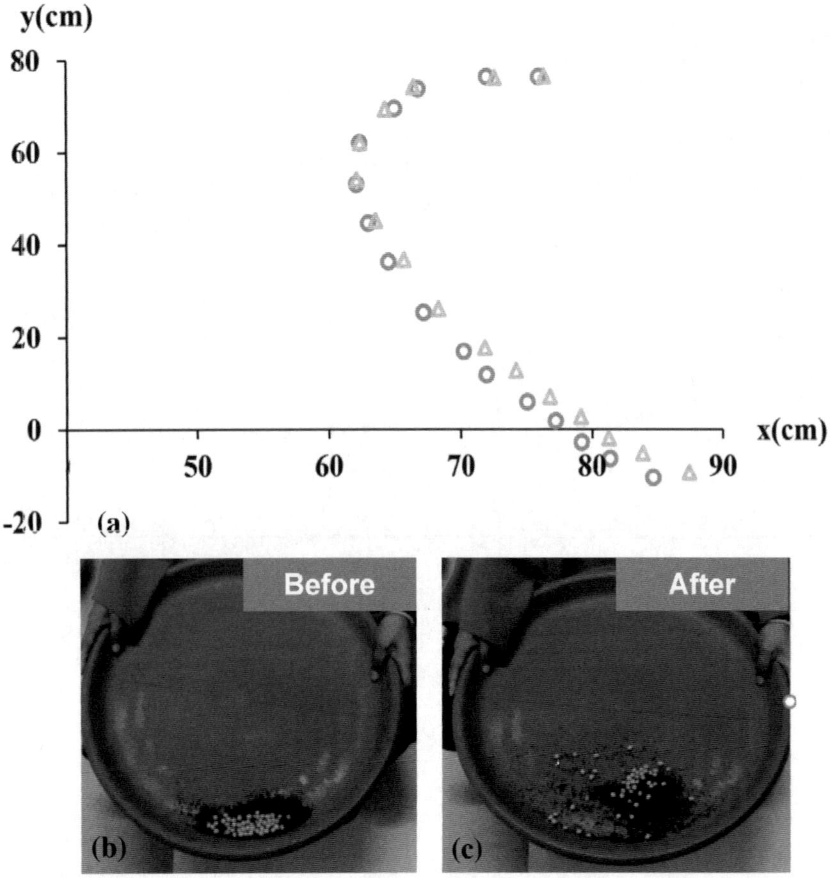

Figure 3. Profiles for falling motion of the mixture granulars by manually winnowing. (a) The trajectories of mixture granulars: rice (circle symbol, green) and basil seeds (triangle symbols, yellow), (b) before winnowing and (c) after winnowing by 5 times.

Figure 3 shows a similarity pattern of granular motions that occur during the winnowing process using different samples. Figure 3 (a) shows x-y position chart for larger samples of soybean grain trajectory pattern represented by a green dot while the pattern of basil seed with smaller size represented by a yellow triangle. Figure 3 (c) shows the experimental results where granules position were change in two types of sample. The position of the basil seeds which smaller are pushed in front of larger sized soybean granules.

The manual winnowing only produces qualitative data in an attempt to explain segregation process.Then, we designed equipment that can produce an airflow mimicking that of produced by manual flapping to obtain more quantitative results. This result is similar to a formation in a riverbed where the top of a riverbed is typically lined with larger cobbles, while finer sand and small gravel particles make up the deeper layers. This process could be related to the Brazil nut effect [9,10]. One of the main concerns within this experiment was the hand movement speed when winnowing cannot be controlled. Therefore, a tool that resembles human hand movements is designed when manually winnowing.

IOP Conf. Series: Materials Science and Engineering **599** (2019) 012019 doi:10.1088/1757-899X/599/1/012019

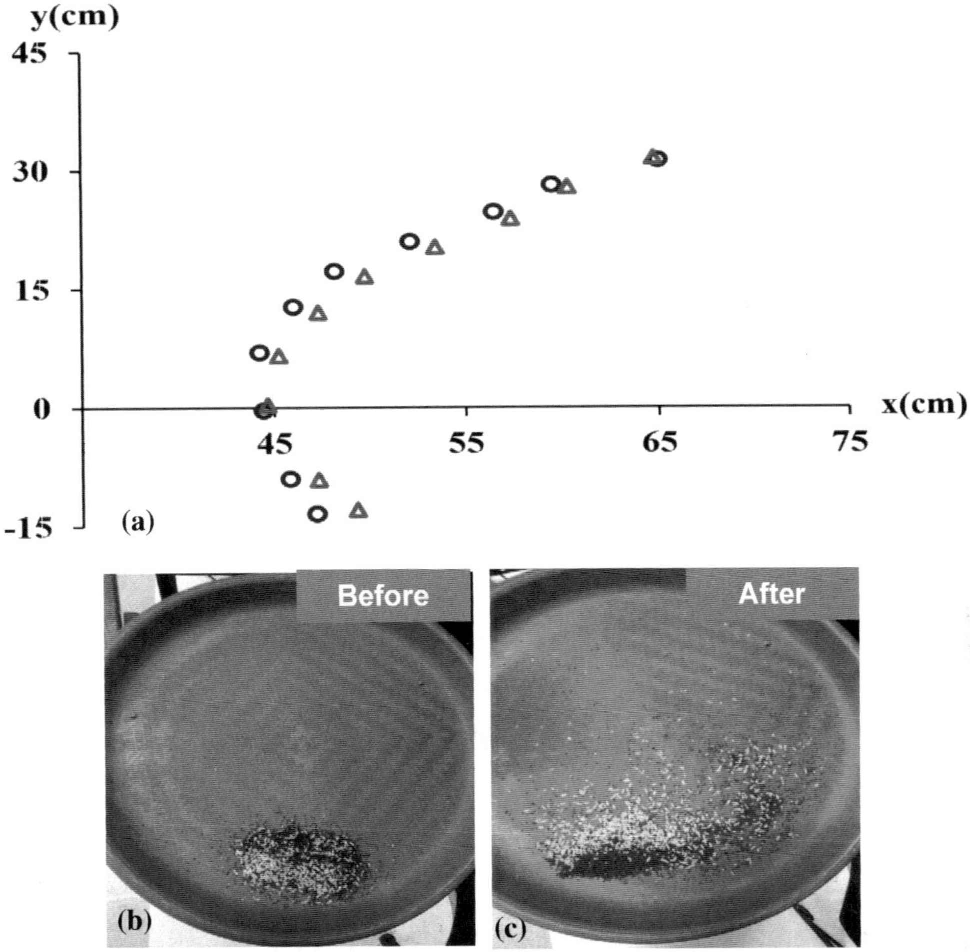

Figure 4. Profiles for falling motion of the mixture granulars using artificial arm instrument. (a) The trajectories of mixture granulars: rice (circle symbol, blue) and basil seeds (triangle symbols, read), (b) before winnowing and (c) after winnowing by 5 times.

Figure 4 shows the profile of the granular mixture motion when winnowing using artificial arm instrument from the initial position at maximum height then falls to the winnower surface. Large rice grain trajectories marked with blue circles and small basil seeds grain are marked with red triangle (see Figure 4 (a)). The Figure 4 (b) shows the condition of the granular mixture before winnowing process. The Figure show that both grains are mixed well. Figure 4 (c) shows the final position of the two types of granules after the swing movement execute within winnowing process. There was a change in each type of grains position, where smaller grains are pushed in front of the bigger grains. these results were achieved after 5 winnowing to attempt.

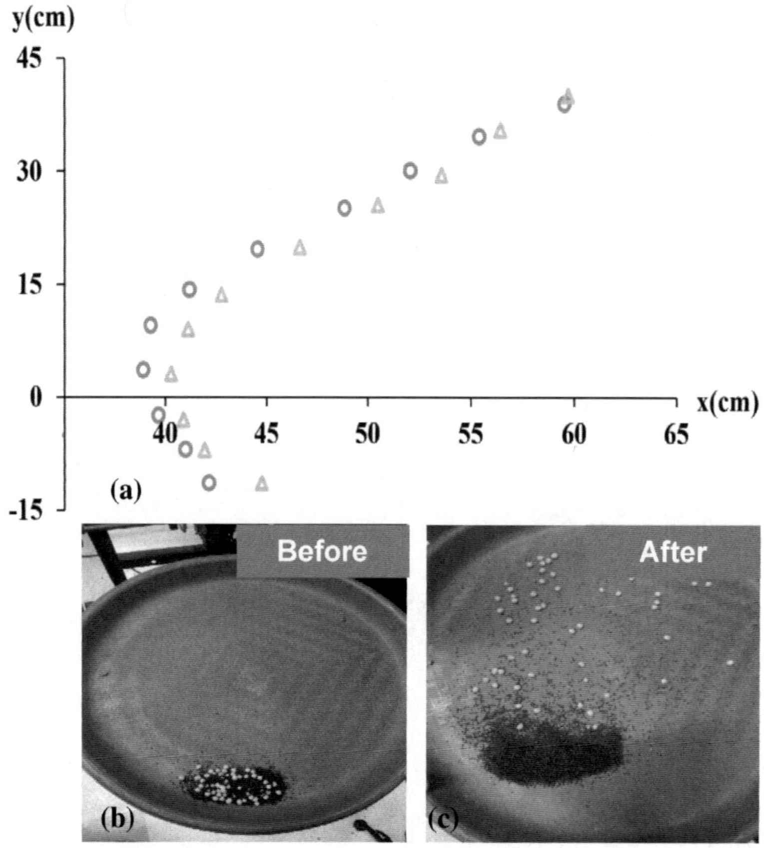

Figure 5. Profiles for falling motion of the mixture granulars using artificial arm instrument. (a) The trajectories of mixture granulars: rice (circle symbol, green) and basil seeds (triangle symbols, yellow), (b) before winnowing and (c) after winnowing by 5 times.

Figure 5 shows a similarity pattern of granular motions that occur during the winnowing process using different samples. Figure 5 (a) shows x-y position chart for larger samples of soybean grain trajectory patterns represented by a green dot while the pattern of basil seed with smaller size represented by a yellow triangle. Figure 5 (b) shows the condition of the granular mixture before winnowing process. The Figure show that both grains are mixed well. Figure 5 (c) shows the final position of the two types of granules after the swing movement execute within winnowing process. There was a change in each type of grains position, where bigger grains are landing far behind the smaller grains.

From the experiments results, we found that there are differences in profile granular trajectories between manual winnowing movements compared to winnowing movements using tools. It appears that the maximum grain height is only 32 cm (Figure 4 (a)) and 38 cm (Figure 5 (a)). While the maximum height of the mixture position is around 75 cm (Figure 2 (a)) and 70 cm (Figure 3 (a)). This result occurs due to the differences of flexibility between human arm and the instrument. But it still produces the same mixture separation pattern. Based on the experiment using granular with variations of different sizes and densities indicate that smaller grains were pushed and landing in front of the large grains in the tray surface.

5. Simulation

The iterations of equations (6) and (7) were solved numerically to simulate the trajectory of the granular mixture.

$$v_x(i+1) = v_x(i) + a_x(i)\Delta t \quad (8)$$

$$v_y(i+1) = v_y(i) + a_y(i)\Delta t \quad (9)$$

$$x(i+1) = x(i) + v_x(i)\Delta t \quad (10)$$

$$y(i+1) = y(i) + v_y(i)\Delta t \quad (11)$$

Equations (10) and (11) are used to obtain the trajectory profile of the mixed particles. By substituting parameters in the form of rice density (1.065 gr / cm3), basil seeds (2.192 gr / cm3), soybeans (1.44 gr / cm3), green beans density (0.546 gr / cm3) and air density ρa≈ 1 kg / m3, the simulation results obtained as shown in Figure 6.

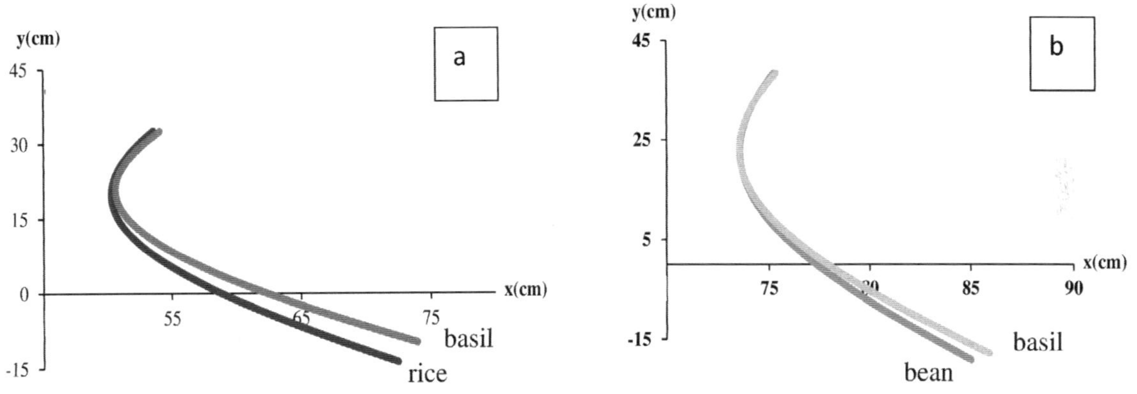

Figure 6. Profile the path of granulars of different sizes and densities. Granulars mixtures (a) rice (blue) and basil (red): (b) Bean (green) and basil (yellow).

Figure 6 is a graph of the profile profile of granular mixtures of different sizes and densities based on simulation using equation 10 and 11. Figure 6 (a) shows profile profile of rice granules, large grains (lines, blue), (b) trajectory of basil seeds, small grains (lines, red).It appears that at the initial position of the track at the maximum height, initially it is still mixed and after reaching the surface of the surface, the mixture undergoes separation. This is indicated by the difference in the final position of the mixture with blue and red coinciding lines. The same is also seen in the profile of the granular mixture (Figure 6 (b)). Based on the above simulations results (Figure 6), profiles for falling motion of the mixture granulars similar to the experiment results. Therefore, these are indicated that granulars segregation due to particle motion in vortex.

6. Conclusion

We have designed a simple granular segregation instrument and studied the profile of particle motion with video processing using a video tracker. The granular motion profile of the experimental results resembled simulation model of particle motion equations in vortex. This vortex arises from the motion of a swinging arm or instrument. This phenomenon shows that granular segregation due to the motion of vortex particles.

7. References

[1] John Mark, NTG 2018 Particle Segregation in Dense Granular Flow, *Annual Review Mechanics* **50** 407 – 433

[2] Behrooz, F., Carlos, P.O., Morgane, H. Dan Douglas, J.J. 2017 River – bed Armouring as a Granular Segregation Phenomenon *Nature Communications* **8**(1363)

[3] Tunuguntla, DR., Bohove, O. Dan Thornton, AR.2014 A Mixture Theory for Size and Density Segregation in Shallow Granular Free – Surface Flows, *J. Fluid Mech* **749** 99 – 112

[4] Xiao, H., McDonald, D., Unbanhoar, PB., Fan, Y., Ottino, J.M dan Lueptow, RM 2017 Controlling Granular Segregation Using Modulated Flow, *Powder Tecknology* ***312***

[5] Shalev, ST. Dan Emily, EB. 2018 Gravity – Independent Grain Size Segregation in Experimental Granular Shear Flows as a Mechanism of Layer Formation, Geophysical Research Letters **45**(16) 8136-8144

[6] Sultanbawa, F.M., Owens, W.G. and Pandiella, S.S. 2001 A new Approach to the prediction of Particle Separation by Sieving in Flour Milling *Food and Bioproducts Processing*, **79**(4) 211-218

[7] Munir, R., Rahmayanti, H.D., Murniati, R., Rahman, D.W., Viridi, D. and Abdullah, M.2018 Experiment and Modeling of Rice Winnowing: Granular Segregation Method in Ancient Traditions, https://arxiv.org/abs/1810.11076

[8] Accetta, K., Fellinger, R., Goncalves, P.L., Musselw, H.S., Pelt,W.P.V. 2013 Current Research in Egyptology 2013, in Proceedings of the Fourteen Annual Symposium, United Kingdom

[9] Kudrolli, A. 2004 Size separation in vibrated granular matter, *Rep. Prog. Phys.* **67** 209–247

[10] Shimoska, A., Nousou, I., Shirakawa, Y., and Hidaka, J. 2013 Effect of particle shape on size segregation of particles, *Chem. Eng. Trans.* **32**, 2143-2148

Acknowledgment

This study was supported by the Ministry of Research, Technology and Higher Education, Republic of Indonesia through the Research Doctoral Dissertation Grant (PDD) 2019.

IOP Conf. Series: Materials Science and Engineering **599** (2019) 012020 doi:10.1088/1757-899X/599/1/012020

Tin-Empty Palm Bunch Ash Impregnated Zeolite as Suitable Catalyst for Simultaneous Transesterification-Esterification Reaction of Palm Oil

Thamrin Usman, Rudiyansyah, Nelly Wahyuni, Ismail Astar, Yudi Yustira, Andi Hairil Alimuddin, Winda Rahmalia*

Department of Chemistry, Mathematics and Natural Science, Universitas Tanjungpura, Jl. Prof. Dr. H. Hadari Nawawi, Pontianak 78124, West Kalimantan, Indonesia

*winda.rahmalia@chemistry.untan.ac.id

Abstract. A single-step method was developed for methyl ester production from palm oil using tin-empty palm bunch ash impregnated zeolite (Sn-PBA-Zeolite) as heterogeneous catalyst. The weight ratios between Sn solution and zeolite 1:4, while the weight of the PBA was varied from 5, 10, 15, 20, 25, and 30 g in 60 mL of distilled water. Effects of reaction time, molar ratio of reactants, and catalyst concentration on the yield of methyl ester were studied. Characteristics of methyl ester produced were also investigated. The optimal conditions for transesterification reaction carried out by using a Sn-PBA-Zeolite of 1:4:25 weight ratio were a palm oil/methanol molar ratio 1:6, catalyst concentration (3% wt/wt), and 3 hours reaction time, whereas the maximum yield of methyl ester reached 76.21% wt/wt. The optimal conditions for esterification reaction carried out by using a Sn-PBA-Zeolite of 1:4:5 weight ratio were a palm oil/methanol molar ratio 1:12, catalyst concentration (3% wt/wt), and 3 hours reaction time, and the maximum yield of methyl ester reached 94.76% wt/wt. The methyl ester resulted has a density of 0.89 g/mL, viscosity of 7.82 cSt, and dominated by a methyl oleate, methyl palmitate, and methyl stearate.

1. Introduction

Methyl esters are type of fatty acid ester that are derived by transesterification of fats with methanol. They are used to produce detergents and biodiesel. The process feedstock can be derived from edible oils such as palm, soybean, groundnut, sunflower, sesame, safflower, waste cooking oil or non-edible oils such as *Jatropha curcas* and *Pongomia pinnata* [1-6]. One of the largest oil sources in Indonesia is palm (*Elais guineensis* Jack.) oil. The oil extracted from the palm fruit mesocarp and it is the largest source of vegetable oil. However, the oil palm contains high lipase activity that increases free fatty acids (FFA).

A conventional transesterification reaction usually uses strong alkali (i.e. NaOH or KOH) or acid (i.e. H_2SO_4 or H_3PO_4) solutions. These homogeneous catalysts are quite sensitive to FFA in the oil feed stocks. FFA can react with the basic catalyst and form soaps. This soap formation complicates the glycerol separation, and drastically reduces the methyl ester yield [7, 8]. For conventional processes using homogenous catalysts, the FFA content in the feedstock must be lower than 0.50 %wt. Instead, the two-step conversion process: an acid-catalyzed esterification pretreatment to lower the FFA content followed by the alkali-catalyzed transesterification, are widely used in both industry and laboratory [9]. The strong acid or alkali catalysts used are corrosive. It must be removed from the methyl ester product by multiple washing. Thus, a significant amount of waste water is generated, together with lost of catalyst.

Based compounds of Sn are one of the compounds which are potentially to be developed as heterogeneous catalysts in methyl ester production. This metal presents higher Lewis acidity in comparison to other transition metal such as zinc (Zn), lead (Pb), and mercury (Hg) [10]. Cardoso et al. [11] reported that $SnCl_2$ used as a catalyst in the esterification reaction of oleic acid and showed catalytic activity which was very similar to H_2SO_4. Therefore, it is advantageous to develop a new class of heterogeneous catalysts, which has a higher tolerance to FFA in oils, and can simultaneously catalyze both of esterification and transesterification reactions.

Xie et al. [12] has mentioned that the SnO_2/SiO_2 materials with various Sn loadings ranging from 1 to 16 wt.% had been prepared and used as heterogeneous acid catalyst for soybean oil transesterification to produce biodiesel. The dispersed amorphous SnO_2 species on the silica surface are considered to be active sites for the transesterification reaction. Several catalyst based on tin compound (tin(II) acetate, tin(II) chloride, tin(II)2-ethylhexanoate and tin(II) stearate) has been tested by Casas et al. [13] for esterification and transesterification of acid vegetable oils. However, Sn was soluble in methanol and glycerol. To overcome this, Sn metal needs to be entrusted to the supporting matrix, one of them by impregnating Sn metal in the zeolite [10].

Zeolite is microporous crystalline alumino silicates, chemically similar to clay minerals, but they are different in their well-defined three dimensional microporous structure. Silicon, aluminum, and oxygen are arranged in a regular structure of $[SiO_4]$ and $[AlO_4]$ tetrahedral units that form a framework with regular pores form of channels, tunnels, or cavities of about 0.12 nm diameter running through the material [14]. Zeolite has been used as ion exchange and molecular sieves in the separation and removal of gasses and solvents. They also have the ability to act as a catalyst for chemical reactions which take place within the internal cavities [15]. Zeolite has been widely used as industrial heterogeneous catalysts because they are inexpensive and environmentally benign. They offer generous surface area and high porosity [16].

Since Sn-Zeolite is acidic, it is favorable for low-qualified oil feedstock with high FFA. In present study, to develop a simultaneous esterification-transesterification reaction, PBA from palm buch ash was used for increasing alkali sites of Sn-Zeolite. Usman et al. [17] has investigated that PBA contains more than 40% of alkali based potassium. From practical point of view, the use of PBA is advantageous since it is highly available as palm oil waste and renewable.

2. Materials and Methods

2.1 Materials

Palm oil and PBA were acquired in PTPN 13, West Kalimantan, Indonesia. Zeolite was purchased from one of the markets in West Kalimantan, Indonesia. The following chemicals, all analytical grades: oxalic acid ($C_2H_2O_4$), sulfuric acid (H_2SO_4), dichloromethane (CH_2Cl_2), ethanol (C_2H_5OH, 95%), phenolphthalein indicator, potassium hydroxide (KOH), anhydrous magnesium sulfate ($MgSO_4$), methanol (CH_3OH), n-hexane (C_6H_{14}), sodium chloride (NaCl), tin (II) chloride ($SnCl_2$), and G F254 silica gel were supplied by Sigma-Aldrich, Germany.

2.2. Methods

2.2.1 Materials Preparation

Zeolite particles were washed 3 times with distilled water, and dried in an oven at 110°C. Dry zeolite was then crushed using a mortar until smooth shape with a size of 100 mesh. PBA was also crushed using a mortar until smooth shape with a size of 100 mesh, and activated in a furnace at a temperature of 600°C for 2 hours. Preparation of palm oil was conducted by determining its water content, acid values, and FFA content.

2.2.2 Synthesis of Cayalysts

Six series of PBA solutions, while the weight of the PBA was varied from 5, 10, 15, 20, 25, and 30 g in 60 mL of distilled water, were soaked and stirred for 6 hours. The resulting filtrate was mixed with 5 grams of pre-prepared of Sn impregnated zeolite (Sn-Zeolite) as has been done by Alimuddin et al. [10]. The preparation process of Sn-PBA-Zeolite was carried out in a three necked flask (250 mL) equipped with a condenser, a thermometer and a magnetic stirrer bar. Impregnation of PBA with Sn-Zeolite was performed at 60°C for 2 hours. After impregnation process had completed, the Sn-PBA-Zeolite was separated from solutions using vacuum filtration. Products were dried in an oven at 105°C for 24 hours to remove the water content followed by calcination in a furnace at 450°C for 4 hours. The final products obtained are referred as to Sn-PBA5-Zeolite, Sn-PBA10-Zeolite, Sn-PBA15-Zeolite, Sn-PBA20-Zeolite, Sn-PBA25-Zeolite, and Sn-PBA30-Zeolite for using 5, 10, 15, 20, 25, and 30 g of weight of PBA respectively. Catalysts obtained were characterized by XRD and XRF.

2.2.3 Transesterification and Esterification Reactions of Palm Oil by using Sn-PBA Zeolite Catalyst

Production of methyl ester was carried out in a three necked flask (250 mL) equipped with a condenser, a thermometer and a magnetic stirrer bar by mixing palm oil, methanol, and catalyst at 65°C. The molar ratio of the reactants was varied from 1:2 to 1:16, Sn-PBA-Zeolite catalysts concentration was varied from 1 to 9 weight percent, time reaction was varied from 1 to 5 hours. After reaction was completed, mixture was centrifuged, and product was washed with warm distilled water. Anhydrous $MgSO_4$ was added to remove water content and filtered. Methyl ester yield and methyl ester product conversion percentage were calculated by following equation 1 and 2.

$$\text{Methyl ester yield} = \text{product mass/sample mass x } 100\% \quad (1)$$

$$\text{Conversion percentage } (\%) = (FFA_{intial} - FFA_{end})/FFA_{initial} \text{ x } 100\% \quad (2)$$

Physical properties (i.e. density, solubility, and viscosity) of solvent-free methyl ester product and residual catalyst were determined. Their chemical properties were also identifying by using GC-MS.

3. Results and Discussion

3.1. Characterization of Catalyst

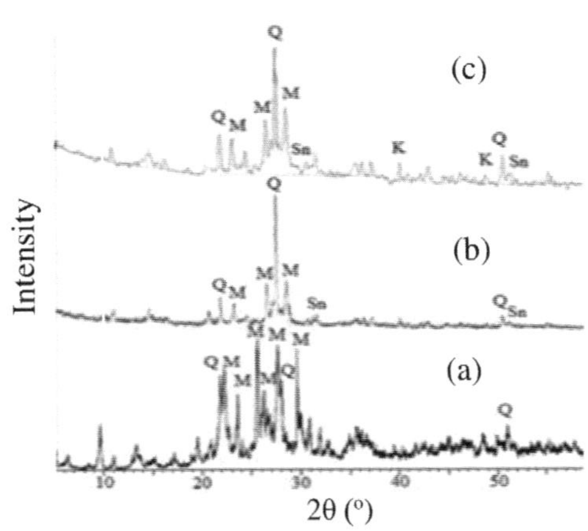

Figure 1. *XRD patterns of natural zeolite (a), Sn-Zeolite (b), and Sn-PBA-Zeolite (c)*

Since catalytic properties of zeolites is dependent on crystallite size and structure, materials used in this study were characterized by using XRD. Figure 1 illustrates XRD patterns of natural zeolite, Sn-Zeolite, and Sn-PBA-Zeolite. As can be seen in Figure 1a, the major mineral found in the natural zeolite is mordenite with a small content of quartz. Mordenite is one of the highest silica zeolites. It has two pore channels, i.e. 6.5 × 7.0 Å, which are parallel to c axis are connected by tortuous pores of 2.6×5.7 Å, and also to b axis that form side pockets. Since the latter are too small for most molecules to enter, mordenite is generally regarded as a mono-dimensional zeolite, which is widely used in catalysis because of its uniform, small pore size, high internal surface area, flexible framework, and controlled chemistry [18,19]. The presence of quartz in the sample, since the zeolite used originated from nature and in this study was not purified further.

Figure 1b and 1c show the loading of Sn and PBA into natural zeolite decrease the crystallinity as indicated by a decrease in the intensity of some characteristic peaks of modernite. Potassium was detected on the XRD pattern after loading of PBA. This proves that potassium is a major element in PBA and can play a role in providing base sites to zeolite. Impregnation is also selective. The presence of potassium (K) as the largest element in PBA is confirmed by the results of XRF analysis (Table 1).

Table 1. XRF Analysis Date

Sample	Number of Compositions (% w/w)						
	Si	Al	Sn	K	Mg	Ca	Fe
Natural zeolite**	33.09	7.68	-	3.85	0.68	1.87	1.89
Sn/zeolite**	31.71	6.83	5.52	3.46	0.44	1.49	1.65
PEFB ash	10.14	1.67	-	26.0	8.43	7.41	2.45
Sn-PBA5-Zeolite	64.61	11.91	7.72	7.97	1.32	3.28	1.64
Sn-PBA10-Zeolite	54.87	8.34	11.2	11.08	2.28	5.67	2.89
Sn-PBA15-Zeolite	62.91	12.24	7.98	8.69	0.86	3.35	1.58
Sn-PBA20-Zeolite	61.38	11.34	8.55	9.28	1.07	3.69	1.74
Sn-PBA25-Zeolite	54.84	8.84	10.4	13.3	1.45	4.64	2.38
Sn-PBA30-Zeolite	57.99	8.91	9.79	12.9	1.50	4.34	2.35

** Source: Alimuddin et al. [12]

XRF data analysis confirms that the Si/Al ratio of natural zeolite is 4.32. Si/Al ratios of zeolites slightly increase in Sn-Zeolite (4.64), but increase significantly when PBA had been impregnated (5.42, 6.57, 5.13, 5.41 for Sn-PBA5-Zeolite, Sn-PBA10-Zeolite, Sn-PBA15-Zeolite, and Sn-PBA20-Zeolite respectively). Changes in cation composition, in particular alkali and alkaline earth cations indicate that there has been a cation exchange to neutralize the zeolite charge and increase the base site of zeolite.

3.2. Transesterification and Esterification Reaction of Palm Oil by Using Sn-PBA-Zeolite as a Catalyst

3.2.1 Effects of Reaction Time, Molar Ratio of Reactants, and Catalyst Concentration on the Yield of Methyl Ester

Effects of reaction time, molar ratio of reactants, and catalyst concentration on the yield of methyl ester were studied by using Sn-PBA5-Zeolite as a catalyst. The transesterification reaction was carried out by using palm oil that have a water content of 0.09%, acid number of 9.27 g/mol, and FFA level of 4.23%. Meanwhile, the esterification reaction was carried out by using palm oil that had a water content of 1.98%, acid number of 145.36 g/mol, and FFA level of 66.32%.

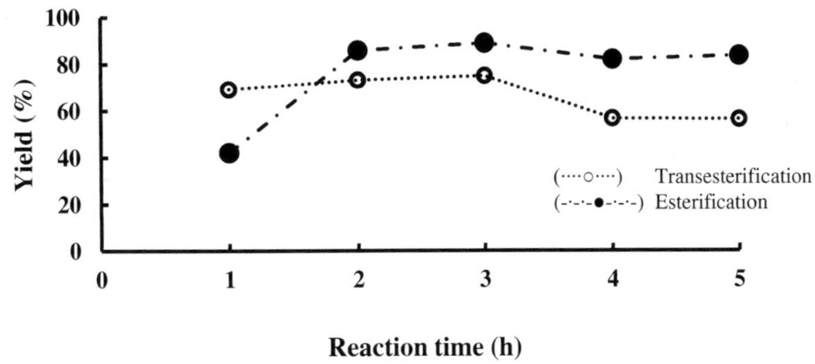

Figure 2. Methyl Ester Resulted at Varied Reaction Time of Transesterification and Esterification.

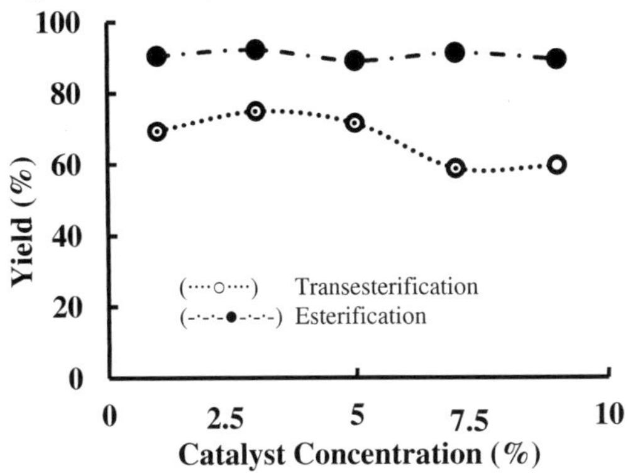

Figure 3. Methyl Ester Resulted at Varied Catalyst Concentration for Transesterification and Esterification.

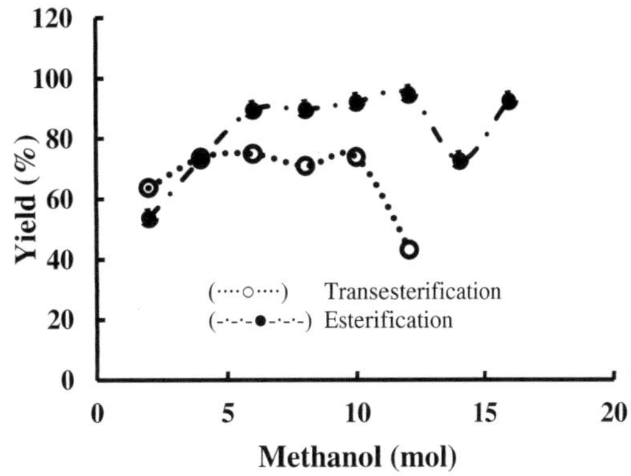

Figure 4. Methyl Ester Resulted Varied Molar Ratio of Reactants for Transesterification and Esterification.

Figure 2 shows the optimum reaction time reached at 3 hours, for both transesterification and esterification reactions. The maximum yield of methyl ester obtained in this time was 74.91% and 88.89% for transesterification and esterification reactions respectively. The longer reaction time provides a great opportunity for the reactant molecules to collide and react to one another to form methyl esters. However, since transesterification and esterification reactions are a reversible reaction, a longer reaction time lead to decreasing of methyl ester yield.

The optimum catalyst concentration in transesterification reaction was at 3% weight percent which was able to produce 74.91% of methyl ester and 92.17% when esterification reaction was applied (Figure 3). Generally, increasing of catalyst concentration increase the reaction rate where the number of molecules and chance to collide will increase [20]. The use of a larger catalyst concentration causes a decrease in the resulting methyl ester yield. This is due to the excess reaction of the catalyst with triglycerides in palm oil to produce a more glycerol by-product. In addition, this may also be due to the increasing amount of methanol that acts as a catalyst solvent rather than as a reactant thereby decreasing the conversion of the product because methanol may act as a reactant as well as a protic solvent.

Figure 4 shows that the optimum molar ratio of reactats in transesterification reaction is 1:6 which was able to produce 74.91% of methyl ester. The optimum molar ratio of reactats in esterification reaction was 1:12 which was able to produce 94.76% of methyl ester. Increase in methanol molar ratio: palm oil actually decreases percent yield. This occurs because of excessive use of methanol will cause the difficulty of separating the glycerol from the reaction product.

3.2.2 Effects of concentration of PBA in Sn-PBA-Zeolite Catalyst

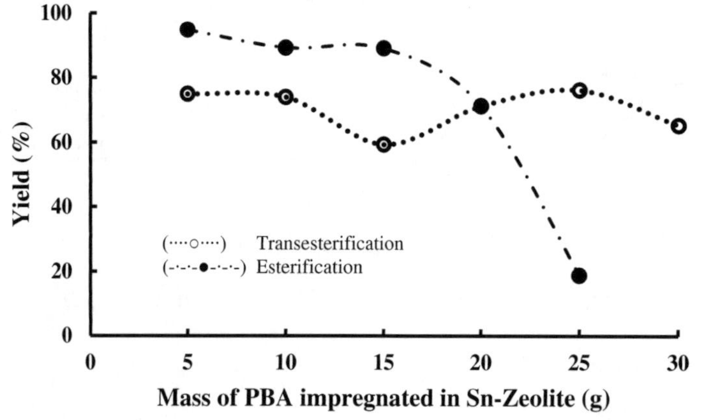

Figure 5. Methyl Ester Resulted Varied of Catalysts for Transesterification and Esterification.

As described above, the optimal conditions for transesterification reaction carried out by using a Sn-PBA5-Zeolite were a palm oil/methanol molar ratio 1:6, catalyst concentration (3% wt/wt), and 3 hours reaction time. Meanwhile, the optimal conditions for esterification reaction carried out by using a Sn-PBA5-Zeolite were a palm oil/methanol molar ratio 1:12, catalyst concentration (3% wt/wt), and 3 hours reaction time. To determine the optimum concentration of PBA to be impregnated as base sites in Sn-Zeolite, weight of PBA used were varied.

Figure 5 shows that use of Sn-PBA25-Zeolite in esterification reaction produced maximum percentage of methyl ester yield, and use of Sn-PBA5-Zeolite for that the esterification reaction. These results were influenced by the concentration of potassium impregnated in Sn-Zeolite. This proves that the presence of base sites on Sn-Zeolite-based catalyst greatly affects the type of reaction would be applied. Transesterification reaction requires a catalyst that has a higher alkalinity than esterification reaction.

3.3. Characteristics of Methyl Ester

Characteristics of methyl ester obtained in this study was determined its physical and chemical properties including density, viscosity, and composition. The density and viscosity of methyl ester at 40°C is 0.89 g/mL and 7.82 Cst respectively. The result of GC-MS analysis (Table 2) shows that the percentage of methyl ester obtained up to 91.86%. The major component of methyl ester on the transesterification-esterification reaction of palm oil with Sn-PBA-Zeolite are methyl oleate (38.92%), methyl palmitate (20.59%), and methyl stearate (14.67%).

Table 2. Chemical Composition of Methyl Esters in Products of Transesterification and Esterification Reactions of Palm Oil by Using Sn-PBA-Zeolite as a Catalyst

Methyl esters	Area (%)
Methyl caprylate	0.10
Methyl pelargonate	0.06
Methyl stearate	14.69
Methyl azelaaldehidat	0.06
Methyl laurate	0.15
Methyl myristate	0.92
Methyl palmitate	20.59
Methyl margarine	0.44

Methyl linoleate	6.94
Methyl oleate	38.92
Methyl linoleate	0.35
Methyl ricinoleate	1.49
Methyl trans-9,10-epoxystearate	0.49
Methyl Cis-9,10-epoxystearate	0.22
Methyl 10-ketostearate	1.60
Methyl arachidate	2.84
Methyl 10-hydroxy-hexadecanoate	0.58
Methyl behenat	0.55
Methyl tricosanoate	0.11
Methyl Lignocerate	0.50
Amount	91.60

4. Conclusions

The presence of base sites on Sn-Zeolite-based catalyst greatly affects the type of reaction would be applied. Transesterification reaction requires a catalyst that has a higher alkalinity than esterification reaction. Empty palm bunch ash has the potential to be increased in useful value, one of them in Sn-Zeolite-based catalyst. The optimal conditions for transesterification reaction carried out by using a Sn-PBA-Zeolite of 1:4:25 weight ratio were a palm oil/methanol molar ratio 1:6, catalyst concentration (3% wt/wt), and 3 hours reaction time. The optimal conditions for esterification reaction carried out by using a Sn-PBA-Zeolite of 1:4:5 weight ratio were a palm oil/methanol molar ratio 1:12, catalyst concentration (3% wt/wt), and 3 hours reaction time. The methyl ester resulted has a density of 0.89 g/mL, viscosity of 7.82 cSt, and dominated by methyl oleate, methyl palmitate, and methyl stearate.

5. References

[1] Hameed B.H., Chin C.S.G.L.H. 2009 Process optimization for methyl ester production from waste cooking oil using activated carbon supported potassium fluoride *Fuel processing Technology* **90** 1532-1537

[2] Satyanarayana M, Muraleedharan C 2011 A comparative study of vegetable oil methyl esters (biodiesels) *Energy 36* 2129-2137

[3] Babu R, Godiganur S 2015 Use of vegetable edible oil methyl esters as fuel in internal combustion (CI) engines-a review, *International Journal of Latest Research in Science and technology* **4** 85-89

[4] Mofijur M, Masjuki H.H, Kalam M.A, Rasul M.G, Atabani A.E, Hazrat Mahmudul, M.A, H.M. 2015 Effect of biodiesel-diesel blending on physico-chemical properties of biodiesel produced from Moringa oleifera *Procedia Engineering* **105** 665-669

[5] Thunyaratchatanon C, Luengnaruemitchai A, Chollacoop N, Yoshimura Y 2016 Catalytic upgrading of soybean oil methyl esters by partial hydrogenation using Pd catalysts, *Fuel* **163** 8-16

[6] Abdullah, R.N.R. Sianipar, D. Ariyani, I.F. Nata 2017 Conversion of palm oil sludge to biodiesel using alum and KOH as catalyst, *Sustainable Environment Research* **27** 291-295

[7] Yan S, Salley S.O, Simon K.Y 2009 Simultaneous transesterification and esterification of unrefined or waste oils over $ZnO-La_2O_3$ catalysts, *Applies Catalysis A: General* **353** 203-212

[8] Olutoye M.A, Lee S.C, Hameed B.H. 2011 Synthesis of fatty acid methyl ester from palm oil (Elais guineensis) with Ky(MgCa)2xO3 as heterogeneus catalyst, *Bioresource Technology* **102** 10777-10783

[9] Thoai D.N, Tongurai C, Prasertsit K, Kumar A 2017 A novel two-step transesterification process catalyzed by homogeneous base catalyst in the first step and heterogeneous acid catalyst in the second step *Fuel Processing Technology 168* 99-104

[10] Alimuddin A.H, Usman T, N. Wahyuni, Rudiyansyah, Y.E. Prawatya, I. Astar, Y. Yustira 2017 Synthesis and characterization of Sn/zeolite and catalytic activity test in the esterification reaction of sludge oil, *AIP Conference Proceedings* **1823** 020052-1-5

[11] Cardoso A.L, Neves S.C.G, Silva M.J 2008 Esterification of oleic acid for biodiesel production catalyzed by $SnCl_2$: a kinetic investigation, *Energies* **1**

[12] Xie W., Wang H., Li H. 2012 Silica-supported tin oxides as heterogeneous acid catalysts for transesterification of soybean oil with methanol *Industrial and Engineering Chemistry Research* **51** 225–231

[13] Casas A, Ramos M.J, Rodríguez, Pérez Á, 2013 Tin compounds as Lewis acid catalysts for esterification and transesterification of acid vegetable oils, *Fuel Processing Technology* **106** 321-325

[14] Zaarour M. Dong B, Naydenova I, Retoux R., Mintova S 2014 Progress in zeolite synthesis promotes advanced applications *Microporous and Mesoporous Materials* **189** 11-21

[15] Manadee S, Sophiphun O, Osakoo N, Supamathanon N, Kidkhunthod P, Chalenk N, Wittayakun J, Prayoonpokarach S 2017 Identification of potassium phase in catalysts supported on zeolite NaX and performance in transesterification of Jatropha seed oil *Fuel Processing Technology* **156** 62-67

[16] Hassani M, Najafpour G.D, Mohammadi M, Rabiee M 2014 Preparation, characterization and application of zeolite-based catalyst for production of biodiesel from waste cooking oil, Journal of Scientific & Industrial Research **73** 129-133

[17] Usman T, Ariani L, Rahmalia W, Advant R 2009 Esterifikasi asam lemak dari limbah kelapa sawit (*sludge oil*) menggunakan katalis tawas, *Indonesian Journal Chemistry* **9** 474-478

[18] Bhadauria J, Singh B.K, Tomar A, Tomar R 2011 Synthesis and characterization of analogue of mordenite and its role as a catalyst for Friedel-Crafts acylation of anisole **3**(2):245-257

[19] Smail H, Shareef K, Ramli Z 2017 Synthesis of mesoporous mordenite zeolite by different natural raw materials *Australian Journal of Basic and Applied Sciences* **11** (2017) 27-34

[20] A.B.M.S. Hossain, M.A. Mazen 2010 Effects of catalyst types and concentrations on biodiesel production from waste soybean oil biomass as renewable energy and environmental recycling process, *Australian Journal of Crop Science* **4** 550-555

Acknowledgment
This study received Financial support from The Ministry of Research, Technology, & Higher Education (Pascasarjana research grant).

Zeolites as Adsorbent Materials for Decolorization of Crude Terpineol

A Indarto[1,2,*], I A Putra[1], Riyano[3], S Noersalim[3], Y Hartanto[3], L Handojo[1]

[1]Department of Chemical Engineering, Institut Teknologi Bandung, 40132 Bandung, Indonesia
[2]Department of Bioenergy Engineering and Chemurgy, Institut Teknologi Bandung, 40132 Bandung, Indonesia
[3]Department of Chemical Engineering, Parahyangan Catholic University, 40141 Bandung, Indonesia

*E-mail: antonius.indarto@che.itb.ac.id

Abstract. Zeolites are microporous crystalline aluminosilicates extensively used as adsorbents, catalysts, and ion exchange beds. The adsorption property of zeolite is expected to be applied as a decolorization agent for terpineol. Terpineol is an alcohol compound derived from turpentine oil and widely used as an active ingredient for disinfectants, cleansers, perfumes, and pharmaceutical purposes. Industrial-scale of terpineol production used two stages of reaction: the first stage is hydration of turpentine into terpin hydrate and the second stage is dehydration of terpin hydrate to terpineol. However, as terpin hydrate is easily oxidized, the product color will change from clear into dark solution and make the price of product cheaper. This study aims to increase terpineol selling price by removing unreacted terpin hydrate (decolorization) from crude terpineol. Zeolite is the main absorbent used in this study along with other similar materials i.e. activated carbon and silica gel. Performances of the adsorbent are indicated by its adsorption capacity and color clarity analysis. The results would be analyzed using GC-MS. Based on the experiment, it can be concluded that the best adsorbent is zeolite activated by acetic acid with 8-15 mg/g adsorption capacity and producing a terpineol purity of 83.5%. This activated zeolite was able to remove terpin hydrate from 9.41% to 0% and other impurities from 14.06% to 6.78%.

1. Introduction

Zeolites are microporous crystalline aluminosilicates composed by silica, alumina, and oxygen atoms connecting and forming into a tetrahedral structure. These materials are extensively used as adsorbents, catalysts, and ion exchange beds. Zeolites can be found in nature but commonly produced in large scale industry to improve its substantial properties like surface area, selectivity, porosity, and thermal stability. Adsorbent applications of zeolites are highly implemented in purification and separation process of the chemical industry i.e. CO_2 removal of natural gas, sweetening of petroleum gas, and sulfur compound removal [1]. This adsorption property is expected to be applied as a decolorization agent for terpineol.

Terpineol ($C_{10}H_{18}O$) is an alcohol compound that can be processed from turpentine (α-pinene). Turpentine can be obtained from merkusii pine that contains 65-85% of α-pinene [2]. Terpineol is one of the most important derivative products of turpentine and commonly used as perfume, anti-fungal, disinfectant, and insect repellant [3]. The reaction of terpineol begins with the formation of terpin hydrate and then followed by the formation of terpineol from terpin hydrate. Terpin hydrate can be formed by reacting dilute acids, such as hydrochloric acid, nitric acid, and phosphoric acid. The reaction of terpin hydrate formation is shown in Figure 1.

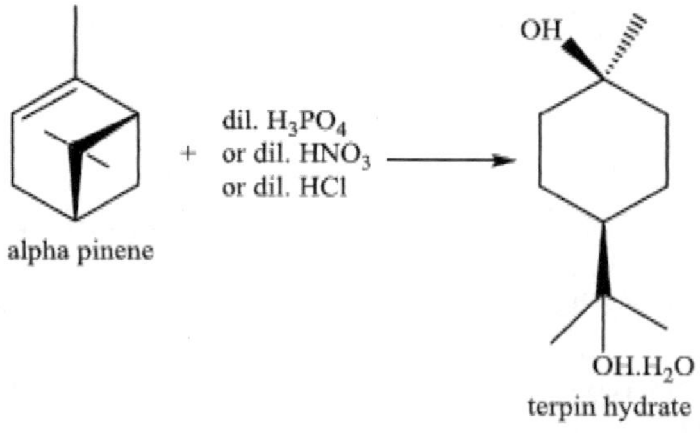

Figure 1. The reaction of terpin hydrate formation.

The process is then followed by the separation of dilute acid and purification of terpin hydrate by steam distillation and neutralization. Another method for separation of terpin hydrate is by washing terpin hydrate with water repeatedly. Terpin hydrate purification is very important to produce pure terpin hydrate product. Pure terpin hydrate is then processed into terpineol by dehydration. The dehydration process requires less mineral acids or organic acids such as sulfuric acid, oxalic acid, or phosphoric acid to take H atoms from terpin hydrate molecules and thus terpineol is formed [4].

This process produces terpineol with dark color which is due to the presence of terpin hydrate. Terpin hydrate has the properties of oxidizing easily which can cause the dark color of terpineol. In addition, terpin hydrate also has a stinging odor that can limit the uses of terpineol as a perfume. Therefore, the terpin hydrate component needs to be removed to increase the selling price of terpineol and increase the usefulness of the product itself [5].

Adsorption is chosen to decolorize crude terpineol because it is cheaper, simpler, and more energy saving. Adsorption has been applied in large scale of industry for years. Adsorption is usually affected by the type of adsorbent and adsorbate. Besides that, adsorption is also affected by adsorbent activation. Therefore, a suitable adsorbent and adsorbate selection is required in the adsorption process by varying the adsorbent composition. In addition, a suitable selection of chemical activation is also required by varying the chemical compound.

2. Research Method

2.1. Material

The materials used in this research were crude terpineol, zeolite, activated carbon, silica gel, nitric acid, acetic acid, hydrochloric acid, sodium hydroxide, and demineralized water. The crude terpineol was obtained from Perhutani Pine Chemical Industry (PPCI), Pemalang, Indonesia and used without further pretreatment.

2.2. Experiment Variation

In this experiment, the decolorization was first performed by using zeolite, activated carbon, silica gel, a combination of zeolite-activated carbon, and zeolite-silica gel with a ratio of 1 to 1. These non-zeolite powders were used for screening material to find the best adsorbent. The best decolorization results will be used further in the variation of chemical activation. Chemical activation was performed to improve the adsorbent properties by acidic and base treatment. Nitric acid was used as the acid compound, while sodium hydroxide was used as the base compound. In the case that acid compound gave a better result, then the next experiment was continued by varying the acid compound and vice versa.

2.3. Adsorbent Preparation

First, the adsorbents were crushed with pestle and mortar, then filtered using mesh size of 100. The powder was then washed with aquadest and heated at 120°C for 24 hours then stored in the desiccator. The dried adsorbents were soaked with acid/base compound for 6 hours at 80°C. After that, the adsorbent was filtered and heated at 120°C for 24 hours then stored in the desiccator. Treated adsorbent will be used in the decolorization experiment.

During the experiment, 5 grams of adsorbent were fed into the adsorption column and crude terpineol was dropped slowly from the top of the column. The treated terpineol that came out from the column was collected on the watch glass. The mass per drop was weighed and the clarity was measured using a light lux meter. Data retrieval is done until the intensity was below 70%.

2.4. Sample Analysis

In order to measure the clarity of the product, 3 drops of demineralized water was put on the watch glass. Then, the intensity of demineralized water was measured with a lux meter as a reference intensity. A color clarity was done by using below equation. The data for the equation was obtained by measuring the light intensity of the sample every certain period. Color clarity analysis was performed to determine the level clarity of terpineol.

$$Sample\ clarity\ (\%) = \frac{sample\ intensity}{reference\ intensity} \times 100\%$$

Terpin hydrate removal analysis was carried out using Gas Chromatography-Mass Spectrophotometry (GC-MS) (Shimadzu GCMS-QP-2010 with Rtx-5MS capillary column). By evaluation of peak area, the concentration of the component could be calculated [6,7]. This analysis was performed to determine the adsorbent ability and efficiency in removing terpin hydrate as the impurities in crude terpineol.

3. Results and Discussion

3.1. Decolorization using Various Adsorbents

The result of decolorization process with the variation of adsorbent materials is shown in Figure 2 and Table 1. From these results, it shows that zeolite adsorbent produces the clearest color of terpineol product.

(a) (b) (c) (d) (e)

Figure 2. Results of crude terpineol decolorization with (a) activated carbon; (b) silica gel; (c) zeolite; (d) zeolite-activated carbon; (e) zeolite-silica gel.

Table 1. Results of curde terpineol decolorization using various adsorbents composition.

Adsorbent	Clarity Percentage
Activated carbon	low
Silica gel	19
Zeolite	66.5
Zeolite-activated carbon	45
Zeolite-Silica gel	55

Decolorization results can be seen from the characteristic of adsorbent and its adsorbate. Terpin hydrate which acts as an adsorbate is a polar organic and hydrophilic compound because the synthesis reaction is done in acidic condition [2]. From these 3 adsorbents (activated carbon, silica gel, and zeolite), activated carbon is the only a hydrophobic adsorbent [8]. Thus, activated carbon could not adsorb terpin hydrate that has a hydrophilic characteristic and the color remained dark-colored terpineol. The clarity of dark colored-terpineol is unknown because its intensity could not be identified by lux meter because the value was too low.

Both silica gel and zeolite adsorbent were able to adsorb terpin hydrate in crude terpineol. But, from Figure 2 and Table 1, zeolite adsorbent produced clearer terpineol than silica gel adsorbent. These results are due to the fact that silica gel is an inorganic adsorbent, while zeolite is an organic adsorbent. Thus, zeolite will perform better because terpin hydrate is also an organic compound [9].

Zeolite adsorbent that has been combined, either with activated carbon or silica gel, produced less clear terpineol than single zeolite adsorbent. This was due to the amount of zeolite that was being reduced. When zeolite was not combined, the total mass of zeolite that was 5 grams, while after the combination, the mass was reduced to only 2.5 grams (the combination is done with a ratio of 1 to 1). By reducing zeolite mass, the performance of adsorption was also reducing.

3.2. Decolorization with the Variation of Physical and Chemical Activation Method

From the previous results, it can be concluded that adsorption with zeolite produced the clearest terpineol. Thus, only zeolite adsorbent was used for the decolorization with the variation of activation type. The yield of decolorization is shown in Figure 3 and Table 2.

(a) (b)

Figure 3. Results of crude terpineol decolorization using treated zeolite (a) physically activated; (b) Physically-chemically activated.

Table 2. Results of crude terpineol decolorization with the variation of activation type.

Zeolite Adsorbent	Clarity Percentage
Physically activated	66.5
Physically-chemically activated	81.5

From Figure 3 and Table 2, physically-chemically activated zeolite produced clearer terpineol than physically activated zeolite. Physical activation was used to evaporate water that is trapped in the pores of zeolite crystal. Thus, the number of zeolite pores and the surface area will significantly increase. Chemical activation was used to clean the pore surface and to arrange the position of the atom that can be exchanged to enhance zeolite ion exchange-ability. By combining physical and chemical activation, zeolite's surface area will increase, and its adsorption ability will also increase.

3.3. Decolorization with the Variation of Acid-Base Chemical Activation
From the previous results, it can be concluded that adsorption with physically-chemically activated zeolite produced the clearest terpineol. Thus, variation is only done by a variating compound that is used in chemical activation. The yield of decolorization is shown in Figure 4 and Table 3.

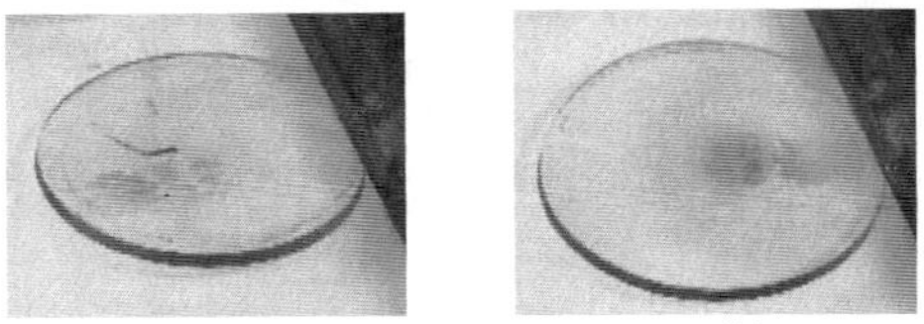

(a) (b)

Figure 4. Results of crude terpineol decolorization with chemically treated zeolite (a) Nitric acid activated; (b) Sodium hydroxide base.

Table 3. Results of crude terpineol decolorization with the variation of acid-base chemical activation.

Zeolite Adsorbent	Clarity Percentage
Nitiric acid activated	81.5
Sodium hydroxide base activated	59

From Figure 4 and Table 3, the adsorption process using nitric acid activated zeolite produced clearer terpineol than sodium hydroxide base activated zeolite. Zeolite is an adsorbent that is composed of tetrahedral's units of AlO_4 and SiO_4 which is connected by O atom. Acid-activated zeolite will cause zeolite's surface to be surrounded by H^+ ion because zeolite functional groups are protonated. In the acidic condition, the zeolite surface becomes positive [10]. Thus, resulting in an exchange between H^+ ion and Si element that has positive charges in zeolite [11]. The ratio of Si/Al in zeolite will decrease and zeolite will have hydrophobic characteristic [12]. This causes zeolite to adsorb terpin hydrate better because both zeolite and terpin hydrate has the same hydrophilic characteristic.

On the contrary, base activated zeolite will cause zeolite's surface to be surrounded by OH^-. Zeolite surface will become negative resulting in an exchange between OH^- ion and Al element [10,11]. Thus, the ratio of Si/Al in zeolite will increase. Base-activated zeolite produced zeolite with a hydrophobic characteristic that caused zeolite to have the worse ability in adsorbing terpin hydrate that has hydrophilic characteristic [12].

3.4. Type Decolorization with the Variation of Acid Chemical Activation
From the previous results, it can be concluded that adsorption with acid-activated zeolite produced the clearest terpineol. Thus, variation is only done by variating acid compound. The yield of decolorization is shown in Figure 5 dan Table 4.

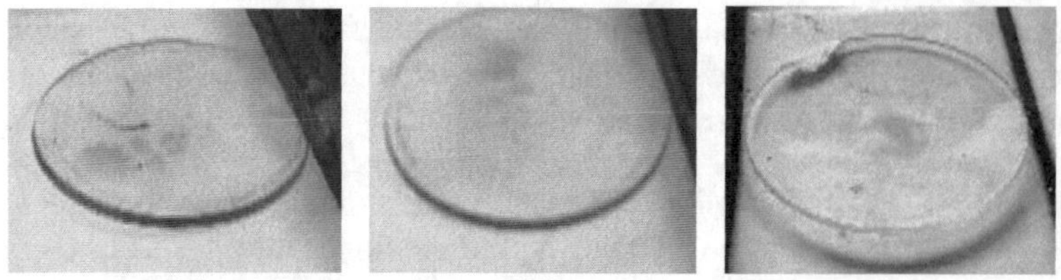

(a) (b) (c)

Figure 5. Results of crude terpineol decolorization with acid-treated zeolite (a) Nitiric acid activated; (b) Acetic acid activated; (c) Hydrochloric acid activated.

Table 4. Results of crude terpineol decolorization with the variation of acid chemical activation.

Zeolite adsorbent	Clarity Percentage
Nitric acid activated	81.5
Acetic acid activated	83.5
Hydrochloric acid activated	71.5

From Figure 5 and Table 4, acetic acid activated zeolite produced the clearest terpineol. These results are due to acetic acid that has the characteristic of polar organic acid, while nitric acid and hydrochloric acid has the characteristic of an inorganic acid. Impurities in terpineol was a polar organic compound because terpineol is synthesized from α-pinene in an acidic condition that form carbocation [13]. A carbocation is a carbon atom that has three bonds, positive charges and more favorable in polar organic condition [14]. Thus, terpineol impurities were easier to be adsorbed using organic material. In this case, acetic acid will give the best result since acetic acid is the only organic acid.

Even though both nitric acid and hydrochloric acid are inorganic acids, but nitric acid activated zeolite produces better terpineol than hydrochloric acid activated zeolite. This result is due to nitric acid that is more polar compared to hydrochloric acid. The level of polarity is determined by dipole moment value. The dipole moment is a quantity to describe two different charges that are separated by distance. The greater the dipole moment value of a compound, the greater its polarity. Nitric acid has 2.17 D, while hydrochloric acid has 1.07 D [15]. Thus, nitric acid is more polar than hydrochloric acid and nitric acid will adsorb terpin hydrate better.

3.5. Removal of Terpin Hydrate

From the previous results, it can be concluded that adsorption with acetic acid and nitric acid produce clearer terpineol. The results of GC-MS analysis are shown in Table 5.

Table 5. Results of GC-MS analysis.

Compound	Composition (%)		
	Crude	Acetic Acid Activated	Nitric Acid Activated
Terpineol	58.18	81.42	78.98
Terpin hydrate	9.41	0	0.19
α-pinene	1,29	0.58	1.05
Limonene	8.24	9.85	13.46
Camphene	8.82	1.39	0.81
Undefined residue	14.06	6.78	5.52

From this analysis result, it can be concluded that both adsorbents can remove terpin hydrate. Terpin hydrate is a by-product that is produced from terpineol's synthesis reaction. Terpin hydrate is wished to be removed from terpineol because terpin hydrate is easily oxidized that causes terpineol to have a dark color. This dark-colored terpineol will cause terpineol selling price to decrease.

Even though both adsorbents can remove terpin hydrate, but acetic acid activated zeolite perform better by removing terpin hydrate perfectly. This is due to terpin hydrate is an organic compound [13]. Thus, terpin hydrate can be adsorbed easier by acetic acid activated zeolite because acetic acid is also an organic compound. On the contrary, nitric acid is an inorganic compound. Thus, nitric acid activated zeolite performs worse than nitric acid zeolite.

3.6. Adsorption Capacity

The adsorption capacity was determined by comparing the percentage clarity of the solution product. The adsorption capacity was calculated at the clarity limit of 70%. This was because 70% of clarity is the minimum standard from the industry. The trend of terpineol clarity by acid activated zeolite is shown in Figure 6. Adsorption capacity calculation result is shown in Table 6.

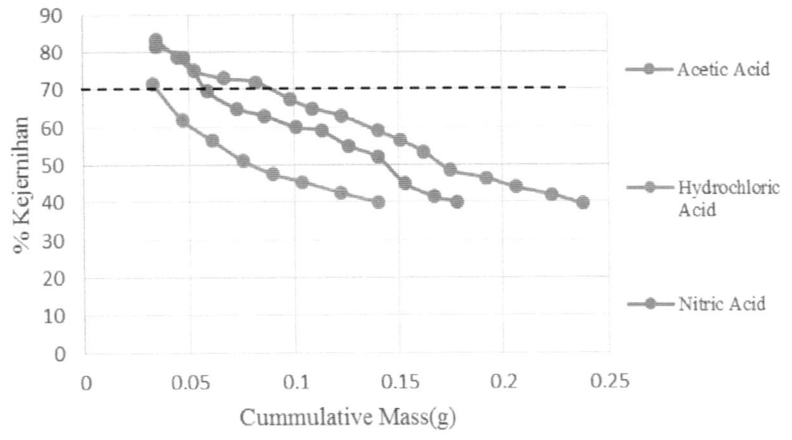

Figure 6. The trend of product clarity as a function of liquid mass for 5 grams of adsorbent.

Table 6. Adsorption capacity of acid-activated zeolite adsorbent.

Zeolite Adsorbent	Capacity (mg/g)
Nitric acid activated	6.6-9
Acetic acid activated	8-15
Hydrochloric acid activated	2.6

It shows that acetic acid zeolite has the greatest adsorption capacity. This is due to terpin hydrate is a polar organic compound and since acetic acid is the only organic compound, acetic acid activated zeolite was able to adsorb terpin hydrate better [14].

Even though both nitric acid and hydrochloric acid were inorganic acid, nitric acid activated zeolite has greater adsorption capacity than hydrochloric acid activated zeolite. This is due to nitric acid's polarity is greater than hydrochloric acid's polarity. The level of polarity can be determined by each compound's dipole moment. Nitric acid has 2.17 D, while hydrochloric acid has only 1.07 D [15]. Thus, since terpin hydrate is a polar organic compound, nitric acid that has more polar characteristic will also have a greater adsorption capacity.

4. Conclusions

A decolorization process of crude terpineol has been conducted by using zeolites and other non-zeolites powder, such as activated carbon and silica gel. Adsorption with acetic acid activated zeolite produced the clearest terpineol product. Adsorption process using acetic acid activated zeolite was able to remove terpin hydrate perfectly from crude terpineol. Among other acid compounds, acetic acid activated zeolite has the best performance that has the greatest adsorption capacity of 8-15 mg/g.

5. References

[1] Payra, P. and Dutta, P.K. Handbook of Zeolites Sciences and Technology. Marcell Decker The Ohio State University, Columbus, Ohio, U.S.A

[2] Handojo, L., Putra, I.A., Azis, M.M., Prakoso, T., Soerawidjaja, T.H. and Indarto, A 2019 Isomerization of Turpentine using various heterogeneous and homogeneous acid catalysts. AIP Conf. Proc. 2085, 020056

[3] Prakoso, T., Hanley, J., Soebianta, M.N., Soerawidjaja, T.H. and Indarto, A. 2018 Synthesis of terpineol from α-pinene using low-price acid catalyst *Catal. Lett.* **148** 725-731

[4] Herrlinger, R. 1959 Production of Alpha-Terpineol. US Patent No. 2,898,380

[5] Fraise, A.P., Lambert, P.A. and Maillard, J.Y. 2008 Russel, Hugo and Ayliffe's Principles and Practice of Disinfection, Preservation and Sterilization. Massachusetts, John Wiley & Sons, pp. 18-19

[6] Indarto, A., Choi, J.W., Lee, H. and Song, H.K. 2006 Treatment of CCl_4 and $CHCl_3$ emission in a gliding-arc plasma. Plasma Dev. Oper. **14** 1-14

[7] Indarto, A., Yang, D.R., Choi, J.W., Lee, H. and Song, H.K. 2007 CCl_4 decomposition by gliding arc plasma: role of C_2 compounds on products distribution. *Chem. Eng. Comm.* **194** 1111-1125

[8] Allport, H.B. 1997 Activated Carbon. In: Mc-Graw-Hill Encyclopedia of Science and Technology. Mc Graw Hill Book Company, New York, pp. 94

[9] Payra, P. and Dutta, P. 2003 Zeolites: A Primer, in Auerbach, S.M., Carrado, K.A., Dutta, P.K. Handbook of Zeolite Science and Technology. Marcel Dekker, New York, pp. 1-25

[10] Bodner, G.M. & Purdue, H.L. 1989 Chemistry an Experimental Science. John Wiley & Sons, New York

[11] Ozkan, F.C. and Ulku, S. 2008 Diffusion mechanism of water vapour in a zeolitic tuff rich in clinoptilolite. *J. Therm. Anal. Calorim* **94**: 669-702

[12] Ackley M.W., Rege, S.U. and Saxena, H. 2003 Application of natural zeolites in the purification and separation of gases *Microporous Mesoporous Mater* **61** 25-42

[13] Hartanto, Y., Indarto, A., Yaswari, Y., Zunita, M. and Soerawidjaja, T.H. 2017 Decolorization of crude terpineol by adsorption *Sep. Sci. Technol* **52** 1967-1972

[14] Handojo, L., Mulyadi, Y., Suryajaya, J., Hartanto, Y. and Indarto A. 2018 α-pinene conversion to terpineol and other derivatives: molecular modeling and infrared mechanistic study *Curr. Phys. Chem* **8** 211-221

[15] Sutresna, N. (2008). Kimia. Grasindo Media Pratama, Bandung

Hydrogen Gas Production using Electro-activated Carbon Catalyst from Coconut Shells

C E Rustana[1*], D S Khaerudini[2*], D Junia[2], F D Dirgantara[2], A Badruzzaman[2], V Marsya[2]

[1]Renewable Energy and Environment Research Group, Department of Physics, Faculty of Mathematics and Sciences, State University of Jakarta, Rawamangun Muka Street, East Jakarta 13220 Special Capital Region of Jakarta Indonesia
[2]Fuel Cell and Hydrogen Technology Group, Research Center for Physics, Indonesian, Institute of Sciences (LIPI), Gd.440 Kawasan Puspiptek Serpong Tangerang Selatan 15314 Banten Indonesia

E-mail: crustana@unj.ac.id, deni.shidqi.khaerudini@lipi.go.id

Abstract. The synthesis of heterogeneous carbon catalyst using electro-activation method has been carried out to assess the effect of temperature on surface morphology of hydrogen production and the surface area of the carbon catalyst as well as characterization of hydrogen production that was resulted. The electrolyte solution of NaOH with a concentration of 2.7 M which was produced from 2 grams of carbon and 25 ml of distilled water was used for electro-activation process. The activation process was taken place with a current of 3 A for 120 minutes and the variation of carbonization temperature of 1000, 1100 and 1200 °C. The electrode system that was used in the process comprised a carbon cathode and copper anode. The results of carbon mass yield testing indicated that purification of charcoal applied at high temperature. The optimum amount of gas with an average rate of 9.19 ml/min/gr Al for a time of 1 hour was produced from carbonized catalyst at carbonization temperature of 1200 °C. The SEM-EDS examination indicated that the reacted product consisted of a mixture of $Al(OH)_3$ and carbon catalysts which would be reprocessed through the re-electro-activation. The characterization of Brunauer-Emmet-Teller (BET) of the surface area of catalyzed carbon shown a value of 394.456 m^2/g; while the result test of Scanning Electron Microscope (SEM) shown that the morphology of surface layer tends to be more pores after carbon catalyzed was used with the concentration of NaOH electrolyte solution of 2.7 M. Results of Fourier Transform Infra-Red (FTIR) analysis also shown the formation of new carboxyl functional groups (O-H) in the wavelength range 464.41 cm^{-1}.

Keywords: Hydrogen, Electro-activated carbon, Catalyst, Coconut Shells

1. Introduction

Indonesia is one of countries which have a big number of coconut plantations in the world. The area is about 3.712 million hectares (for about 31.4% of the world's coconut plantation area) with coconut production of approximately 12.915 billion grains (24.4% of world's production). Coconut shell weight reaches 12% of the weight of coconuts as whole. With an average weight of a coconut for about 1.5 kg, the potential of Indonesia's coconut shell reaches 2324.7 million tons / year [1].

Coconut shell is a suitable material for charcoal or carbon, because it has high hardness, content of silica (SiO_2), a high carbon as well as low mineral ash. Coconut shell charcoal is the main product of the coconut shell pyrolysis process which also contains volatile matters, water and ash [2]. Meanwhile, activated carbon was used as an absorbent material and purifiers, but small amounts of activated carbon can also be used as a catalyst [3].

A catalyst is a chemical compound that causes the reaction to be faster for reaching its equilibrium without experiencing chemical changes at the end of the reaction. Effective catalysts must have at least sufficient selectivity, activity and durability. It can also produce the desired product in not too extreme operating conditions. To produce an effective catalyst, the role of the buffer as a place for the spread of the active nucleus can increase the effectiveness of the entire catalyst. Thus, the catalyst support must have a structure that can disperse the active region, so that it can multiply the number of active areas on the surface of the catalyst. As the surface area of the active area increases, activity will increase, thus overall durability will also increase compared to the active region without support [4].

Carbon can electrochemically be activated and used in chemical reactions with water and fuel, such as aluminum, to produce hydrogen by electro-activation of carbon, where the by-products are the non-harmful aluminum oxide or aluminum hydroxide. Control the reaction temperature, and the amount of aluminum to carbon ratio and electro-activation process can provide amount of hydrogen according to the desired hydrogen rate [5,6]. Therefore, in this study a new approach of electro-activated carbon catalyst based on hydrolysis (water and aluminum reaction) with the aim of accelerating the production process of hydrogen with relatively abundant and inexpensive raw materials is proposed.

2. Experimental

In the present study, carbon was produced from coconut shells and carried out using electro-activation methods. In the electro-activation process, Natrium hydroxide (NaOH) with 2.7 M of concentration, 2 grams of carbon and 25 ml of distilled water was used as electrolyte. The electro-activation process was carried out for 2 hours with 3 amperes of DC current. The electro-activation system consists of a copper plate as a cathode and active graphite carbon from a waste battery as an anode. Carbon catalyst with a minimum temperature at 1000 °C and a maximum at 1200 °C were applied in the process of pyrolysis-carbonization. In the production produced hydrogen and by-product materials. The produced hydrogen was measured through an addition of the circumference of balloon for every 105 seconds or 1.75 minutes. The carbon catalyst surface was characterized and analyzed by using BET (Quantachrome Nova 4200e) and FTIR (Thermoscientific Nicolet iS-10). The sample of product was analyzed by scanning electron microscope SEM (Hitachi SU-3500) and with energy dispersive x-ray spectroscopy (EDX Horiba).

3. Results and Discussion

3.1 Carbonized catalyst production

The temperature of 1000 °C was chosen as the minimum temperature in the variation of the carbonization temperature. The carbonization process increased the carbon content of coconut shell, and opens the width of the pore surface of the carbon. It triggered more OH^- ions entering and binding to the carbon surface [7]. The process of carbonization powder was carried out for 1 hour under inert atmosphere with nitrogen gas was flowed during the process. Nitrogen gas was utilized to prevent the air entering the pyrolysis system as well as a carrier of the molecules that were evaporated and decomposed during heating process not accumulated on the furnace.

Table 1. Carbon yield of coconut shell after carbonization at 1000, 1100 and 1200 °C

Holding Temperature (°C)	Coconut Shell Mass (gram)	Carbon Mass (gram)	Reduction Mass (gram)	% Reduction Mass	Carbon Yield (wt.%)
1000	19.53	4.57	14.96	76.60010241	23.3999
1000	18.44	4.39	14.05	76.19305857	23.8069
1100	22.83	5.421	17.409	76.25492773	23.7451

1100	11.95	2.74	9.21	77.07112971	22.9289
1200	11.56	2.83	8.73	75.51903114	24.481
1200	11.71	2.69	9.02	77.02818104	22.9718

Table 1 shows that the carbon powder which was produced after the carbonization process had experienced mass reduction. It was caused by the process of material decomposition occurring during the carbonization process. The result indicated that the highest carbon yield was obtained on carbonization process with a temperature of 1200 °C which was as much as 24.48 wt%. This shows the level of purity of carbon at high temperatures to remove other minerals from the previous stage.

3.2 Surface Coconut Shell Carbon Area Before and After Catalysis (Electro-Activation)

The surface area of the carbon catalyst support is an important aspect in the function of carbon as a buffer for OH^- reacting ions. Carbon catalyst with a high surface area is a potential adsorbent for its use in the OH^- ion adsorption process. The surface area of the carbon catalyst in this study was measured by the BET method (Brunauer-Emmett-Teller). The graph of adsorption and desorption of N_2 gas isotherm is shown in Figure 1.

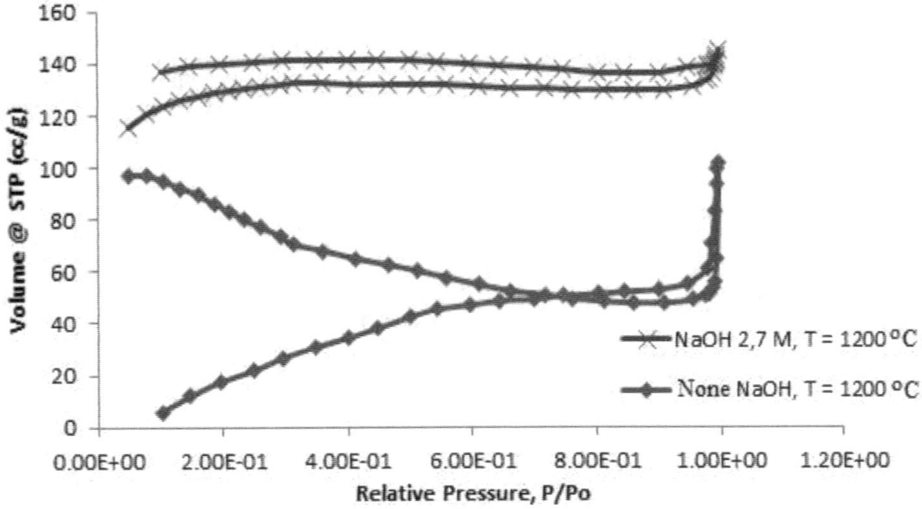

Figure 1. Adsorption and desorption of isotherm N_2 gas on carbon before catalysis and carbonized at 1200 °C (without NaOH) and after catalysis (with NaOH 2.7 M).

Figure 1 shows the volume of N_2 gas adsorption in the highest isotherm condition on the electro-activated carbon. It was caused by a large surface area so that more N_2 gas is absorbed. According to the definition by IUPAC, the adsorbent pores are classified into three groups: micropore (diameter < 2 nm), mesopore (2–50 nm), and macropore (>50 nm). In general, micropores usually account for over 95% of the total surface area for common activated carbons [8]. Thus, Electro-activated carbons using NaOH are considered to be microporous solids with a relatively small external area. The surface area and pore diameter for the two types of carbons above is shown in the following Table 2.

Table 2. Surface area and average pore diameter of carbon catalyst

Carbon Condition	Surface Area (m^2/g)	Average pore diameter (Å)
Before Catalysis	206.669	15.2
After Catalysis	394.456	21. 9

Table 2 shows that the surface area with electro-activation of NaOH was higher compared to before electro-activation of NaOH where the surface area after catalysis was 394.456 m^2/g. With the addition of NaOH, it will oxidize and erode the surface walls of the carbon to form more pores and increase the carbon surface area [9]. After the electro-activation process was complete, activated carbon was then filtered by using filter paper. It was found that carbon results from electro-activation were lesser than before electro-activation. This is because of a reduction in carbon yield. The decrease in the yield for activated carbon was most probably due to the dehydrating reagent (NaOH), which provided elimination and dehydration reactions; thus breaking the C–O–C and C–C bonds of the raw material [10]. Tseng and co-workers [11] reported that the activation mechanism with NaOH could be described according to the following reaction:

$$6NaOH + 2C \rightleftharpoons 2Na + 2Na_2CO_3 + 3H_2 \quad (1)$$

3.3 Hydrogen gas production

Data collecting of the produced hydrogen gas had been done by measuring the circumference of the balloon by a piece of threads taking at every 105 seconds (1.75 minute) measurement. The results shown on the following graphs.

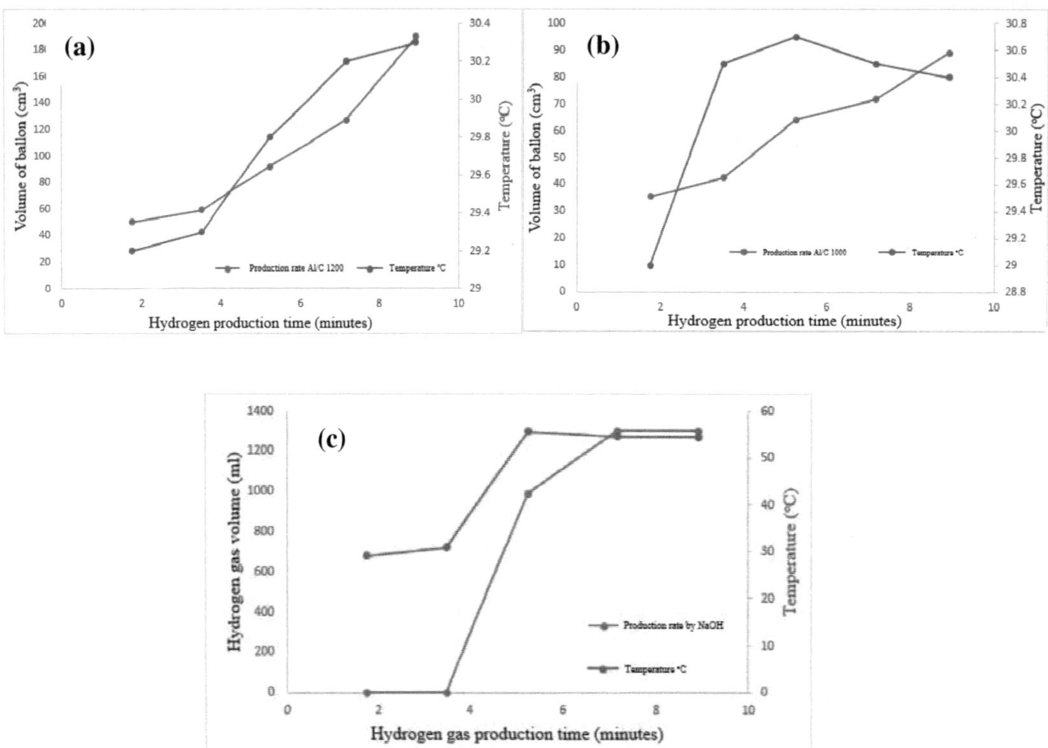

Figure 2. Hydrogen production of Al-water reaction with catalyst of (a) C carbonized at 1000 °C, (b) Carbonized at 1200 °C, and (c) NaOH.

3.3.1 Reaction time or induction time or delay time

In the production of hydrogen gas using a carbonized catalyst, it seems did not require such induction time (the delay time of the reaction to be started). Since the beginning of the reaction until the circumference of the balloon circle, it had perfectly been measured only at 1.75 minutes, as indicated in Figure 2(a) and 2(b). While, the hydrogen production process by using NaOH catalyst, the reaction needed more than 3 minutes to start producing hydrogen gas as shown in Figure 2(c).

The induction time did not occur in the reaction using carbon catalyst since the carbon catalyst directly works by splitting the hydrogen atoms from water molecules. While, in the reaction by using a catalyst of NaOH, it required time to be able to directly contact with aluminum until OH^- ions to finish bonding with aluminate that covered aluminum called the corrosion process of aluminum. Deng et al [7] speculated that the pressure difference between hydrogen bubbles trapped with the environment caused failure of the hydrated oxide layer, so that water can contact aluminum instead of the alumina layer. The method of producing hydrogen gas using a NaOH catalyst only relies on aluminum corrosion methods, so that it takes a long time for water to directly penetrate and contact aluminum.

Carbon catalysts contact directly with water molecules and attract positively charged the hydrogen atoms since the surface of carbon becomes negatively charged after the electro-activation process. It is similar to the mirror charge produced on parallel capacitor plates. The potential of produced electricity is homogeneous for all carbon surfaces, since carbon is a conductor. As part of a water molecule, a hydrogen atom (which some electrons are depleted because of O^{2-} ion that binds to carbon and forms CO and CO_2 as gases, and has a positive single-proton nucleus) will be attracted to the negative carbon surface. The reaction between hydrogen and carbon will help carbon attracting hydrogen atoms from water molecules [12].

As an experiment conducted by Lee [14] using $Al(OH)_3$, precursor of $Al(NO_3)_3$ and without the use of carbonized catalyst concluded that $Al(OH)_3$ effectively reacted with an oxide layer that envelops pure aluminum. The most optimum combination of Al, $Al(OH)_3$, and water to produce hydrogen to produce hydrogen were respectively 3, 15, and 50 with reaction temperature of 95 °C. However, the result of hydrogen production is almost 100% for 6 minutes with a specific volume of hydrogen of 1360 ml H_2 gr^{-1} Al. Whereas, the experiment conducted by Phillips [5] which used a wet carbon catalyst, the production of hydrogen gas was taken place at temperatures of around 65.6 to 87.8 °C. While our research indicated that the reaction in producing hydrogen gas was faster and taken place with lower temperature that was around 30.4 °C. It could be due to differences in the size of the catalyst used. Gai [13] revealed that the induction time of Al particles is mainly related to the H diffusion in bulk of Al and the critical gas pressure of H_2 bubbles at Al: Al_2O_3 interface. Moreover, our research used aluminum flakes and powder carbon. Accordingly, the activation energy increased with increasing the average size of carbon particles which probably originates from the fact that larger-size of Al having a wider particle size distribution. While, the small surface area of the carbon catalyst supported the function of carbon as a buffer for OH^- ions to react with H particle from the water or Al bulk.

3.3.2 *Hydrogen yield*

Hydrogen yield or produced hydrogen was obtained through the difference in the calculated volume of balloon subtracted by the initial calculated volume of balloon and divided by the total volume of measured hydrogen and multiplied by 100%, as described in equation (2).

$$Hydrogen\,Yield = \frac{Calculated\ volume - Initial\ calculated\ volume}{Calculated\ volume\ total} x100\% \qquad (2)$$

The highest yield of hydrogen gas was obtained from the reaction using the result of carbon catalyze of 1200 °C, It had increased to 61.1% when left for 1 hour with volume was 489.1 ml and be predicted to continually increase for more than three days (based on preliminary research). However, it is because of the carbon content formed at 1200 °C in carbonization process was larger compared to the carbon content of 1000 °C. Increasing carbon content will cause more OH^- ion concentrations that can stick to the carbon surface [14].

3.3.3 *The production rate of hydrogen gas*

The rate of hydrogen gas production can be seen clearly in Figures 2 (a), 2(b) and 2(c). Figure 2(a) reveals a graph of the production rate of hydrogen gas that produced in the reaction of hydrogen production using carbon catalysts resulted from carbonization temperatures of 1200 °C. The production rate of hydrogen gas increased with increasing temperature; it was predicted to continually increase in certain time and temperature which indicated by the saturated of curve.

The obtained data indicated that there was a strong influence between the temperature and the rate of hydrogen gas production. As the temperature increases, it was also followed by increasing the production rate of hydrogen gas. Likewise, the production rate of hydrogen gas decreased as the reaction temperature decreased. Figure 2(b) shows the production rate of hydrogen gas using the carbonized carbon at temperature of 1000 °C decreased when it reached 7.17 minutes; while the production temperature decreased to 30.4 °C however the volume was still experienced a little increasing. It also occurred in the reaction using a catalyst NaOH (Figure 2(c)) which indicated by the reaction only ended for about 1.75 minutes and temperatures starting to decline. The calculation model was done by Chen et al. [15] described that the reaction of aluminum and water should increase from room temperature to reach a temperature of 95 °C, while in an experiment conducted by Gai [13] proved that the reaction could stop at 60 °C.

3.4 By-product Analysis

Further evaluation was carried out in this study to understand the microstructure of aluminum hydroxide ($Al(OH)_3$) sample by using SEM/EDX micrograph analysis as shown in Figure 3. From the SEM observation, indeed some void still can be observed as shown in the surface top view. Figure 3 shows the morphology of mixture of $Al(OH)_3$ and carbon after the first hydrogen production between aluminum - water by using a carbon catalyst that was resulted in a carbonization temperature of 1200 °C. Figure 3(a), 3(b) and 3(c) show the appearance of the reaction between aluminum that has reacted with water which then becomes $Al(OH)_3$ flakes (mark in red circle). Figure 3a shows that the content of $Al(OH)_3$ was still consist of carbon (mark in blue circle). The mixture of carbon and $Al(OH)_3$ was confirmed by the EDX test results, as indicated in Figure 5.

Figure 4 shows the morphology of $Al(OH)_3$ and carbon mixture from the test results of the second reaction of hydrogen production between aluminum - water using a carbon catalyst resulted from a carbonization temperature of 1200 °C. When, it is compared with the SEM result of the first reaction (Figure 3), it is only $Al(OH)_3$ that was detected. Figure 4(b) and 4(c) shows the appearance of the reaction between aluminum used in conjunction with water which then becomes flakes of $Al(OH)_3$. At this condition, it can be seen that the content of $Al(OH)_3$ is indicated more compared with the previous result. Its by-product of carbon and $Al(OH)_3$ were also confirmed by EDX analysis, as shown in Figure 6.

Figure 3. Characterization result of Al(OH3) morphology mixture from first reaction, (a) 5.000x magnification, (b) 10.000x magnification, (c) 20.000x magnification.

Figure 4. Characterization result of Al(OH3) morphology mixture from second reaction, (a) 5.000x magnification, (b) 10.000x magnification, (c)10.000x magnification.

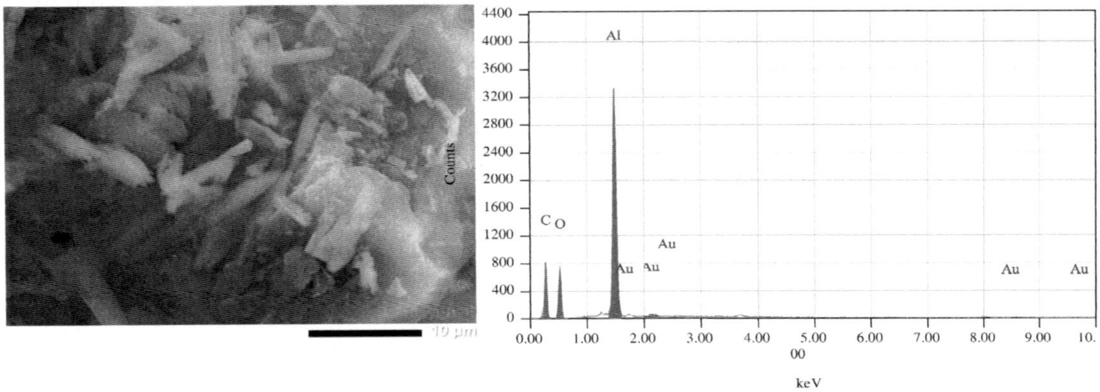

Figure 5. Characterization result of Al(OH3) mixture content from first reaction.

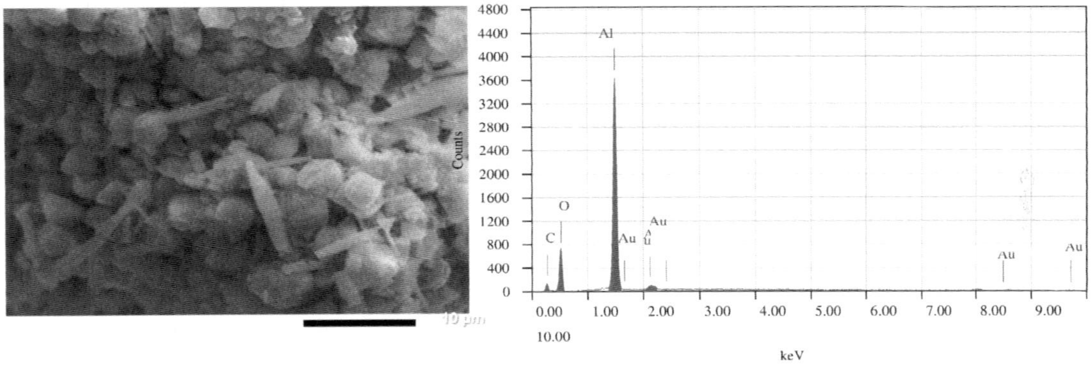

Figure 6. Characterization result of Al(OH3) mixture content from second reaction.

Figure 5 and 6 shows the location of EDX that shows a mixture of $Al(OH)_3$ and carbon in the sample. It also shows a graph based on the mass value of aluminum and oxygen detected from the EDX results. The comparison of the masses will be applied to determine the formula that can be calculated using Proust's law of mass comparison. Based on the results of calculations that can obtained by comparing with the result of EDX analysis. It is known that the phase of $Al(OH)_3$ consist of aluminum and oxygen with mass ratio of 23: 48.

4. Conclusions

Based on the data analysis and discussion, it can be concluded that catalysts based on electro-activated carbon from coconut shell can be utilized to produce hydrogen gases from reaction between aluminum and water. The carbon catalyst that was used to produce hydrogen gas was produced in the reaction of aluminum and water using carbonization temperature of 1200 °C. FTIR analysis showed that carbon catalyst contained OH^- ions which entered the O-H functional group in the wavelength region of 464.41 cm^{-1}, 1035.36 cm^{-1}, and 3431.43 cm^{-1}. The production rate of hydrogen gas using carbon catalysts with carbonization results of 1200 °C was more stable with the average production rate of 9.19 ml/minute/gram. However, the rate of hydrogen gas production by using NaOH was much greater compared to previous one, but the reaction taken place for only 7.17 minutes and unstable.

5. References

[1] Wijaya Y 2012 *Dampak Teknologi Pirolisis Pembakaran Tertutup terhadap Kualitas Lingkungan dan Nilai Tambah Produk Tempurung Kelapa* (Lampung: Universitas Lampung)
[2] Wang Q and Sarkar J 2018 *Int. J. of Energy Prod. & Mgmt.* **3** (1) 34

[3] Cheremisinoff P N and Morresi A C 1978 *Carbon Adsorption Application,* In: Carbon Adsorption Handbook ed P N Cheremisinoff and F Ellerbusch (New York: An Arbor Science) p 1-53
[4] Viswanathan B, Neel P I and Varadarajan T K 2009 *Methods of Activation and Specific Applications of Carbon Materials* (Chennai: Indian Institute of Technology Madras)
[5] Phillips D H 2015 *Methods and Systems for Producing Hydrogen* US 9,102,529 B2 (United States: United States Patent)
[6] Walther H J 1980 *Carbon Adsorption Handbook* ed P N Cheremisinoff and F Ellerbusch (New York: Ann Arbor Science Publishers)
[7] Deng Z Y, Ferreira J M F and Sakka Y 2008 *J. Am. Ceram. Soc.* **91** (12) 3825
[8] Chrysoula M, George G S, Costas P, 2008 *Bioresour. Technol.* **99** 6400–6408
[9] Shofa 2012 *Pembuatan Karbon Aktif Berbahan Baku Ampas Tebu dengan Aktivasi Kalium Hidroksida* (Depok: Universitas Indonesia)
[10] Basta A H, Fierro V, El-Saied H, and Celzard A 2009 *Bioresource Technology* **100** 3941–3947
[11] Wu F C, Tseng R L, Juang R S 2005 *Sep. Purif. Technol.* **47** 10–19
[12] Onda A, Ochi T and Yanagisawa K 2009 *Topic in Catalyst* **52** (6-7) 801
[13] Gai W Z, Liu W H, Deng Z Y, and Zhou J G 2012 *Int. J. Hydrogen Energy* **37** 13132
[14] Alfatah T 2015 *Pembuatan Karbon Aktif dari Tempurung Kelapa dengan Aktivator NaOH serta Aplikasi Adsorpsi Merkuri menggunakan Kolom Unggun Tetap.* (Aceh: Universitas Syiah Kuala)
[15] Chen Y K, Teng HT, Lee TY, et al, 2014 *Int J Energy Environ Eng* **5** 87

Authorship Contribution

D S Khaerudini conceived and directed the project. D Junia and A Badruzzaman prepared and pre-treated the raw material, calcined and sintered sample. F D Dirgantara carried out XRD and SEM/EDS experiments and data analysis. D S Khaerudini and C E Rustana supervised and controlled the preparation of coconut shell into activated carbon/carbon catalyst material as a proper carbon source. F D Dirgantara carried out hydrogen gas production preparation and mass of balance calculations. D S Khaerudini and C E Rustana supervised, provided consultation and problem solving as well as help in analyzing data for the hydrogen gas production phenomena, XRD and SEM/EDS characterizations. C E Rustana, D Junia, and D S Khaerudini drafted and edited the manuscript (with the help of F D Dirgantara, A Badruzzaman, and V Marsya). All the authors discussed the data and contributed in the manuscript.

Acknowledgements

The authors would like to thank to Research Center for Physics – LIPI for technical and facilities support.

Study on The Effect of Para-hydroxybenzoic Acid as A Reduction Agent on Stability, Shape, and Size of Silver Nanoparticles

Fadliah

Petroleum Engineering Department, Faculty of Earth Technology and Energy, Universitas Trisakti
Jalan Kyai Tapa No.1 Grogol, Jakarta Barat 11440, Indonesia

E-mail: fadliah@trisakti.ac.id

Abstract. Research of the study on the effect of para-hydroxybenzoic acid as a reducing agent on the stability, size, and shape of silver nanoparticles has been done. In this study, pH variations were carried out to examine the effect on silver nanoparticles growth. Besides, various concentrations of AgNO3 precursors were also performed to see the effect on the number of silver nanoparticles formed. Nanoparticles tested for their stability, shape, and size using UV/Vis spectrophotometer. The Transmission Electron Microscope (TEM) was used to characterize the shape and size of the formed silver nanoparticles. The results showed that the conditions of silver nanoparticles formation was at pH 9 with uniform morphology and the diameter of 34 ± 1.78 nm. Furthermore, from the less decreasing and small shift of the 6 weeks of age silver nanoparticles UV spectrum, it has been proven that the nanoparticles have good stability.

Keywords: Para-hydroxybenzoic, Silver Nanoparticles, Stability, Size, Shape

1. Introduction

Silver is one of the constituent elements of the earth that was abundant and has a flexible nature. This causes silver widely used in various fields such as photography, cosmetics, jewelry, health, industry, etc. Silver is generally used in bulk phase [1, 2]. In this phase, silver has a small surface area which causes the performance of silver less effective. Therefore to improve the performance of silver, the size of silver must be converted into nanoparticles size [3].

Nanoparticles are particles that have size 1-100 nm8. Decreasing the size of silver particles into nanoscale causes the effectiveness of silver increase due to the increased surface area of silver. The changes of silver into nano form can generally be done in two ways, chemically and physically [4, 5]. One chemical method was to reduce the size of silver by adding organic compounds containing hydroxy groups as reducing agents [6, 7].

In the process of silver nanoparticles synthesis, agglomeration often occurs and colloidal system damage due to precipitation and flocculation. This problem can be solved by using stabilizers and reducers that are suitable so that they can form a system that protects silver particles and prevents aggregation. Stabilizers were used to protect particles from the aggregation process and reduce the occurrence of collisions between silver particles so that they get stable and monodispersed silver nanoparticles [8, 9].

Previous studies used reducing agents and stabilizers with two different compounds so that they were seen less efficient. Therefore in this study used reducing agents in the form of para-hydroxy benzoic acid which has an OH group so that it is expected to reduce Ag^+ to Ag^0 and with the carboxylic group (-COOH) there can be electrostatic interactions of carboxylic groups with Ag^+ ions in solution. This interaction can prevent the aggregation of Ag^+ ions from settling.

2. Problems

The problem that often arises in the research of silver nanoparticle synthesis was the lack of stability of silver nanoparticles that characterized by the formation of aggregates or agglomeration of silver nanoparticles. Therefore, this study used benzoic acid derivative (para-hydroxybenzoic acid) which is expected to be able to maintain the stability of the formed nanoparticles. This research was carried out by a chemical reduction method using $AgNO_3$ as the precursor and para hydroxy benzoate acid as a reduction. Observations were made on variations in pH, variations in concentration and effect of storage time on silver nanoparticles stability and size.

3. Materials and Methods

3.1 Determination of potential reduction of para-hydroxybenzoic Acid

Measurement of potential reduction using an electrochemical analyzer with measurement speed 0,01 Vs-1 and the potential is 0.0 to -0.8 V. The data obtained from the electrochemical analyzer in the form of a voltammogram were used to determine the magnitude of the reduction potential energy of the hydroxybenzoic acid. $E_{1/2}$ is the magnitude of the potential energy of the ortho hydroxy benzoate acid reducing agent, Epa was the magnitude of the potential value measured at the cathode and Epc is the potential value at the anode.

3.2 Synthesis of silver nanoparticles

The synthesis of silver nanoparticles was carried out by reducing $AgNO_3$ with reducing agents at a pH that varied from 1 to 12 by adding a solution of HNO_3 and NaOH. The color of the solution formed was then observed visually and if a yellow color was formed indicates silver nanoparticles have been formed. All solutions obtained were analyzed using a UV / Vis spectrophotometer at a wavelength of 200-800 nm. The maximum absorption of visible light in the wavelength range of 400-500 nm indicates the formation of silver nanoparticles [10].

3.3 Optimization of $AgNO_3$ concentration

$AgNO_3$ with a concentration of 1×10^{-4} to 5×10^{-4} mol L^{-1} compared to the solution without the addition of $AgNO_3$ precursors, then in each of the test tubes was added a reducing agent which has been set at maximum pH. All solutions were heated in the water bath. The heated solution was analyzed using a UV/Vis spectrophotometer at a wavelength of 200-800 nm with fast measurement speed.

3.4 Characterization of silver nanoparticle morphology

Silver nanoparticle colloids formed were characterized using TEM to determine their shape and size and to prove the presence of nano-sized particles.

3.5 Study of the stability of silver nanoparticles

Observations were made on changes in absorbance or shifts in absorption of the maximum wavelength of silver nanoparticles during the interval of 0 hours, 1 hour, 2 hours, 3 hours, 4 hours, 5 hours, 1 day, 2 days, 3 days, and 5 days, 1 week, 2 weeks , 3 weeks, 4 weeks, 5 weeks, and 6 weeks.

4. Results and discussion

4.1 Determination of potential reduction of para-hydroxybenzoic acid

The potential reduction of a reducing agent was important because this was related to the ability of the reducing agent that we will use in reducing the precursor of $AgNO_3$. The potential reduction of para hydroxy benzoate acid was measured using cyclic voltammetry. The measurement results were shown in Figure 1.

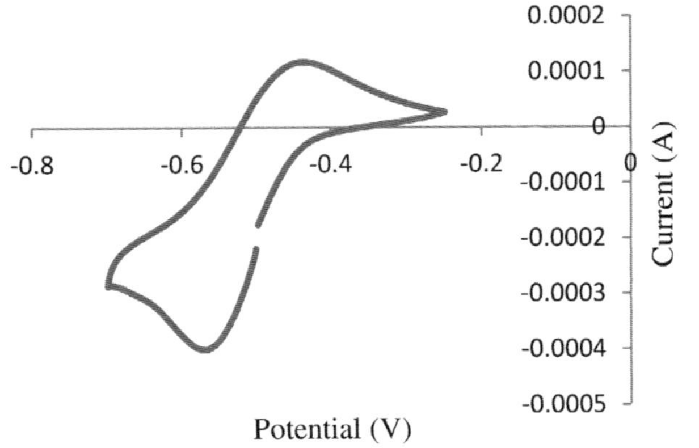

Figure 1. The cyclic voltamogram of para hydroxy benzoate acid.

Figure 1 shows that para hydroxy benzoate acid can reduce silver ions because the reduction potential possessed was lower than silver ions. Besides, the E value of the reaction between para hydroxy benzoate acid and silver ion can be calculated by calculating the potential energy difference. If the difference is positive then the reaction takes place spontaneously, while if the difference is negative then the reaction is not spontaneous[11].

4.2 Effect of pH on the system on the absorbance of silver nanoparticle colloids

PH measurement was done to see the condition of silver nanoparticles formed at acidic or basic pH. The pH variation was carried out on the precursor $AgNO_3$ by using NaOH to adjust the atmosphere of alkaline and HCl to regulate the acidic atmosphere. Silver nanoparticles were not formed at acidic pH but at alkaline pH and maximum in pH 9 marked by changing the color of the clear solution to yellow as seen in figure 3.

Figure 2 shows that the higher the pH of the solution, the more Ag + reduced to Ag0 at pH 8-9. The pH increases, the color of the solution will change from clear to yellow. This shows in absorbance that measurement using UV/Vis spectrophotometer [12-15].

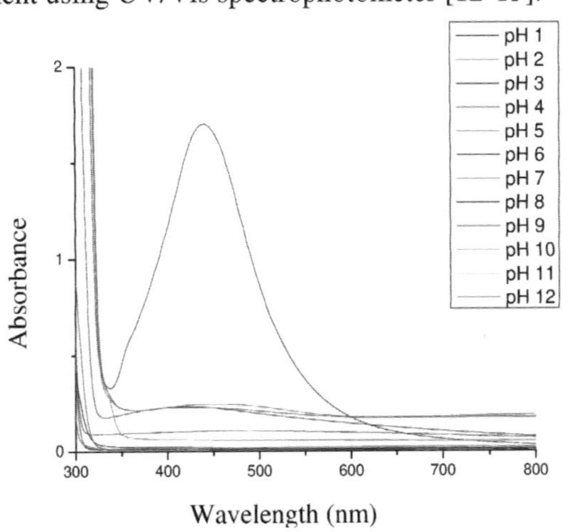

Figure 2. The spectrum of silver nanoparticles synthesized using $AgNO_3$ 3×10^{-4} mol L^{-1} and para-hydroxybenzoic acid pH 1 to pH 12.

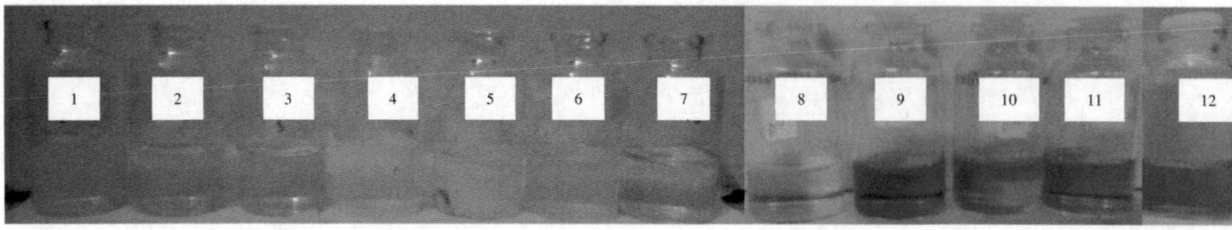

Figure 3. Colors of silver nanoparticles synthesized using $AgNO_3$ 3×10^{-4} mol L^{-1} and para-hydroxybenzoic acid pH 1 to pH 12.

The formation of silver nanoparticles was characterized by the presence of maximum absorption of visible light in a wavelength range of 400-500 nm followed by a change in color to pale yellow [16].

4.3 The effect of the concentration of $AgNO_3$ on the absorbance of silver nanoparticle colloids

The effect of the concentration of $AgNO_3$ can be seen from the absorbance of the measured solution using a UV/Vis spectrophotometer when the concentration of $AgNO_3$ used increases, the absorbance value at the maximum absorption wavelength that is almost close or equal will increase, because more silver nanoparticles are produced [17].

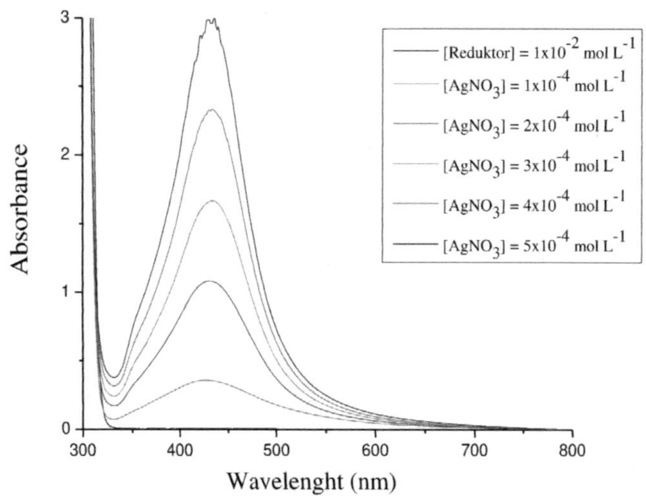

Figure 4. The spectrum of silver nanoparticles synthesized using $AgNO_3$ 1×10^{-4} to 5×10^{-4} mol L^{-1}.

4.4 Characterization of silver nanoparticles morphology

The size of the silver nanoparticles produced was 34 ± 1.78 nm.

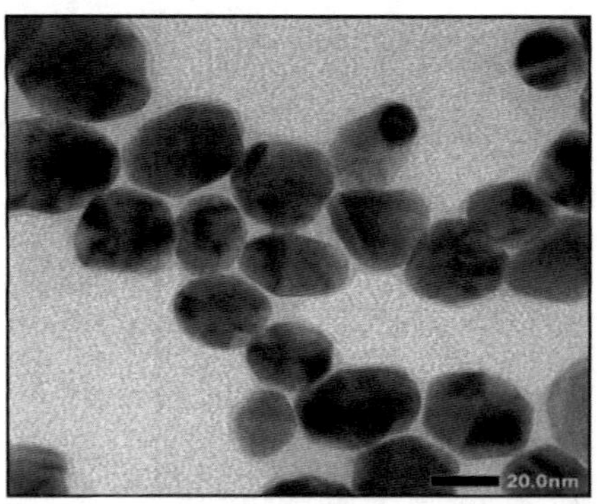

Figure 5. Shape of silver nanoparticles formed.

Characterization results of nanoparticles shape and size with TEM showed that the silver nanoparticles produced had a uniform morphological shape with a uniform size. Besides, from TEM images, no aggregate was formed which indicates the para-hydroxy benzoate acid reducing agent besides having the ability to reduce silver ions also can maintain the stability of silver nanoparticles as indicated by the absence of agglomeration in the silver nanoparticles.

4.5 Study of the stability of silver nanoparticles

The stability of silver nanoparticles was observed by looking at the absorbance of nanoparticles formed at alkaline pH and the shift in their wavelengths. The results of absorbance measurements showed that the silver nanoparticles produced were stable for 6 weeks of storage. This can be seen from the shift in wavelength that was not too significant, which is only 4 nm from the wavelength of 427 nm to 431 nm.

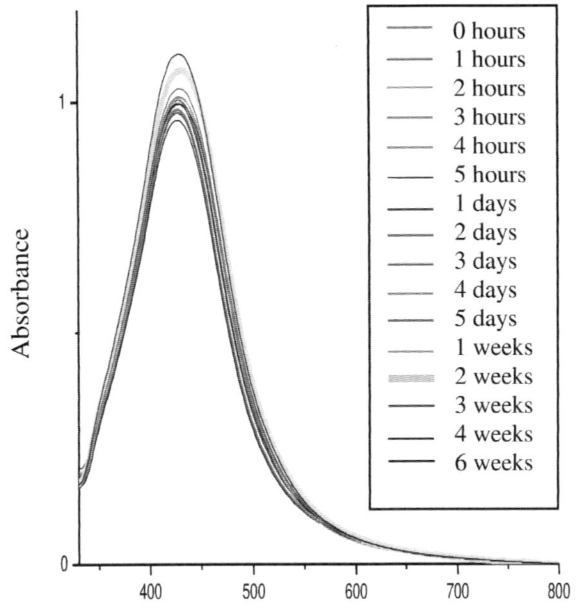

Figure 6. The spectrum of silver nanoparticles stability.

5. Conclusions

Reduction of Ag (I) ions with para-hydroxy benzoic acid which also as a stabilizer can produce silver nanoparticles that formed at alkaline pH. The silver nanoparticles produced proved to be stable for 6 weeks storage, which was characterized by a small wavelength shift as 4 nm. The morphological form of the silver nanoparticles produced has a spherical morphology with a uniform size as 34 ± 1.78 nm.

6. References

[1] Wijnhoven S W P, Peijnenburg W J G M, Herberts C A, Hagens W, Oomen A G and Heugens, E H W 2009 *Nanotoxicol.* **3** 109

[2] Khan Z, Al-Thabaiti S A, Obaid A Y and Al-Youbi A O 2011 *Colloids Surf. B.* **82** 513-517

[3] Kirk R E and Othmer D F 1981 *Encyclopedia of Chemical Engineering Technology* (New York: John Wiley and Sons Inc)

[4] Liu J, Lee J B, Kim D H, and Kim Y 2007 *Collods Surf. A.* **302** 276-279

[5] Chen H and Schluesener H J 2008 *ToxitolLett.* **176** 1-12

[6] Lu Y C and Chou K S 2008 *J. Chin. Ins. Chem. Eng.* **39** 673-678
[7] Dabbagh H A, Teimour A, Chermahini A N and Shahraki M 2008 *Spectrochim. Acta. Part A.* **69** 449-459
[8] Indumathy R, Sreeram K J, Sriranjani M, Aby C P and Nair B U 2010 *Mater. Sci. Appl.* **1** 271-278
[9] Ahmad M B, Lim J J, Shameli K, Ibrahim N A and Tay M Y 2011 *Molecules.* **16** 7237-7248
[10] Oldenburg S J 2011 Silver Nanoparticles: Properties and Applications (USA: Sigma-Aldrich)
[11] Vanýsek P 2010 Electrochemical Series, CRC Handbook of Chemistry and Physics, 90th Edition (Bora Raton: CRC Press/Taylor and Francis)
[12] Patil R S, Kokate M R, Jambhale C L, Pawar S M, Han S H and Kolekar S S 2012 *Adv Nat. Sc. Nanosci Nanotechnol.* **3** 1-7
[13] Ma H Y, Yin B S, Wang S Y, Jiaou Y L, Pan W, Huang S X, Chen S H and Meng F j 2004 C*hem Phys Chem.* **5** 68-75
[14] Dong F, Heinbuch S, Xie Y, Bernstein E R, Rocca J J, Wang Z, Ding X L and He S G 2009 *J. Am. Chem. Soc.* **131** 1057-1066
[15] Qin Y, Ji X, Jing J, Liu H, Wu H and Yang W 2010 *Colloids. Surf. A.* **375** 172
[16] Zielinska A, Ewa S, Adriana Z, Maria G and Jan H 2009 *Procedia Chemistry.* **1** 1560–156
[17] Fadliah 2017 *Jurnal Petro.* **6** 38-42

IOP Conf. Series: Materials Science and Engineering **599** (2019) 012024 doi:10.1088/1757-899X/599/1/012024

Characterization of Nano Chitosan-Cucumber Suri (*Cucumis melo* L.) Seeds with Sodium Tripolyphosphate as Crosslinker

***Rissa Laila Vifta[1], Fania Putri Luhurningtyas[2], Febi Indah Fajarwati[3]**

[1,2] Department of Pharmacy, Faculty of Health Science, Universitas Ngudi Waluyo, Indonesia

[3] Department of Chemistry, Faculty of Mathematics and Natural Science, Universitas Islam Indonesia, Indonesia

*rissa_laila@yahoo.co.id. (Corresponding author 1)
*febi.indah@uii.ac.id. (Corresponding author 2)

Abstract. Cucumber suri (*Cucumis melo* L.) is one of natural materials which has many bioactivities, such as antioxidant, antifungal, antibacterial, anti-cholesterol, antidiabetic, and several other activities. Eventough, Cucumber suri (*Cucumis melo* L.) has a fairly low flavonoid content. In addition, bioactive compounds in cucumber suri seeds tend to be difficult to dissolve in water, unstable to pH, and even have a low bioavailability. Nanoparticles can be used to increase the activity of natural materials by using chitosan and sodium tripolyphosphate as crosslinking agents. This study aims to determine the characteristics of nano chitosan-cucumber suri seeds (*Cucumis melo* L.). First, Cucumber Suri seeds weas previously extracted using maceration method and identified theirs bioactive compounds. And then, Cucumber Suri seed extract was encapsulated with Chitosan: STPP (5: 1). Characterization test was carried out using UV-Vis Spectrophotometry, Infrared Spectrophotometry, and Particle Size Analyzer (PSA). The encapsulation process of Cucumber Suri seed extract produces a clear, homogeneous and stable supernatant. The values of characterization with UV-Vis Spectrophotometry in 0 to 60 minutes was >90%, that indicated nanosized particleswas succefully formed. The IR characterization values indicate a shift in the nano chitosan-cucumber suri seeds wavelength compare with chitosan. The shift of –OH stretching vibration at wavelength 3403 cm^{-1} and stretching vibration of C=O in 1642 cm^{-1} showed there was an interaction between chitosan and Cucumber Suri seed extract. Analysis using Particle Size Analyzer (PSA) showed that chitosan-cucumber suri seeds had a normal polydispersion index (PI) of 0.677 and average particle diameter size was 360 nm.

1. Introduction

Cucumber suri (*Cucumis melo* L.) is an annual plant that produces fruit, speradly growth, and has yellow and bell-shaped flowers [1]. Cucumber suri contains secondary metabolites which generally have a low bioavailability and they are very sensitive to some processing factors. Because it is difficult to dissolve in water, it becomes difficult to absorb. In addition, many compound of the active ingredients has a bitter taste that limits their use in oral form [2]. To resolve the insolubility in water and also low bioavailaibility, another technology needs to be applied [3]. One of many technologies that can be applied is nanotechnology. The synthesis of nanoparticles was designed as a delivery system to control the particle size, the character of the surface and drug releasing system to achieve specific drug action to optimize of therapeutic effects [4]. The application of nanotechnology can be a solution to the main problem of active compounds that have low bioavailability, because nano-sized particles have a very large surface area that allows it to be more effective and easy to pass through the intestinal wall [5].

Chitosan is kind of polymers that often used as encapsulant. The advantages of chitosan as encapsulant has been widely used to improve the drug release and other subtances [6,7]. The use of chitosan as a nanopolymer material needs the addition of Sodium tripolyphosphate (STTP) as a crosslinker to stabilize the nanoparticles polymers [8]. This can be occur because of the positive charge of chitosan will bind to the negative charge of STPP (Sodium tripolyphosphate) [9]. Tripolyphosphate (TPP) its self is a polyanion which can be used as crosslinker in nanoparticles formation [10].

In the previous study, chitosan as an encapsulant had been used to develop the mechanical and physicochemical properties of *Yerba mate* exctract [7]. In another research group, chitosan and tripolyphosphate as crosslinkers was found to be useful to increase the antiulcerogenic activities of *Arrabidaea chica* exctract [11]. In another hand, polymers such as chitosan can be used to deliver the active compounds contained in natural product. El-Aziz et *al.*, 2018 reported that combination of Chitosan and *Mentha longifolia* extract showed the highest effect in antifungal activities between bulk chitosan and chitosan nanoparticles. Chitosan was temporary used in deliver the active compounds because of its abilities to provide protection and stability [12].

In this study will be carried out the synthesis and characterization of the combination between nanochitosan-*Cucumis melo* L. with Sodium Tripolyphosphate (STPP) as the crosslinker. The result is expected to provide a preliminary study of the formation of nanoparticles between chitosan and *Cucumis melo* L. This research can be used as an alternative way in phytopharmaca, especially about the functionalization and optimization of natural product as an alternative for complementary treatment through the nanoparticles formation.

2. Method

2.1. Plant material
Cucumber suri (*Cucumis melo* L.) seeds was obtained from Salatiga, Central Java and previously determined in Laboratory of Ecology and Biosystematics, Universitas Diponegoro.

2.2. Testing material
Testing materials were Ferri-chloride 1%, dragendorff reagent, Mayer reagent, sitroborate reagent, HCl pro analysis (Merck®), Chloride acid 2N, H_2SO_4 pro analysis, n-hexane, ethyl acetate, ethanol 96%, Chitosan, Sodium Tripolyphosphate (STPP) from PT. Brataco, and Aquades from CV. Bratachem.

2.3. Extraction procedure
Making Cucumber Suri (*Cucumis melo* L.) seed extract was made by maceration method using 1500 grams of cucumber suri seed powder that added by ethanol 96% (1:10), and then macerated for 24 hours. The maserate was collected and the pulp was remecerated for 24 hours until the maserate was obtained. The collected maserate was evaporated using a rotary evaporator at 60°C until the thick extracts was obtained.

2.4. Phytochemical screening
Phytochemical screening using the thin layer chromatography (TLC) method was carried out to determine the secondary metabolites contained in the extract. The mobile phase used was n-hexan and ethyl acetate with a ratio of 8: 2. Identification of the compounds was carried out using several spray reagents, namely identification of the flavonoids (using H_2SO_4), identification of alkaloids (using Dragendorff reagent), and identification of tannins (using $FeCl_3$ 1%).

2.5. Synthesis of nano chitosan-cucumber suri
One milligram of Cucumber Suri extract was diluted in 10 ml of ethanol pro-analysis and stirred for 20 minutes. After that, the sediment and the solution were separated. The solution was collected and ready to be encapsulated. 0.069 grams of chitosan was diluted in 11 ml of glacial acetic acid 2.5% (pH 4). The chitosan solution that was obtained then added with the Cucumber Suri seed extract. The next step, the solution was stirred at 400 rpm for 20 minutes. Then, the STPP solution with a concentration of 0.25% as much as 300μl was added and stirred at 400 rpm for 20 minutes. The result of the formulations that was obtained, was centrifuged for 15 minutes.

2.6. Characterization of nano chitosan-cucumber suri
The Characterization of Chitosan-Cucumber Suri Seed nanoparticles was carried out using UV-Vis Spectrophotometer (Shimadzu, UVmini-1240), IR Spectrophotometer (Perkin Elmer, Frontier FT-IR 96772), and Particle Size Analyzer (PSA) (Malvern, Zetasizer Nano-S90). The initial characterization was done by observing the transmittance at the maximum wavelength. The scanning of wavelengths was carried in the range of 190-900 nm according to the research conducted by Yasin *et al*., 2013. Identification of specific groups of nano chitosan-cucumber suri seed was carried out using Infrared Spectrophotometer at the wave number 450-4000 cm^{-1}. The particle size and polydispersion index tests were carried out using Particle Size Analyzer.

3. Result and Disscussion
The synthesis of nano chitosan-cucumber suri seeds was began by determining the material. The determination was aimed to find out the truth of the plants and avoid errors in the material collection [1]. The result of the determination of cucumber suri seeds conducted at the Laboratory of Ecology and Biosystematics, Department of Biology, Faculty of Science and Mathematics, Diponegoro University, Semarang showed that the sample that used in this research was Cucumber Suri seeds (*Cucumis melo* L.) which included in the family of Cucurbitaceae.

The extraction of cucumber suri seed was done using maceration method by ethanol 96% as the solvent. The maceration method was chosen because the process is easy and does not use high temperatures which might damage the secondary metabolites contained in the Cucumber Suri seeds. The extraction was carried out using two processes, maceration and remaseration for five days in total it produced of 135.335 grams extract with a percentage of the rendement obtained 9.022% b/b. The rendement of the extract was obtained based on comparison between constant weight of the thick extract and the initial weight of the powder used.

Table 1. The analysis of rendement cucumber suri seed extract

Powder mass (gr)	Constant mass (gr)	Rendement (% b/b)	Characteristic of the Extract		
			Form	Colour	Odor
1500	135.335	9.022	Thick	Brown	Typical

Phytochemical screening was carried out using Thin Layer Chromatography (TLC) that used n-hexane: ethyl acetate (8: 2) as mobile phases and silica GF_{254} as the stationary phase. The solvent had been choosen related with the properties of the compound and the stationary phase used. The solvents should not be carcinogenic and also does not react with samples during the test. The stationary phase that was used is silica GF_{254} (Merck®) because its can be separated the phenolic compounds, alcaloids, fatty acids, sterols, and terpenoids widely. The result was shown in Table 2 which indicated that the active compounds contained in Cucumber Suri seeds were flavonoids, alkaloids, and tannins.

Tabel 2. Thin layer chromatograpy of cucumber suri seed extract

Secondary metabolits compounds	Reagent for Identfication	Visualization	UV_{254} nm	UV_{366} nm	Result
Alcaloids	Dragendorff	Yellowish-White	Yellowish-Brown	Yellow	+
Flavonoids	Sitroborate	Yellow	Brown	Green	+
Tannins	Ferri-chloride	Yellowish-White	Blackish-Green	Blackish-Green	+

Based on the TLC result of the Cucumber Suri seed extract reacted with dragendorff reagent, yellowish brown spot was seen on the UV_{254} nm and yellow spot on the UV_{366} nm were showed the alcaloid content. The test using sitroborate reagents was showed brown spots on the UV_{254} nm and green spots on the UV_{366} nm which indicated the flavonoid content. The test using ferri-chloride reagent showed in blackish-green spots on the UV_{254} nm and blackish-green on the UV_{366} nm which indicated the tannin content.

The synthesis of nanochitosan-cucumber suri was carried using ionic gelation method by adding chitosan and STPP to form the nanoparticles. After that, the characterization of particle had been seen using the spectrophotometry to know the nanoparticles formed. The principles of using UV-Vis Spectophotometry method was to measure the transmittance values from nanoparticles. Based on the result shown in Table 3, the transmittance was >90%, which indicated that the nanoparticles has been formed quite well.

Table 3. The Formulation of nanochitosan-cucumber suri seeds result

Chitosan: STPP ratio	Result	
	Visual	Transmittance (%)
5:1	Clear-Transparant	99.5

Chitosan was chosen as the encapsulant material with several advantages, such as chitosan is a polymer that has bioactive properties, biocompatible and biodegradable, but chitosan has several disadvantages which are, it can absorb water quickly, so it is easy to be swollen which will decrease the ability in delivering and releasing system of the drug. Therefore, it is necessary to add a crosslinking material, such as Sodium Tripolyphosphate (STPP) which can reduce the swelling degrees and improve the biocompatibility. The use of STTP as a low-dose crosslinker was intended to lower the bond between the polionic TPP and the amine group on chitosan [9].

The characterization of nano chitosan-cucumber suri using UV-Vis spectrophotometer was carried out by observing the transmittance for 60 minutes with 5 minutes time interval. The purpose of this is to determine the stability of nanoparticles formation. The result showed that Nanochitosan-cucumber suri had a good stability which was identified by looking transmittance values that not exceeding the initial interval (>90%). The transmittance >90% means that the smaller particle size and larger surface area are make it easier to read the absorbance. The small size of particle resulted in the faster brown motion, prevent the sedimentation process and cause the supernatant becomes clearer [13].

Table 4.The analysis of transmittance measurement in nanochitosan-cucumber suri seeds

Time (minutes)	Visual	%T (average)	K*%T
0-60	Clear	99.50±0.066	99.640

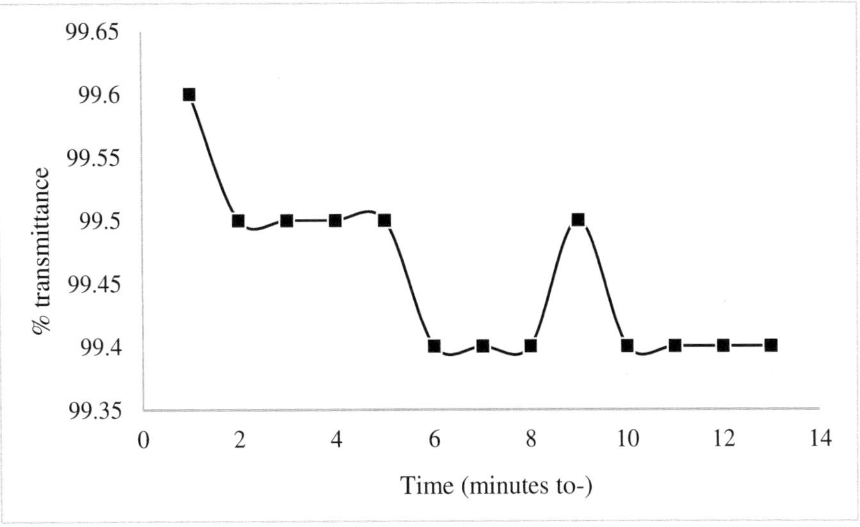

Figure 1. The correlation between time and transmittance percentage of Nanochitosan-Cucumber Suri Seeds.

The nanoparticles formation was confirmed by looking at result from the values of the PSA with a value of Polydispersity Index (PI) of 0.677 and an average diameter of 360 nm which shown in the Table 5. Polydispersity Index (PI) is used to estimate the particle size distribution range and determine whether there is an aggregation formation in nanoparticles samples or not. The particle size distribution is expressed uniformly if the PI value is in the range 0.01-0.7 [14]. Polidispersity index (PI) measured was around 0.677 which indicated a good level of particle distribution uniformity. The large size of the particle can be caused by the small number of active substances that bonded with the chitosan, so that the remaining active substances formed larger size particle aggregations [15].

Table 5. Particle size analysis of nanochitosan-cucumber seeds result

Chitosan:STPP ratio	Characterization Result (PSA)	
	PI	Averge diameter (nm)
5:1	0.677	360.00

The characterization of nano chitosan-cucumber suri seeds using infrared spectrophotometry which shown in Figure 1 was carried out to identify the effect of encapsulation between chitosan and cucumber suri seed extract which was identified by the shifting of the absorption wave number. Chitosan has a typical absorption of its -OH stretch vibration in the 3400an cm^{-1} region, -CH stretching vibration in 2800 cm^{-1}, C=O stretching vibration in 1651 cm^{-1}, –NH bending vibration at 1597 cm^{-1}, and –CO stretching vibration in 1080 cm^{-1} [16]. The presence of –OH stretching vibration shift at 3403 cm^{-1}, C–H stretching vibrations at 2928 cm^{-1}, and C=O stretching vibrations at 1642 cm^{-1} showed that there was interaction between encapsulated chitosan and flavonoids contents in the cucumber suri seed extract.

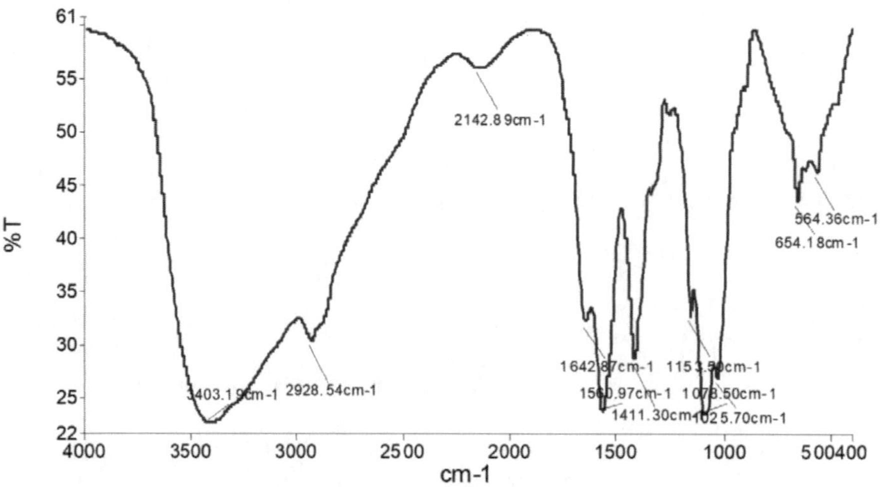

Figure 2. The analysis result using infrared spectroscopy of nanochitosan-cucumber seed.

4. Conclusion

Nanochitosan-Cucumber Suri Seeds had been successfully synthesized using ionic gelation method and Sodium Tripolyphosphate (STPP) as the crosslinker. The characterization result using UV-Vis spectrophotometry showed the presence of nanoparticles formation by looking at the transmittance that was more than 90% and confirmed using Particle size Analyzer with polydispersity index (PI) at 0.677 and the average particle diameter was 360 nm. The analysis using infrared spectrophotometry showed that shift in wavenumber of chitosan which indicated that was interaction between chitosan and cucumber suri seed extract.

5. References

[1] Vifta, R.L., Khotimah, S.K. and Luhurningtyas, F.P. 2018 Uji Aktivitas Antifungi Ekstrak Etanol Biji Timun Suri (*Cucumis melo* L.) terhadap Pertumbuhan Candida albicans secara In Vitro. *Indonesian Journal of Pharmacy and Natural Product*, **1**(1)

[2] Munin, A. And Edwards-lévy, F 2011 Encapsulation of Natural Polyphenolic Compounds; a Review *Pharmaceutics* **3**(4) 793–829

[3] Türk, C. T. Ş., Z. S. Bayindir., and U. Badilli. 2009. Preparation of Polymeric Nanoparticles Using different stabilizing agents. *Journal of Faculty of Pharmacy Ankara*; **38**(1): 257–268

[4] Singh, R. and Lillard, J.W., 2009, Nanoparticle-based Targeted Drug Delivery, *Exp. Mol. Pathol*; **86**(3): 215-223

[5] Studies, R. A. 2010. *Nanotechnology and Food Safety*. Safety Food and Environmental Hygiene Department The Government of the Hong Kong Special Administrative Region. Hongkong

[6] Aranaz, I. Mengíbar, M., Harris, R., Paños, I., Miralles, B., and Acosta, N. 2009. Functional Characterization of Chitin and Chitosan Functional Characterization of Chitin and Chitosan. *Current Chemical Biology* **3**(2) 203–230

[7] Anbinder PS, Deladino L, Navarro AS, Amalvy JI, and Martino MN 2011Yerba mate extract encapsulation with alginate and chitosan systems: interactions between active compound encapsulation polymers *Journal of Encapsulation and Adsorption Sciences* **1**(04) 80-7

[8] Mora-huertas, C. E. Fessi, H., and Elaissari, A 2010 Polymer-Based Nanocapsules For Drug Delivery. *International Journal of Pharmaceutics* **385**(1) 113–142

[9] Shah, B. R., Yan L., Weiping J., W., Yaping A., Lei H., Zhenshun L., Wei X., and Bin L 2016 Preparation and Optimization of Pickering Emulsion Stabilized by Chitosan Tripolyphosphate Nanoparticles for curcumin encapsulation. *Food Hydrocolloids* **52** 369–377

[10] Bhumkar, Devika R., Pokharkar, and Varsha B. 2006 Studies on effect pH on cross-linking of chitosan with sodium tripolyphosphate: a technical note *AAPS PharmSciTech* **7**(2)

[11] Servat-Medina L, González-Gómez A, Reyes-Ortega F, Sousa IM, Queiroz ND, Zago PM, Jorge MP, Monteiro KM, de Carvalho JE, San Román J, and Foglio MA 2015 Chitosan–tripolyphosphate nanoparticles as Arrabidaea chica standardized extract carrier: synthesis, characterization, biocompatibility, and antiulcerogenic activity *International journal of nanomedicine* **10** (3897)

[12] El-Aziz AR, Al-Othman MR, Mahmoud MA, Shehata SM, Abdelazim NS 2018 Chitosan nanoparticles as a carrier for Mentha longifolia extract: synthesis, characterization and antifungal activity *Current Science* **114**(10):2116

[13] Sumardjo, D., 2006. *Pengantar Kimia: Buku Panduan Kuliah Mahasiswa Kedokteran Dan Program Strata 1* Fakultas Bioeksakta. Jakarta: EGC

[14] Binarjo, A., Yuwono, T. and Priyanti, R., 2015 Pengembangan Preparasi Nanopartikel Thymoquinone-Kitosan Dengan Metode Kosolven Menggunakan Isopropil Alkohol. *Pharmaciana*, **5**(2)

[15] Alam, S., Khan, Z.I., Mustafa, G., Kumar, M., Islam, F., Bhatnagar, A. and Ahmad, F.J. 2012 Development and Evaluation Of Thymoquinone-Encapsulated Chitosan Nanoparticles For Nose-To-Brain Targeting: a Pharmacoscintigraphic Study. *International Journal of Nanomedicine* **7** 5705

[16] Pavia, D.L., Lampman, G.M., Kriz, G.S., and Vyvyan, J.R. 2001 *Introduction to Spectroscopy*, Washington: Thomson Learning

Acknowledgment
This study was funded by Ngudi Waluyo University using the 2018 internal research fund to facilitate, so this research can be done as well.

IOP Conf. Series: Materials Science and Engineering **599** (2019) 012025 doi:10.1088/1757-899X/599/1/012025

Adsorption Study of Methylene Blue and Eriochrome Black T Dyes on Activated Carbon and Magnetic Carbon Composite

***F I Fajarwati, M Anugrahwati, I Yanti, R A Safitri, Yeni, E Yuanita**

Department of Chemistry, Faculty of Mathematics and Natural Sciences, Universitas Islam Indonesia

*Email: febi.indah@uii.ac.id

Abstract. Dye is one of the dangerous contaminants because it can interfere with the health of living things and the environment. This research aims to study the comparison of adsorption capacity between activated carbon and its magnetic composite on methylene blue and eriochrome black T, as azo dyes, an organic compound contains a functional group of –N=N-, that are hardly to be degraded in the environment. The activated carbon used in this study was obtained commercially while the composite material was synthesized using activated carbon which was carried by iron species from a solution of Fe^{2+} and Fe^{3+} salts. The variables studied in this research were contact periodic time and dye concentration. The adsorption process was evaluated using a kinetics and isotherm model. Furthermore, the characterization of the two adsorbents was conducted using Fourier-transform Infrared (FTIR) spectrometer. The results of FTIR characterization show a functional group that indicates of the proper of magnetic carbon composite preparation as indicated at wave number from the vibration of Fe-O bonds. Adsorption result from activated carbon and magnetic carbon composite on both dyes followed the kinetics model pseudo-second order. Meanwhile, the isotherm adsorption model of the adsorbents against methylene blue follows the Langmuir and Freundlich equations for eriochrome black T.

Keyword: adsorption, activated carbon, magnetic composite carbon, pseudo-second order, Langmuir

1. Introduction

Nowadays, contamination of dyes becomes one of focuses to be concerned in liquid waste treatment activities. Source of dyestuff waste originates from various industrial processes such as in the textile, paints, polymers and many other industries. The textile industry plays a role as the largest sources of the presence of dyestuffs in the environment, in which most of the dyes have toxic, mutagenic and carcinogenic properties [1].

Methylene blue (MB) and Eriochrome black T (EBT) are included in azo dye compounds which represents more than half of the global dye production. These reactive azo dyes contain one or more azo bonds (–N=N–) that act as chromophore in the molecular structure. These dyes have been identified as the most problematic dyes in effluents due to its resistance to be degraded by the exposure of light, water and chemicals as it has a complex chemical structure [1].

Several methods in processing dyestuff waste continue to be developed in order to minimize the adverse effects prior its disposal in the environment. Adsorption is one method that is widely used in wastewater treatment processes because it is affordable, effective and easy to be operated [2]. Some materials that potentially applied as adsorbents are activated carbon, fly ash, resin, metal oxides, clay materials, and composite magnetic particles [3,4]. Activated carbon is the most commonly adsorbent applied for the adsorption of organic pollutants because of its large availability, low cost and effectiveness. After adsorption process, separation between activated carbon and the treated solution is usually done by filtration or centrifugation which considerably time consuming, therefore, previous research fabricates it with the magnetic properties in order to be more efficient in practical [13, 14]. However, the fabrication brings consequences such as in its physical properties and adsorption ability to certain organic pollutants [15].

In order to access the adsorption ability, it is important to evaluate how much adsorbate can be adsorbed by the adsorbent which can be analyzed using Langmuir and Freundlich isotherm. The Langmuir model is based on the assumption that monolayer adsorption occurs on a homogeneous surface with a finite number of adsorption sites and negligible mutual interactions between the adsorbed molecules [10, 11]. The linear form of Langmuir equation is given as:

$$\frac{1}{q_e} = \frac{1}{q_m K_L C_e} + \frac{1}{q_m} \quad (1)$$

Where : C_e (mg/L) and q_e (mg/g) are equilibrium concentration and adsorption capacity, respectively, q_m (mg/g) and K_L (L/mg) are the maximum adsorption capacity and rate of adsorption (Langmuir constants), respectively.

The Freundlich model assumes that molecules are adsorbed on the heterogeneous surface of adsorbent based on different sites with different adsorption energies [12]. This model takes into account the mutual interaction between adsorbed molecules. The application of the Freundlich equation also suggests that adsorption energy exponentially decreases upon the completion of the sorption centers of the adsorbent. The linear form of Freundlich equation is given as:

$$\ln q_e = \ln K_F + \frac{1}{n} \ln C_e \quad (2)$$

Where: $1/n$ and K_F (mg/g).(L/mg) are Freundlich constants related to the favorability of adsorption process and the adsorption capacity of the adsorbent, respectively. The heterogeneity factor, $1/n$ express adsorption intensity of adsorbent; the adsorption bond is stronger with decreasing $1/n$ value.

From the previous research, adsorption study of EBT using activated carbon (AC) has been conducted [17] however, its study using magnetite carbon composite (MCC) has not been done. Therefore, in this research, determination of adsorption capacity of EBT as well as MB (as the model compound) on the MC was studied in comparison with the adsorption on the AC to observe the adsorption abilities of both adsorbents; the characteristic of the adsorbents, effect of the contact time variation and effect of dye concentration on the adsorption processes. Analysis of adsorption kinetics data from the contact time variations was tested by pseudo first-order and pseudo second-order equations. Furthermore, the equilibrium data was also analyzed based on the Langmuir and Freundlich equations.

2. Experimental

2.1. Materials

Materials used in this research for the preparation of adsorbent were activated carbon, $FeCl_3$, $FeSO_4$, NaOH and ethanol 96%. Meanwhile, for the kinetics study the materials used were filter paper methylene blue, eriochrome black T, pH paper. All of these materials were acquired in pro analysis (p.a.) quality from Merck.

2.2. Equipments
Laboratory equipments used in this study were magnetic stirrer, analytical balance, oven, vacuum filter, mortar and pestle. Moreover, analytical instruments which were used i.e. Fourier Transform Infrared (HITACHI), spectrophotometer UV-Vis (U-2010 Spectrophotometer), shaker (SCILOGEX), glassware and plastic apparatuses. All of these equipments are provided by the research laboratory of Department of Chemistry and the Integrated Laboratory of Universitas Islam Indonesia.

2.3. Adsorbents preparation
Activated carbon (AC) used in this research was obtained commercially that could be used directly after being dried at 105 °C in the oven for 1 h. Meanwhile, the preparation of the composite (MCC) was performed based on the procedure of Mohan et al. [5]. The activated carbon that has been obtained was taken as much as 5 g then being soaked in 50 mL of distilled water. The ferric chloride solution was prepared by adding 1.8 g of $FeCl_3$ to 130 mL of distilled water. At the same time, a ferrous sulphate solution was prepared by adding 2 g of $FeSO_4$ into 150 mL of distilled water. The two solutions were then mixed and stirred at 60-70 °C. The suspension formed was then added into the activated carbon suspension at room temperature while stirring slowly for 30 minutes.

After mixing, 10 M of NaOH solution was added drop wise into the mixture suspension until the pH reached 10-11. During the addition of NaOH, the suspension colour became black at pH 10. After stirring for 60 minutes, the suspension was left at room temperature for 24 hours and repeatedly washed with distilled water, followed by ethanol. Then, the formed composite was filtered with a vacuum filter and dried overnight at 50 °C in the oven. Eventually, both of the adsorbents were characterized by FTIR in order to observe the difference emerged.

2.4. Adsorption study

2.4.1 Effect of the contact time on the adsorption process
The kinetic study was carried out using of initial dye concentration of 1000 mg/L and the adsorbent mass of 0.04 g. Each sample of 10 mL dye solution was agitated in a shaker at 260 rpm at room temperature during the adsorption process. Then, the treated samples were collected at pre-determined time interval from 5 to 75 min. After separated from the adsorbents, the final dye concentration was analyzed using UV-Vis spectrophotometer and the adsorption capacity (q_e) was calculated. Here, the kinetic equation of each dye was determined.

2.4.2 Effect of concentration on the adsorption process
For the isotherm study, the experiment was performed using 10 mL dye solution and adsorbent mass of 0.04 g. Concentration variation of methylene blue was in the range of 200 to 1000 mg/L and eriochrome black T was 50 to 250 mg/L. The dye solutions were agitated in a shaker at 260 rpm at the optimum time of adsorption process that was 15 min. After separation from the adsorbents, the final dye concentration was analyzed using UV-Vis spectrophotometer and the adsorption capacity (q_e) was calculated. Finally, the equations of Freundlich and Langmuir were determined.

3. Results and discussion

3.1 Material characterization
In this study, the preparation of the composite was performed using the same activated carbon obtained commercially and then both characteristics were compared. In the process of preparing MCC, mixtures of Fe^{2+} and Fe^{3+} cations were used to form iron oxides on the carbon surface as a result of the adsorption, hydrolysis and precipitation processes. In this preparation, when NaOH was added to the mixture, the colour of the solution changes with increasing pH, from light brown, dark brown and finally to black at pH of 10 which indicates the formation of magnetite (Fe_3O_4), material which is a mixture of iron oxide from Fe^{2+} and Fe^{3+} [6]. As an adsorbent, AC is commonly separated via filtration from the treated water sample,

IOP Conf. Series: Materials Science and Engineering **599** (2019) 012025 doi:10.1088/1757-899X/599/1/012025

meanwhile, MCC can be separated from the sample using magnetic field as it has a magnetic property resulting more efficient separation process after adsorption process.

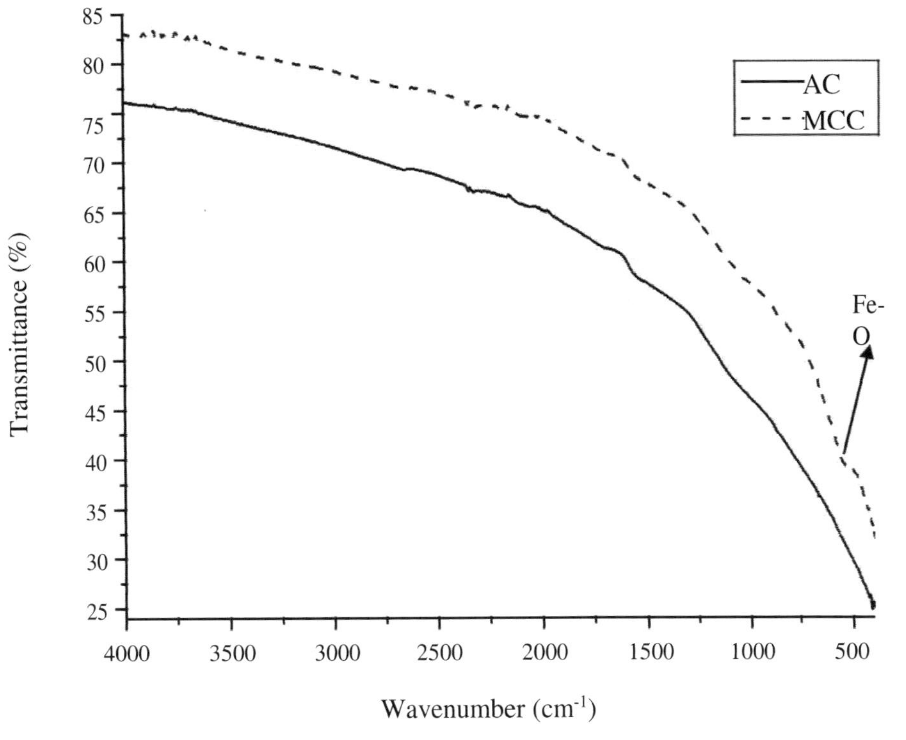

Figure 1. FTIR spectra of AC and MCC.

The material characterization in this study was conducted with FTIR aiming to determine the active groups present in AC and MCC. Through FTIR spectra, it can also be observed the formation of magnetic property from the presence of new absorption peak. Figure 1. shows the FTIR spectra of the two adsorbents used in this study in which each spectrum shows a wave number in the range of 400-4000 cm^{-1}. The two spectra almost depict the same characteristics of the functional group in the activated carbon while the difference that appears in the composite material spectra (MCC) compared with the spectra of AC is the absorption peak at wave number of 533.47 cm^{-1}, that indicates the presence of vibration from Fe-O bond, i.e. characteristic for magnetite [14].

3.2 Adsorption

For the adsorption study, the effect of various contact time was observed to study the kinetics of the adsorption process as well as to find out the optimum contact time which then used in the experiment of the effect of the dyes concentration on the adsorption process. In Figure 2., it is known that contact time affects the adsorption of dyes (both MB and EBT) on the AC and MCC. In the Figure 2.a MB adsorption occurs quickly at the beginning of contact time, i.e. 15 minute with the adsorption capacity of 6.81 mg/g for AC and 12.5 mg/g for MCC, this adsorption capacity decreases with the increasing of the contact time until the equilibrium is reached. The same trend also occurs for EBT adsorption (Figure 2.b), the adsorption capacity is rapid at the beginning, i.e. 15 minutes with adsorption capacity of 255.72 mg/g for AC and 244.27 mg/g for MCC. As time increases, the EBT adsorption capacity decreases significantly and the increase is hardly observed.

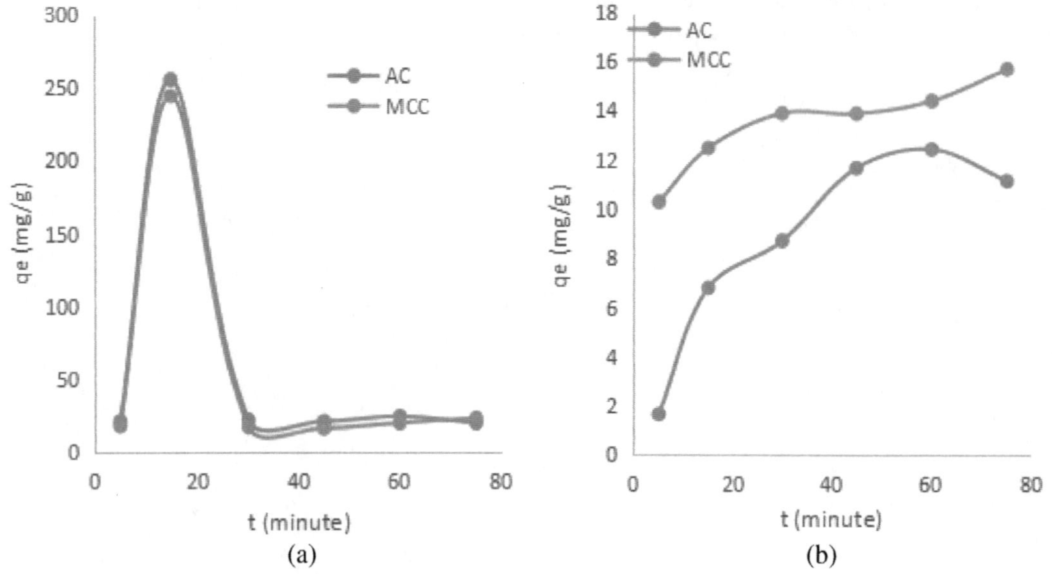

Figure 2. Adsorption study of methylene blue (a) and eriochrome black T (b) using activated carbon (AC) and magnetic carbon composite (MCC) in the variation of time.

From Figure 3. it can be seen that the adsorption capacity of MB and EBT is different in which the capacity increases with the addition of the adsorbate concentration. The adsorption of methylene blue (Figure 3.a) by AC and MCC has a maximum adsorption capacity at a concentration of 600 mg/L with q_e of 158.02 mg/g for AC and 146.18 mg/g for MCC. Whereas in the EBT adsorption (Figure 3.b), the maximum capacity was obtained at a concentration of 100 mg/L with q_e of 3.88 mg/g for AC and 3.98 mg/g for MCC.

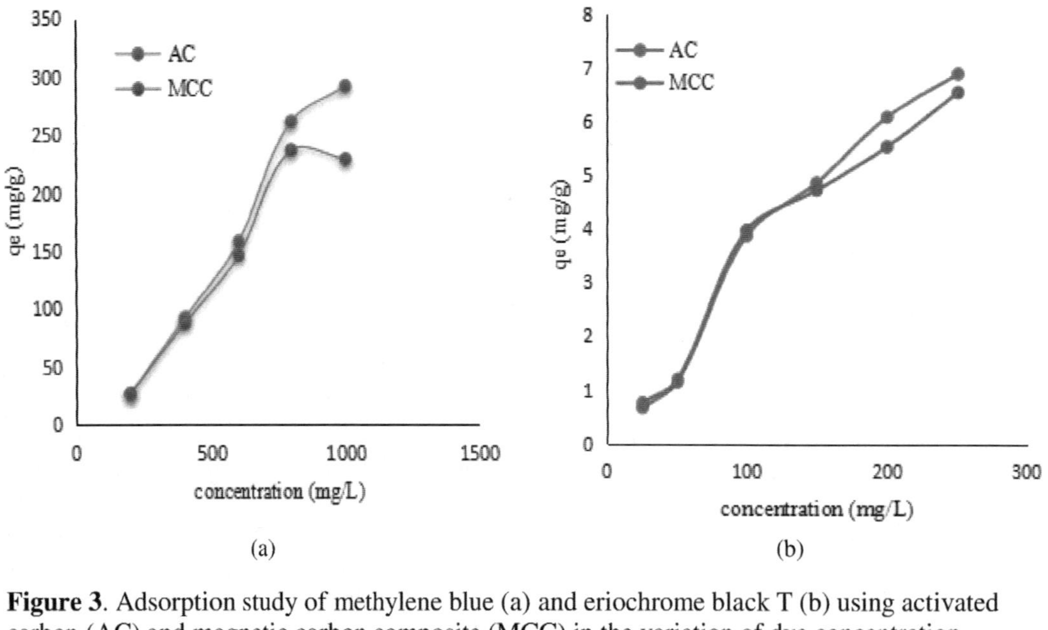

Figure 3. Adsorption study of methylene blue (a) and eriochrome black T (b) using activated carbon (AC) and magnetic carbon composite (MCC) in the variation of dye concentration.

Based on the value of R^2 obtained in Table 1. it was observed that MB adsorption by AC and MCC in this study tended to follow pseudo second order kinetics while EBT adsorption by AC and MCC followed pseudo first-order kinetic equations. If the kinetics model follows a pseudo-first order, the adsorption reaction is likely to occur in the form of physisorption whereas if the kinetics model follows a pseudo second order, the adsorption reaction is likely to occur in the form of chemisorption. According to Hamdaoui [7] the pseudo second-order kinetic equation model shows that adsorption runs in a two-stage process; which is fast at the initial and slower at the second stage. The first stage of the adsorption process begins through physical adsorption via van der waals interaction, i.e. the adsorbate approaches the surface of the adsorbent to form an ionic or covalent bond which then results in a chemical adsorption process which causes the adsorbate to adhere strongly to the adsorbent and the adsorbate is not easily being desorbed. In the adsorption process, most dyes adsorption processes follow the chemisorption process which is influenced by the chemical characteristics of the adsorbent material in the form of biomass (organic) to the dye [8]

Table 1. Kinetics data of methylene blue and eriochrome black T adsorption using activated carbon and magnetic carbon composite

Dyes	Kinetic models	Parameter	Adsorbent	
			Active carbon(AC)	Magnetic carbon composite (MCC)
Methylene blue	Pseudo-first order	R^2	0.1822	0.1797
		k (1/minute)	0.0818	0.0806
	Pseudo-second order	R2	0.9577	0.8972
		qe (cal)(mol/g)	19.8020	20.7039
		k (g/mol.minute)	-0.0086	-0.0499
Eriochrom black T	Pseudo-first order	R^2	0.7526	0.9135
		k (1/minute)	-0.0792	-0.0509
	Pseudo-second order	R^2	0.8531	0.9939
		qe (cal)(mol/g)	18.2482	15.9744
		k (g/mol.minute)	0.0016	0.0147

From the values of some parameters shown in Table 2., the maximum adsorption ability (q_{max}) value is obtained from the adsorbent that shows the limited capacity of the adsorbent surface to adsorb the adsorbate. The maximum adsorption ability of MB by AC and MCC was 333 mg/g and 250 mg/g, respectively. Whereas the maximum adsorption of EBT by AC and MCC was -16.4474 mg/g and 20 mg/g that are smaller than ones in methylene blue.

Table 2. Parameters of Langmuir and Freundlich isotherm models.

Dyes	Model Isotherm	Parameter	Adsorbent	
			Active carbon	Magnetic composite carbon
Methylene Blue	Langmuir	qmax (mg/g)	333.0000	250.0000
		K (L/g)	0.0571	0.0358
		E (kJ/mol)	7.1389	-8.3045
		R^2	0.9819	0.9817
	Freundlich	K (L/g)	-0.0526	-0.0821
		n	0.6030	0.3990
		R^2	0.9238	0.9665
Eriochrom	Langmuir	qmax (mg/g)	-16.4474	20.7900

black T		K (L/g)	-0.0429	0.0500
		E (kJ/mol)	0	-7.4717
		R^2	0.5716	0.4973
	Freundlich	K (L/g)	0.0123	0.0009
		n	1.2297	0.9278
		R^2	0.9898	0.9485

In addition, apart from the q_{max} value, other parameters used to determine the adsorption ability of an adsorbent is from the energy isotherm. According to Gasser [9] chemical adsorption has adsorption energy of 40-480 kJ/mol while physical adsorption requires a lower range of energy. Therefore, according to the data, the adsorption energy of the dyes by the adsorbents is assumed to be categorized as physical adsorption considering its low adsorption energy value.

4. Conclusion

It can be conclude from the study that alteration of the adsorbent characteristic affects the adsorption activity for the same adsorbate as well as the structure of the adsorbate itself. Adsorption of MB and EBT on both adsorbents in accordance with the kinetics equation of pseudo second order, however, the adsorption capacity of EBT on the AC and MCC is lower than that of MB as it is shown from the adsorption energy value.

5. References

[1] Barka N., Abdennouri M and Makhfouk, M.E., 2010, Removal of Methylene Blue and Eriochrome Black T from aqueous solutions by biosorption on Scolymus hispanicus L.: Kinetics, equilibrium and thermo dynamics, *Journal of the Taiwan Institute of Chemical Engineers*, **42**, 320-326

[2] Zahra N.L., Sugiyana D and Notodarmojo S., 2014, Adsorption of Reactive Red 141 Textile Dye Onto Natural Local Clay, *Arena Tekstil*, **29(2),** 68-72.

[3] Zohdi N., Mahdavi F., Abdullah L. C and Choong T.S.Y., 2014, Removal of boron from aqueous solution using magnetic carbon nanotube improved with tartaric acid, *Journal of Enviromental Health Sciences and Engineering*, **12(3)**, 1-12

[4] Raveendra R.S., Prashanth P.A and Malini B.R., Nagabhushana B.M., 2015, Adsorption of Eriochrome black-T azo Dye from Aqueous solution on Low cost Activated Carbon prepared from Tridax procumbens, *Res. J. Chem. Sci.*, **5(3)**, 9-13

[5] Mohan, D., Sarswat, A., Singh, V.K., Alexandre-Franco, M., Pittman Jr., C.U., 2011 Development of magnetic activated carbon from almond shells for trinitrophenol removal from water, *Chemical Engineering Journal* **172**(2-3) 1111-1125

[6] Ahn, T., Kim, J.H., Yang, H., Lee, J.W., dan Kim, J., 2012, Formation Pathways of Magnetite Nanoparticles by Coprecipitation Method, *Journal of Physical Chemistry C*, **116**, 6069-6076

[7] Hamdaoui O. 2006 Batch Study of Liquid-Phase Adsorption of Methylene Blue Using Cedar Sawdust and Crushed Brick, *J. Hazard Mater* **135** 264-273

[8] Eren, Z. and Acar F.N. 2006 Adsorption of Reactive Black 5 from an Aqueous Solution: Equilibrium and Kinetic Studies, *Desalinat ion* **194**, 1-10

[9] Gasser, R.P., 1987, *An Introductionto Chemisorption and Catalysis by Metals*, Oxford Science Publication, New York

[10] Esfandiar, N., B.N., and T.Ebadi 2014 Removal of Mn (II) from groundwater by sugarcane bagasse and activated carbon (a comperative study): Application of response surfase methodology (RSM) , *Journal of Industrial and Engineering Chemistry* **20** 3726-3736

[11] Asfaram, A., Ghaedi, M., Hajati, S, Rezaeinejad, M., Goudarzi, A., and Purkait, M.K. 2015 Rapid Removal of Auramine-O and Methylene blue by ZnS:Cu Nanoparticles

IOP Conf. Series: Materials Science and Engineering **599** (2019) 012025 doi:10.1088/1757-899X/599/1/012025

Loaded on Activated Carbon: A Response Surface Methodology Approach, *Journal of The Taiwan Institute of Chemical Engineers* **53** 80-91

[12] Han, Y., Li, W., Zhang, J., Meng, H., Xu, Y. and Zhang, X., 2015, Adsorption Behavior of Rhodamine B on Nanoporous Polymer, *RSC Adv.* **2015**(5) 104915-104922

[13] Zhu, X., Liu, Y., Zhou, C., Zhang, S., and Chen, J. 2014 Novel and High-Performance Magnetic Carbon Composite Prepared from Waste Hydrochar for Dye Removal, *ACS Sustainable Chem. Eng.* 2 (**4**) 969-977

[14] Cazetta, A.L., Pezoli, O., Bedin, K.C., Silva, T.L., Junior, A.P., Asefa, T., and Almeida, V.C. 2016 Magnetic Activated Carbon Derived from Biomass Waste by Concurrent Synthesis: Efficient Adsorbent for Toxic Dyes, *ACS Sustainable Chem. Eng.*, **4** (3), 1058-1068

[15] Juang, R., Yei, Y., Liao, C., Lin, K., Lu, H., Wang, S., and Sun, A. 2018 Synthesis of Magnetic Fe_3O_4/activated Carbon Nanocomposites with High Surface Area as Recoverable Adsorbents, *Journal of the Taiwan Institute of Chemical Engineers* **90** 51-60

[16] Ahmad AA, Hameed BH.2009 Fixed-bed Adsorption of Reactive Azo Dye Onto Granular Activated Carbon Prepared From Waste, J *Hazard Mater* **175** 298

[17] de Luna, M.D.G., Flores E.D., Genuino, D.A.D., Futalan, C.M., Wan, M. 2013 Adsorption of Eriochrome Black T (EBT) Dye Using Activated Carbon Prepared From Waste Rice Hulls – Optimization, Isotherm and Kinetic Studies, *Journal of the Taiwan Institute of Chemical Engineers* **44** 646-653

IOP Conf. Series: Materials Science and Engineering **599** (2019) 012026 doi:10.1088/1757-899X/599/1/012026

TiO_2 Photocatalytic Degradation of Methylene Blue Using Simple Spray Method

F D Utami[1], D Y Rahman[1], D O Margareta[1], H D Rahmayanti[1], R Munir[1], E Sustini[1] and M Abdullah[1]

[1]Department of Physics, Faculty of Mathematical and Natural Sciences, Bandung Institute of Technology, Jalan Ganeca 10 Bandung 40132, Indonesia

* Email: din@fi.itb.ac.id; fisca.dian.utami1@gmail.com

Abstract. The present work focused on the effects of TiO_2 in degrading organic wastewater. Technical TiO_2 of anatase crystalline phase was used. TiO_2 photocatalyst showed a powerful result in destroying organic effluent. Spray coating was conducted to immobilize the TiO_2 particles onto the plastic buffer followed with heat-treatment process. As a result, 30 mL of 25 mg L^{-1} methylene blue (MB) used in the photocatalyst test was sucessfully degraded after 4 hours of irradiation. Repetitive use of the TiO_2 films still has a great photodecomposition ability of removing 99% of the organic contaminant after 5 times use.

Keywords: Photocatalytic degradation, TiO_2, Spray method.

1. Introduction

Nowadays, the disposal of dyeing process in textile industries contributes to the primary causes of water contamination in the environment. Various kinds of chemicals and dyes are utilized in the textile industry. From the entire dyeing process, it is estimated that about 15 – 20 % dye is released into the environment [1,2]. Dyes are not facilely degraded due to their complex chemical structure and aromatic rings. The presence of textile dyes in water reduces light penetration, aesthetic appearance, may led water to be unsafe for drinking and may caused cancer disease due to its high toxicity. It prompts many scientists to focus on organic dye waste management. The commonly used method in the treatment of organic dye waste is employing a combination of biological oxidation and physico-chemical treatment [3]. This conventional method still leaves some obstacles, such as incomplete waste degradation, the formation of secondary products after waste treatment, and high cost. Because of these obstacles, many scientists consider photocatalyst to be a very potent method to degrade organic contaminants. The term “Photocatalyst” refers to the process that occurs in semiconductor materials when irradiated with light in a certain wavelength range inducing chemical reactions that can destroy the chain of organic contaminant compounds. It produces electron-hole, then reacts with the contaminant compounds [3,4].

Many semiconductor materials are commonly used for photocatalyst applications, such as water treatment, air purification, anti-bacterial, etc. Those semiconductors include TiO_2, ZnO, V_2O_5, Fe_2O_3, CdS, ZnS, ZrO_2 and WO_3 [5,6]. Among these, TiO_2 catalyst is mostly preferred for photocatalyst applications due to interesting properties, such as chemical stability, no secondary product, high reduction-oxidation activity, and low price [6,7]. Besides, TiO_2 also has thermal stability and remains stable after a repetitive catalytic process [6].

There are two types of reactors commonly used in photocatalyst-based waste treatment processes, namely slurry reactors (catalysts directly mixed in pollutant) and immobilization reactors (catalysts immobilized on transparent support surface). Slurry reactors have the advantage of a high contact area with the catalyst, but this photoreactor has several drawbacks, such as the possibility of agglomeration when using high concentration catalysts, the reduction of degradation rate due to scattered light, and high-cost of post-treatment in separating TiO_2 particles after photocatalytic process [8]. Current attention has been focused on immobilizing TiO_2 on transparent substrates to solve such obstacles. When using the immobilized reactor, we can provide reusable catalyst, reduce operating costs due to the absence of post-treatment, and minimize the occurrence of scattered light [8].

The immobilization technique can be carried out on the reactor wall, a buffer material or on the wall of the UV source. Esparza et al. and Stathatos et al. explain the superiority of TiO_2 immobilization, including the expansion of the contact surface between catalysts and pollutants, high adsorption properties and increasing the number of hydroxyl groups on the surface or reducing the risk of recombination of electron-hole pairs [9,10]. Several techniques for immobilizing TiO_2 in various buffers have been reported, including TiO_2 immobilized on a hydrogenation method [11], on Cellulose by dip coating method [12], on NTCs using the sol-gel method [13], on Al_2O_3 ceramic paper by direct spraying method [14], on silica by RF-Magnetron sputtering [15], on Borosilicate glass by spray coating [16], on Zeolite by Photodeposition and Lyophilization [17], on photoanode by heat attachment method [18], on chemically treated metal buffer by sol-gel electrospinning [19], on Polypropylene by thermal milling method [20], on Polyethylene Terephthalate by spray coating method [21-23], and so on.

Mostly, the present methods as mentioned above tend to be complicated and costly. In current work, we propose a simple spray method of immobilizing TiO_2 onto plastic to improve degradation rate. Attaching TiO_2 powder on the surface of plastic with an easier method becomes a high challenge and give distinct advantages of this study.

2. Experimental

2.1. Procedures

Methylene blue (MB) powder from Sakura (Indonesia), a cationic thiazine dye, was directly used as a pollutant model. TiO_2 technical powder of anatase crystalline structure was employed as a catalyst. As mentioned above, we focused on an easy, inexpensive, high degradation rate, and potentially applied method for large-scale application. Here, the solid buffer material used was transparent plastic (local suppliers; Indonesia).

Before the spraying process begins, the transparent substrate was cleaned using 95% ethanol from Sakura (Indonesia). TiO_2 dispersion was achieved by stirring the 5 gr of TiO_2 powder and 100 mL of deionized water at the room temperature. The cleaned transparent plastic was placed on the hot plate then coated by spraying the TiO_2 dispersion solution onto it. The spraying process was conducted using a laboratory spray gun. This spray apparatus has been illustrated in Figure 1.

During the spraying, heat treatment was conducted. The hot plate temperature was set on 27 °C and 100 °C. Each of these samples was labeled with sample 1 and 2, respectively. Sample 1 was dried in room condition (no heat treatment) while sample 2 was put into the electric oven then kept at the temperature of 100 °C for 1 hour. All samples were rinsed in distilled water.

In addition, to determine the effect of spray number to the degradation rate, we varied the spray number during the spraying process, including 10, 20, and 30 times of spraying. The substrates were placed on the hot plate at the temperature of 100 °C. All samples were placed in the electric oven then held at 100 °C for 1 hour, with the same treatment as the sample 2. Samples 1 and 2 were then formed into cylinders and placed on a different photo-reactor. The height of the cylinder is about 5 cm and the diameter is 2 cm.

We performed a photocatalytic test with 30 mL solution with 25 mg L^{-1} MB. It is placed outside the catalyst cylinders in each photoreactor. The photocatalytic tests were investigated under solar exposure. It was performed during 5 hours.

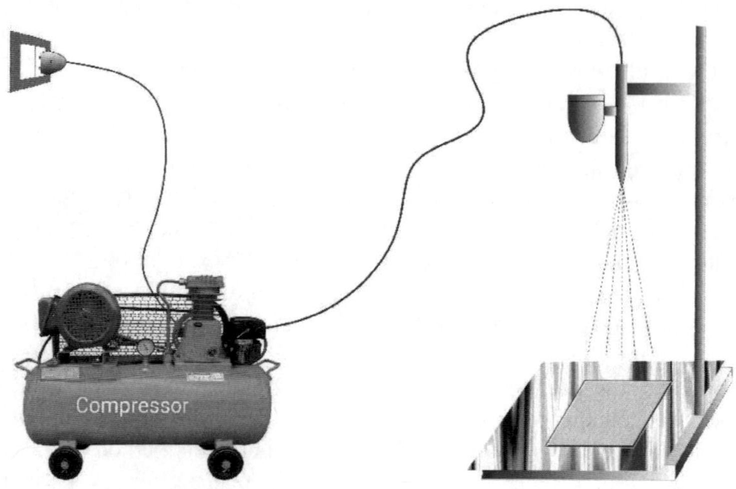

Figure 1. Spray process apparatus set up.

To study the performance of the catalysts, different treatments were applied to each reactor (Table 1). Here, we used 4 labeled reactors: A, B, C and D. Reactor A contained sample 1 which was placed in a dark condition without irradiation. Reactor B contained the pollutant without the use of catalyst. Reactor C contained sample 1 and reactor D contained sample 2. Reactor B, C and D were investigated under solar exposure. The catalyst cylinders were put into the tubular reactor with diameter of 8 cm. The number of catalyst cylinder used were 4 cylinders for each reactor. In addition, to investigate the influence of spray number in catalyst performance toward the degradation rate, three different reactors were employed, each containing the sprayed catalyst of 10 times, 20 times and 30 times. The treatment was the same as in reactor D.

Table 1. The different treatment of each reactor

	Treatment			
Name	Presence of Catalyst	Irradiation	Hot plate temperature (°C)	Annealing temperature (°C)
Reactor A	√	-	100	100
Reactor B	-	√	-	-
Reactor C	√	√	27	27
Reactor D	√	√	100	100

2.2. Analysis

An ocean optic USB2000 spectrometer was employed to measure the UV-Vis absorbance. The shifting and decreasing of MB absorbance peak (UV-Vis) can be used to determine the photoactivity of catalyst in MB degradation. Based on the Beer-Lambert theory related with the light absorption, the MB concentration can be measured since it shows a direct proportion to the MB absorbance. When the monochromatic spectrum surpasses a solution containing a larger number of molecules, the absorbance value also increases; it indicates a higher concentration in substance. The thicker the medium transversed by the beam, the greater the attenuation of the light intensity [24].

3. Result and Discussion

As previously described, the degradation of Methylene blue concentration which is used as the contaminant was determined by UV-Vis absorbance [25]. The maximum peak of UV-Vis absorbance is clearly described in Figure 2 ranging 663-664 nm. We used the wavelength shifting of maximum peak

IOP Conf. Series: Materials Science and Engineering **599** (2019) 012026 doi:10.1088/1757-899X/599/1/012026

to analyze the degradation of contaminants [26,27]. Figure 3 shows a decrease in contaminant concentration after irradiation. The UV-Vis absorbance measurement of MB solution was performed every hour for 5 hours. Line A shows a decrease in MB concentration due to presence TiO_2 but without irradiation (in the dark condition). This decrease is very small compared to other treatments. The degradation process is not a photocatalytic process since there is no source of TiO_2 activating photons. Instead, it is the adsorption of MB particles by TiO_2 surface [28]. This process shows a decrease in Methylene blue concentration due to its adsorption by TiO2 surface without any irradiation effects of the photon source. The corresponding degradation level C/C_0 curve tends to decrease linearly with time. The same result was also reported by Lao et al. [28].

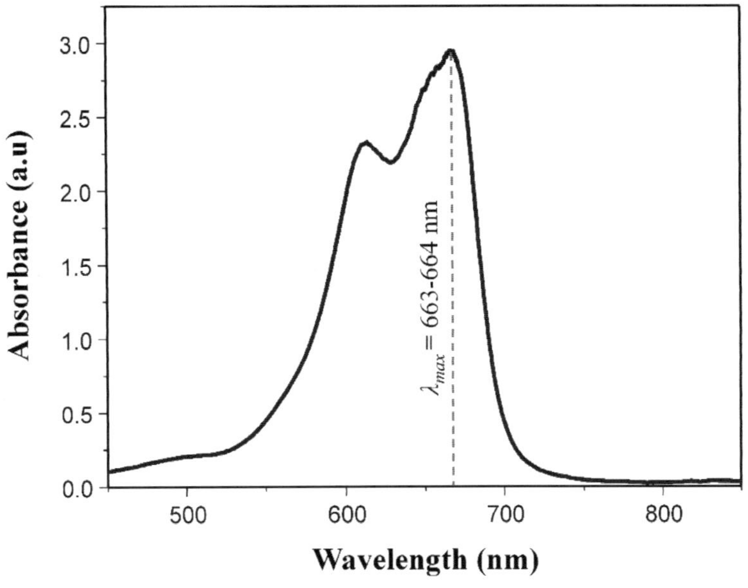

Figure 2. UV-Visible absorbance spectrum of Methylene blue.

Line B represents a decrease in methylene blue concentration due to sun exposure. This process is also not a photocatalytic process since no catalyst (TiO_2) is used. This mechanism is due to the transfer of energy from the solar photons to the catalyst. These photons excite the electrons in the molecules. This process can break the chemical chains of the organic contaminant, in this case methylene blue. This mechanism is called photolysis. It takes longer than the photocatalytic process [29]. Line B also shows a linear decrease in concentration due to the sunlight irradiation without the presence of the catalyst.

Line C and D show reduction in MB concentration due to the photocatalytic process of TiO_2 catalysts. It is activated when illuminated by a source of photons (sunlight) whose energy is higher than its bandgap; it then encourages an electron to travel from VB (valence band) to CB (conduction band) leaving a hole in VB [3,6-10]. It was found that both curves followed the pseudo-first-order decay kinetics toward the time after irradiation (Figure 3).

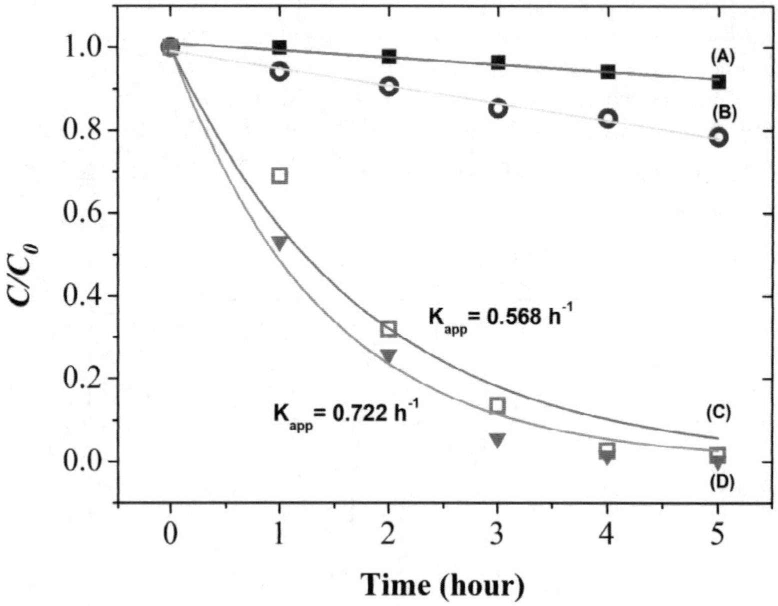

Figure 3. The degradation level of MB after treatment by (A) TiO_2 without irradiation; (B) irradiation without catalyst used; (C) TiO_2-coated plastic using spraying method without heat treatment; and (D) TiO_2-coated plastic using spraying method combined with heat treatment.

The rate of photocatalytic degradation strongly depends on the light radiation and the characteristics of the surface of the photocatalyst. The common model is the kinetic scheme of Langmuir-Hinshelwood or LH model [6]. Here the reaction rate *K* is directly proportional to the surface fraction covered by the catalyst. The adsorption level is determined by the value of the adsorption equilibrium constant, *K*, while *C* indicates the concentration of organic compounds at time *t*. Thus the first-order reaction rate constants are obtained by the equation

$$\frac{C}{C_0} = e^{-K_{app}t} \quad (1)$$

we can determine the reaction rate constants of the photocatalyst by fitting the data using the equation above [6]. It represents a photocatalyst scheme generating a curved characteristic.

There is a difference in reaction rate between lines C and D. The degradation rate of the line C is obtained at 0.568 h^{-1} while the line D is 0.722 h^{-1}. The highest reaction rate is indicated by line D at which TiO_2 was used as catalyst and coated by combined spraying and heat treatment method. It explains why the heating process when the TiO_2 nucleation took place on a plastic substrate influences the TiO_2 coating's strength to the substrate. The strength of the coating increases with the increase of adhesion strength between TiO_2 and the substrate. This is also confirmed by the washing process after the coating process: the released TiO_2 particles are less than the TiO_2 coating process without heating process (conducted at room temperature). It can be stated that the use of the heat treatment-based spraying method produces more coated TiO_2 particles. More TiO_2 particles on the substrate surface allow greater probability that TiO_2 will come into direct contact with organic contaminants, so that the degradation rate of MB is faster.

IOP Conf. Series: Materials Science and Engineering **599** (2019) 012026 doi:10.1088/1757-899X/599/1/012026

Once we see that the catalyst can work effectively, the influence of spray number on the photocatalyst performance was also investigated. The used catalyst was TiO_2-coated plastic using spraying method combined with heat treatment. From figure 4, The degradation rate of blue curve (the sprayed catalyst of 10 times) is obtained at 0.537 h^{-1}, the yellow curve (the sprayed catalyst of 20 times) is 0.723 h^{-1} while the pink curve (the sprayed catalyst of 30 times) is 0.486 h^{-1}. The reactor using the sprayed catalyst of 20 times exhibits the highest photocatalytic activity compared to the other. After 4 hours of sun exposure, 99% of the contaminants were successfully destroyed. The use of catalysts that sprayed 30 times resulted in lower photodegraded performance. When we observe the use of a sprayed catalyst of 10 times compared to the sprayed catalyst of 20 times, the reaction rate of the photocatalyst obviously increases. The increase was caused by the greater probability of catalyst surface contacted directly with the pollutant. More TiO_2 particles were attached on the substrate surface.

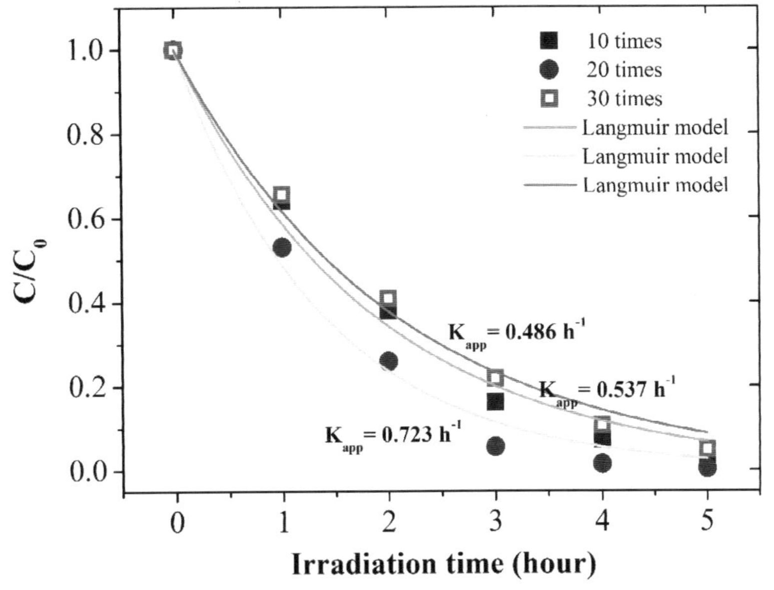

Figure 4. The effect of spray number in catalyst performance toward kinetic rate of photocatalyst activity, blue curve represents the sprayed catalyst of 10 times, yellow curve represents the sprayed catalyst of 20 times, and pink curve represents the sprayed catalyst of 30 times.

The semiconductor was activated by the photons and produced electrons and holes. The electrons will bind to oxygen generating superoxide radicals while the holes will bond to OH in water to form hydroxyl radicals. These radical agents will break the chain of contaminants and promote mineralization in H_2O and CO_2 [3]. The two main parts are playing the role here, i.e. the photon energy and the amount of catalyst present on the surface substrate. Both are the essential parts that play a prominent role in the photocatalytic process and can be applied to the treatment of organic contaminants [30,31]. The higher the light intensity of the photon source, the more surface of TiO_2 generating electron-hole pairs that can be in direct contact with the pollutant so that the organic pollutant degradation process will be faster. These explained why the photo-activation reached the optimum condition with the use of sprayed catalyst of 20 times and returned down when using the sprayed catalyst of 30 times. The coming light can't penetrate effectively when using the sprayed catalyst of more than 20 times, in this case the sprayed catalyst of 30 times, which makes the electron-hole recombined.

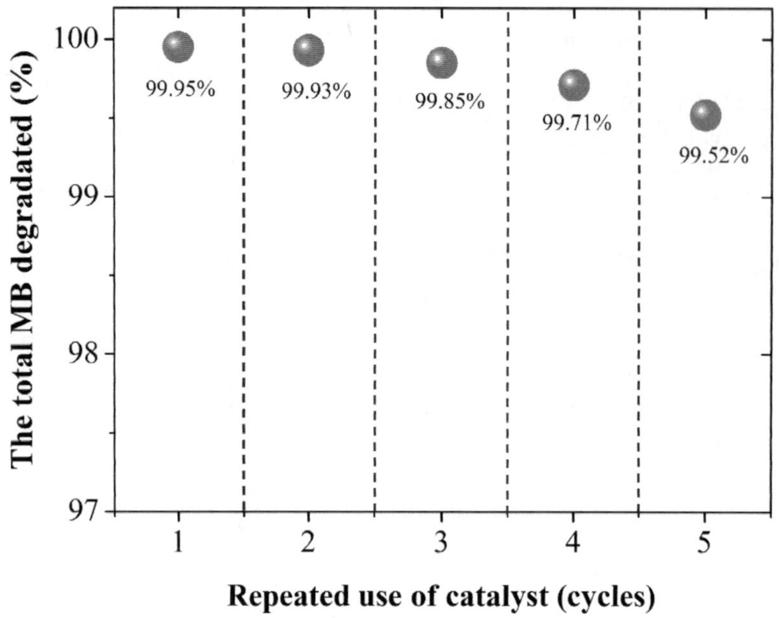

Figure 5. The repetitive usable of catalyst for MB degradation process.

One of the interesting properties of TiO_2 is its stable characteristic after repeated use. In this study, catalyst performance was investigated through repeated use of TiO_2 films. The film was repeatedly used for 5 times sequentially. Each cycle of repetition is carried out for 5 hours of irradiation. After the photocatalytic test, it was found that the film still has a great ability of decomposition by degrading 99% of the MB concentration. Only about 0.5% was reduced from the first use.

4. Conclusion

We have successfully coated TiO_2 on transparent plastic substrate using simple spray method. The excellent performance of the catalyst was achieved with TiO_2 coated plastic made by a combined spraying and heat treatment process. The method produced more coated TiO_2 particles. It allows greater probability that TiO_2 will come into direct contact with organic contaminants, so that the degradation rate of MB is faster. Since this method is adequate easy and affordable price, it is worthy to be promoted as a forthcoming photocatalyst technology.

5. References

[1] Zhu H Y, Xiao L, Jiang R, Zeng G M, and Liu L 2011 *Chemical engineering journal* **172**(2-3) 746-753
[2] Clarke C E, Kielar F, Talbot H M, Jouhnson K L 2010 *Environ. Sci. Technol.* **44** 1116–1122
[3] Fujishima A, Rao T N and Tryk D A 2000 *J. Photochem. Photobiol. C: Photochem. Rev.* **1** 1-21
[4] Ameta R, Solanki M S, Benjamin S, and Ameta S C 2018 Photocatalysis. In Advanced oxidation processes for waste water treatment (pp. 135-175). Academic Press
[5] Tayade R J, Natarajan T S, and Bajaj H C 2009 *Ind. Eng. Chem. Res.* **48** 10262-10267
[6] Chong M N, Jin B, Chow C W K, and Saint C 2010 *Water Res.* **44** 2997-3027
[7] Atout H, Álvarez M G, Chebli D, Bouguettoucha A, Tichit D, Llorca J, and Medina F 2017 *Mater. Res. Bull.* **95** 578-587
[8] Vaiano V, Sacco O, Pisano D, Sannino D, and Ciambelli P 2015 *Chem. Eng. Sci.* **137** 152-160
[9] Esparza P, Borges, M E, Diaz L, Alvarez-Galvan M C, and Fierro J L G 2010 *Appl. Catal. A* **388** 7–14

[10] Stathatos E, Papoulis D, Aggelopoulos C A, Panagiotaras D, and Nikolopoulou A 2012 *J. Hazard. Mater.* **211** 68–76
[11] Wang W, Ni Y, Lu C, and Xu Z 2014 *Appl. Surface Sci.* **290** 125–130
[12] Ortelli S, Blosi M, Albonetti S, Vaccari A, Dondi M, and Costa A L 2013 *J. Photoch. Photobio. A* **276** 58– 64
[13] Dong Y, Tang D, and Li C 2014 *Appl. Surf. Sci.* **296** 1-7
[14] Vereb G, Ambrus Z, Pap Z, Mogyorosi K, Dombi A, and Hernadi K 2014 *React. Kinet. Mech. Catal.* **113** 293-303
[15] Barrocas B, Serio S, Rovisco A, Nunes Y, and Jorge M E M 2016 *Appl. Surf. Sci.* **360** 798-806
[16] Yadini A El, Saufi H, Dunlop P S M, Byrne J A, Azzouzi M El, and Hajjaji S El 2014 *J. Catal.* **2014** 1-8
[17] Jansson I, Yoshiiri K, Hori H, García-García F J, Rojas S, Sanchez B, Ohtani B, Suarez S 2016 *Appl. Catal. A Gen.* **521** 208-219
[18] Lee S L, Ho L N, Ong S A, Wong Y S, Voon C H, Khalik W F, Yusoff N A, and Nordin N 2017 *Chemosphere* **166** 118-125
[19] Ray S, and Lalman J A 2016 *Mater. Today Proc.* **3** 1582-1591
[20] Aliah H, Aji M P, Masturi, Sustini E, Budiman M, and Abdullah M 2012 *Am. J. Environ. Sci.* **8** 280-290
[21] Utami F D, Rahman D Y, Sustini E, and Abdullah M 2019 *J Phys. Conf. Ser.* **1171**(1) 012030
[22] Utami F D, Rahman D Y, Sutisna, Kamirul, Margaretta D O, and Abdullah M 2019 *J Phys. Conf. Ser.* **1204**(2019) 012086
[23] Margaretta D O, Permadi K W, Rahman D Y, Utami F D, Viridi S, and Abdullah M 2019 *J Phys. Conf. Ser.* **1204**(2019) 012051
[24] Aarthi T, Narahari P, and Madras G 2007 *Journal of Hazardous Materials* **149** 725-734
[25] Tseng K H, Chung M Y, and Chang C Y 2017 *Nanomaterials* **7** 133
[26] Slimen H, Houas A and Nogier J P 2011 *J. Photochem. Photobiol. A Chem.* **221** 13-21
[27] Wang F, Min S, Han Y, and Feng L 2010 *Superlattices Microstruct.* **48**(2) 170-180
[28] Thao L, Dang T, Khanitchaidecha W, Channei D, and Nakaruk A 2017 *Materials* **10**(2) 122
[29] Wong C C, and Chu W 2003 *Chemosphere* **50**(8) 981-987
[30] Rajeswari R and Kanmani S 2009 *Iran J. Environ. Health. Sci. Eng.* **6** 61-66
[31] Zangeneh H, Zinatizadeh A A L, Habibi M, Akia M and Isa M H 2015 *J. Ind. Eng. Chem.* **26** 1-36

Acknowledgments
This work was supported by PMDSU fellowship from The Ministry of Research, Technology and Higher Education, Republic of Indonesia No. 328/SP2H/LT/DPRM/II/2016. This work was also supported by KK Research Grant from Bandung Institute of Technology 2019.

Agarose Cleaning Optimization for DGT Devices Specific for Cesium

PH Sadewo[1,2,3a], T Suzuki-Muresan[1b], A Widiharto[2] and B Muresan[4]

[1]SUBATECH (IMT Atlantique, CNRS/IN2P3, Université de Nantes), BP 20722, 44307 Nantes, France
[2] Department of Nuclear Engineering and Engineering Physics, Universitas Gadjah Mada Jl. Grafika No.2, Sendowo, Sinduadi, Mlati, Kabupaten Sleman, Daerah Istimewa Yogyakarta 55281 Yogyakarta, Indonesia
[3] Center for Accelerator Science and Technology, National Nuclear Energy Agency (BATAN), Jl. Babarsari, Yogyakarta 55281, Indonesia
[4] Ifsttar, AME, EASE, F-44341 Bouguenais, France (EU)

[a] prasetyo.haryo@batan.go.id
[b] suzuki@subatech.in2p3.fr

Abstract. Agarose gels commonly used in electrophoresis are regularly checked by its manufacturers for the absence of contaminating nucleases or residual sulfates. It is also tested for several metals and trace elements: Pb, Cd, Co, Cu, Zn, K, Ca, Cr Mg, Mn, Fe, Ni, Na, and Cl. However, no data are available for Cs-impurity measurements. A Diffusive Gradients in Thin films (DGT) device specific for cesium (Cs) is intended to be developed. Therefore, it is important to assess the impurity degree of agarose with Cs. Direct dissolution of agarose in 4 $mol{\cdot}L^{-1}$ HNO_3 gives a concentration of $2.1{\times}10^{-8}$ $mol{\cdot}L^{-1}$ of Cs, which is higher than that is found in natural environments (e.g. $[Cs]<1{\times}10^{-8}$ $mol{\cdot}L^{-1}$ in water). Two methods of Cs impurity removal have been tested, involving acidic cleaning with HCl and extraction by exchanger resin (KCuFC). Agarose powder was mixed with HCl in different concentrations, rinsed and dried. This method shows that the HCl-cleaning method is not quite efficient in removing the Cs impurity compared to untreated agarose gel. The second method, KCuFC resin was mixed with agarose powder in ultra-pure water, heated afterwards, and filtered. This method was able to reduce the amount of Cs by 99% compared to the untreated agarose gel. It appears that acceptable Cs levels (i.e. with respect to ambient Cs concentration in natural waters) are achievable by the removal of preexisting Cs in the agarose powder.

Keywords: Diffusion Gradient in Thin Film, Cesium, Potassium Copper Ferrocyanide, Agarose

1. Introduction

Diffusive gradient in thin film (DGT) is a method used for in-situ preconcentration of trace and ultra-trace labile compounds. It allows estimation of the concentration and the labile fraction of an element. This method does not disturb the dynamics of the particle dissolution, colloid formation or aggregation, and solutes' concentrations in natural water [1]. It has been widely used to probe heavy metals and, more recently, the persistent organic pollutants or radionuclides.

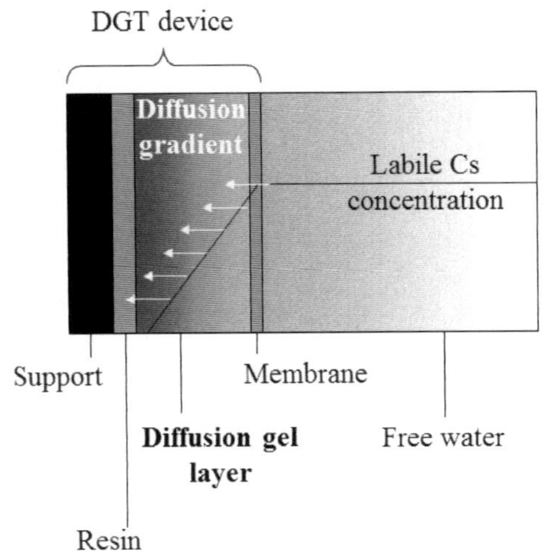

Figure 1. Schematic view of the functioning of a DGT device and concentration evolution during steady state diffusion in water.

Most of the DGT devices include a polyacrylamide as the diffusion gel layer. Polyacrylamide gel may be toxic to the local environment, because monomeric acrylamide is toxic, and polymerization is not complete Polyacrylamide gel for DGT purposes consists of 15% by volume acrylamide monomer, 0.3% by volume agarose based crosslinker, for 10 ml of gel solution 70 µl of ammonium persulphate 10% (APS) initiator and 20 µl TEMED catalyst. In some cases, the crosslinker can be substituted with bis-acrylamide 2% [2].

Cesium (Cs) is a chemical element that displays a high environmental mobility. It travels long distances when suspended in the atmosphere onto particles or dissolved in natural waters. Once deposited on the soil, its mobility vastly depends on local environmental factors including the soil nature and ongoing pedological processes [3]. Clayey soil presents high affinity with Cs, in particular with minerals presenting specific sorption sites such as illite, montmorillonite [3,4].

Cs is a natural chemical element but it can also be anthropogenically produced in the nuclear sector and the military industry. Actually, amounts of radioactive Cs have been released into the atmosphere during nearly all nuclear weapons tests and several nuclear accidents, notably the Chernobyl and Fukushima disaster. In fact, ^{137}Cs was among these isotopes widely distributed into the local and global environment. It thus constituted on of the greatest risk to human health [5]. All these aspects make Cs a useful element for probing the transfer mechanisms and, as a consequence, to monitor the dissemination of radionuclide into pristine and anthropized areas caused by a nuclear accident or the use of nuclear weapons.

In spite of its extreme ecotoxicity, the level of ^{137}Cs may be extremely low. Reportedly, radioactive Cs in the water samples from Fukushima area is less than 0.01 $Bq \cdot L^{-1}$. Comparatively, germanium semiconductor detectors show a limit of detection (LOD) of 0.3 $Bq \cdot L^{-1}$, for 12 hour measurement. Besides, stable Cs is found in free and pore waters at ultra-trace levels (0.004 to 2 $\mu g \cdot l^{-1}$ with an average of 0.05 $\mu g \cdot l^{-1}$ [3]). Direct determination is therefore not always possible [6,7].

Accordingly, a novel Cs-specific DGT device was developed in this work that uses agarose as the diffusion gel layer. Agarose is made from agar by removing the agaropectin which has a lower gelling ability (Figure 2). It is a thermoreversible and ion independent gelling agent. Some types of agarose are sufficiently devoid of charged residues so it can be considered neutral. This makes agarose suitable for electrophoresis as well as for investigating the most charge-sensitive environmental processes [8] (e.g. such as ionic Cs-Illite interactions).

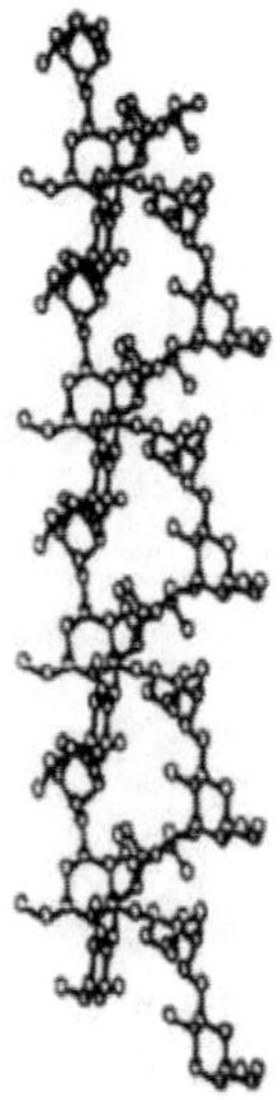

Figure 2. Agarose double helix viewed perpendicular to the helix axis [9].

Ultimately, because one of the most common use of agarose gels is electrophoresis, the impurities that the manufacturers check are the absence of contaminating nucleases or residual sulfates. It is also tested for several metals and trace elements: Pb, Cd, Co, Cu, Zn, K, Ca, Cr Mg, Mn, Fe, Ni, Na, and Cl. However, to our knowledge, there is no available information about the Cs content. Here we ask the question if commercial agarose can be sufficiently cleaned to allow the assessment of natural and radioactive Cs levels in the environment by means of agarose gel DGT devices. To answer this question, it is important to first assess the impurity degree of agarose with Cs and propose a cleaning procedure that gives access to the determination of environmental Cs concentrations.

The information available in the bibliography is given first. Then, the methodology and materials used for agarose gel cleaning are presented. The principal results are then displayed and discussed with respect to Cs concentration in collected water samples.

2. Overview on previous researches on Cs specific DGT devices

There exist several studies about Cs-specific DGT development. In most of these studies, authors use polyacrylamide as a diffusive gel according to the procedure from Zhang and Davison [2]. The difference in all the studies are the type of resin types of resin are tested in order to bind ionic Cs.

For instance, in the research by Chang et al. (1998), the binding agent consisted of AG50WX8 cation exchange resin. The resin is dispersed into an acrylamide solution [10]. Authors observed that the resin is readily saturated due to the continuous uptake of major cations present in the solution. Utilizing AG50WX8, they proved their hypothesis where for Cs and Sr, there was no evidence for the presence of inert complexes which would dissociate too slowly to be accumulated in DGT. There could be situations, however, where colloidal forms are significant and, in these cases, DGT would discriminate by measuring only labile solution species. Furthermore, it can easily be deployed in situ so that the selective binding of labile species would be assessed [10]. Chang et al. concluded that DGT with a general cation exchanger will be suitable for applications that measure multiple cations at low concentrations [10].

Later, in the research of Murdock et al. (2001) [11], they used Ammonium Molybdophosphate (AMP) as binding agent. They found that using known concentrations of ^{134}Cs, measurements with DGT devices with AMP binding agent closely reproduced the actual concentration, with a good degree of accuracy and precision for sampling periods up to 1 day. The DGT deployment period was limited to <1 month because of the significant progressive reduction in performance over longer periods [11]. The hypothesis was to consider that this was due to the growth of biofilms of algae/bacteria, which were seen to color the filter surface after deployment. Another explanation lies in the progressive degradation of the AMP resin. Hence, authors recommended another Cs specific binding agents: namely, potassium cobalt ferrocyanide (KCuFC) that may extend deployment periods [11]. They found a nonlinear effect on the diffusion coefficient of variations in water temperature during the DGT sampling period. With unchanging solution concentration, the error will depend on the accuracy of the temperature record obtained and can become negligible with the use of a temperature logger. When there is variation in both temperature and solution concentration during the sampling period, the error is unquantifiable, although within definable limits. The worst case would occur if all the activity was discharged as a spike when the temperature was either at a maximum or minimum [11].

Finally, in the work of Li et al. (2009) [12], the authors used Copper Ferrocyanide (CuFCN) as the binding agent is immobilized on Chelex 100 resin gel (Chelex-CuFCN) and on polyacrylic acid gel (PAA-CuFCN). They found that both binding phases are suitable to be used as binding phase for Cs accumulated in DGT. Both are capable to selectively bind Cs ions in a high alkali and alkaline earth metal ions environment. But the labile concentration in PAA-CuFCN is closer to the dissolved Cs concentration compared to those in Chelex-CuFCN. The difference is caused by the different binding kinetics of Cs ions on the gel. DGT coupled with a gamma detector can be used for direct determination of radioactive Cs or with ICP-MS for stable Cs measurements [12].

Following this work, a specific DGT device for Cs was developed by using agarose as diffusive gel and copper-based potassium hexacyanoferrate (KCuFC) as binding agent. However, commercial agarose gel contains impurities in Cs higher than the environmental concentration, thus, requiring specific purification treatment. The elaboration of the treatment is the main objective of this study.

3. Materials and Methods

The study of the Cs impurity in agarose and the elaboration of an impurity removal process were conducted due to the trace concentration of Cs in natural waters and the high uncertainty on the initial Cs content of commercial agarose. Due to the trace amounts of Cs in natural water, the containers and flasks are also tested for its Cs content and its interaction with Cs. All containers used are cleaned using HCl and rinsed using ultrapure water.

3.1 Methodology

Because there little to no available information on the commercial agarose contamination with Cs, an initial determination of its Cs content is required. According to the measured content, the agarose samples to be used in DGT devices may undergo either one or several cleaning steps. Additionally, a determination of the remaining Cs content takes place between two successive cleaning steps until the required purification degree is reached. The methodology followed in this research is presented in the figure 3.

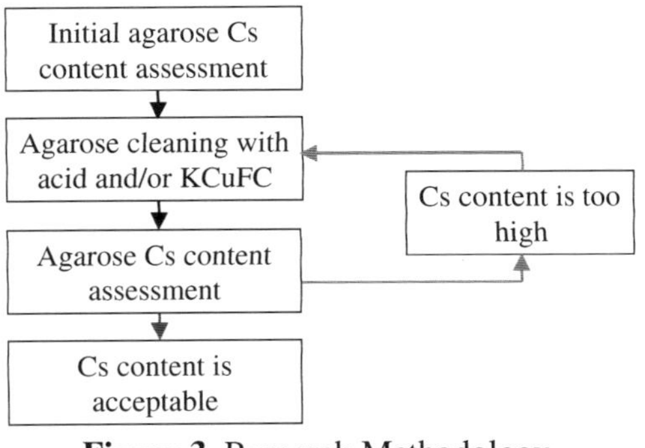

Figure 3. Research Methodology.

3.2 Equipment and materials

In order to prepare the diffusion gel layer of DGT devices, high purity electrophoresis grade agarose was purchased from Alfa Aesar Company. Cs concentration was measured using Induced Coupled Plasma-Mass Spectrometer (ICP-MS) (Xseries 2, THERMOELECTRON). The detection limit in Cs is $(1.5\pm0.2)\times10^{-10}$ mol·L^{-1}. The acid used to digest was ultrapure HNO_3 10 mol·L^{-1}. The two types of acids used in the course of the cleaning steps are taken from ultrapure HCl (37%) or HNO_3 (70%) solutions. The liquid gel and KCuFC exchanger resin were separated using a filter syringe membrane supplied by Sartorius with a pore size of 0.22 µm.

3.3 Agarose digestion

To assess the quantity of Cs in agarose powder, agarose powder was directly digested in HNO_3 4 mol·L^{-1} at room temperature. 0.5 g of agarose powder was digested in 5 ml of HNO_3 4 mol·L^{-1}. The duration of the digestion process is 18 hours.

3.4 Agarose cleaning with Acid

Initially, the cleaning steps consisted in mixing the agarose powder with HCl solutions by using a ratio of agarose mass to acid volume of 62.5 g·L^{-1}. The concentrations of HCl were 0.1 mol·L^{-1}, 1.0 mol·L^{-1} and 2.5 mol·L^{-1}. The contact time was from 1 to up to 18 hours at room temperature. After the cleaning, all trace of acidity was removed by rinsing several times (usually 5-7 times) with ultrapure water. The acid-washed agarose powder was ultimately dried at 35° C in an oven. The cleaned agarose was digested in HNO_3 4 mol·L^{-1} at room temperature for a duration of 18 hours. The solution of digestion is afterward analyzed by ICP-MS.

3.5 Agarose cleaning with KCuFC

According to the targeted degree of purity, the acid-cleaned agarose powder or raw commercial agarose powder may undergo an extra cleaning step. This polishing step pertains to the removal of Cs by adsorption onto KCuFC resin. This is partly because, unlike with Chelex-100, KCuFC resin efficiency is almost pH independent. In practice, the agarose powder is mixed with KCuFC and 20 mL of ultra-pure water (contact time of 1 week). After 1 week, the mixture is heated and gently agitated until the agarose dissolution is complete (at 80 °C). The KCuFC and the agarose solution are then separated by either allowing the resin to settle at the bottom of the vial or by filtration. In the latter case, a syringe filter of 0.22 µm filter was used. The filtrates are after analyzed by ICP-MS. It is important to keep the filtration rate and the temperature of agarose solution high enough to avoid gel formation.

4. Results

4.1 Agarose digestion

The digestion of raw commercial agarose powder using HNO_3 4 mol·L^{-1} gives 2.1×10^{-8} mol·L^{-1} of Cs. This value doesn't seem high but considering the trace amount of Cs in the environment which is below 1×10^{-9} mol·L^{-1}, pretreatment is required before the utilization of the gel. As a matter of fact, the quantity of Cs found in the agarose powder is 56 times higher compared to the average environmental level. Cleaning process is therefore required to be able to distinguish natural Cs in the environment from the Cs impurity in the agarose powder.

4.2 The purity degree of agarose after acidic treatments

The results obtained for agarose cleaning with HCl are displayed in the figure 4 where the x axis represents the ratio of Cs amount in the digested cleaned agarose powder. The contact time is equal to 1 hour because longer durations resulted in pronounced degradation of the agarose samples. As a matter of fact, these turned into a dark mass or black powder when being exposed to concentrated HCl solutions (i.e. 2.5 mol·L^{-1} or 5.0 mol·L^{-1}) for longer periods. The overall decrease in the Cs content is therefore limited: i.e. less than 30%. Furthermore, it appears that the decrease is almost the same for all tested HCl concentrations.

This result shows that the HCl-cleaning method is not quite efficient and calls for longer contact durations, which is susceptible to alter the agarose gelling properties and to promote the formation of charged groups inside the gel matrix. The latter situation may render the interpretation of Cs-DGT measurements complex and thus would call for an exhaustive calibration work of DGT devices. Therefore, we decided to no longer pursue Cs removal from agarose utilizing acids.

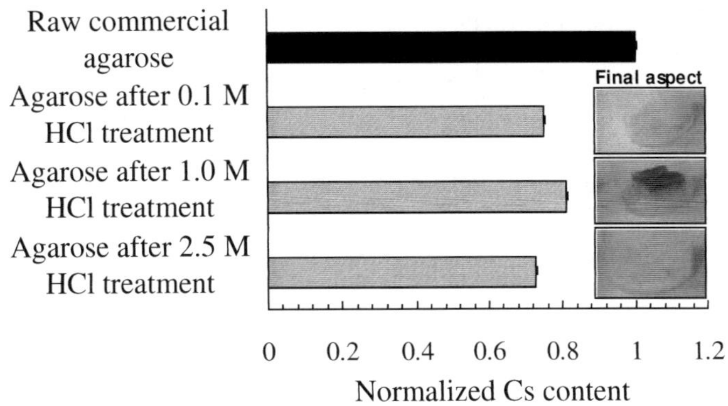

Figure 4. Normalized Cs content of agarose powder in the absence (i.e. raw) or after various treatments with HCl (1-hour cleaning) and subsequent drying. The Cs content of raw commercial agarose is set equal to the unity. The final visual aspect of treated powders is also given as pictures.

Extra cleaning tests conducted with HNO_3 (5 mol·L^{-1}) instead of HCl resulted in complete digestion of the agarose powder. This nevertheless allowed calculating its total Cs content: i.e. 0.1-0.3×10^{-9} mol·g^{-1}. Such low values are encouraging with respect to stable Cs level determination in natural waters by means of agarose gel-containing DGT devices. As a matter of fact, the related Cs contents of the agarose gel formed by using the raw commercial or the HCl cleaned agarose powder are 0.16-0.51×10^{-9} mol·L^{-1} and 0.12-0.36×10^{-9} mol·L^{-1}, respectively. This result also indicates that agarose powder cleaning is feasible providing an adapted washing agent is used.

4.3 Agarose cleaning using KCuFC

Agarose powder was mixed with KCuFC resin in ultra-pure water the suspension was heated afterwards, the KCuFC resin and the agarose powder were separated by filtration. The results (Figure 5.c) shows that the amount of Cs is reduced by 99% compared to untreated agarose gel (Fig 5.a). Using this method, a significantly higher purity was obtained compared to cleaning with HCl 2.5 mol·L^{-1} (around 6.4 times less Cs). Handling operations are also much more convenient because no acids are used and there is no risk of improper rinsing that can compromise the results.

The cleaning duration is another parameter assessed. The effectiveness of the cleaning process increases with time. Variations of 30 minutes and one week has been tested with the results shown in figure 5.c and 5.d respectively. The effectiveness of the cleaning process only stepped up by a factor of 4 when increasing the cleaning duration by a factor close to 340.

Furthermore, using gravity to separate the KCuFC resin from the agarose solution is not feasible due to the viscosity of the agarose solution (figure 5.b). It also resulted in a higher amount of Cs in the gel, which may indicate the contamination of Cs from a previous test or a memory effect of the vials used. Nevertheless, this result also shows the high selectivity of the exchange resin to uptake the Cs and its high sensitivity.

The effect of cleaning process reveals that the combination of time duration, heat, the use of KCuFC resin and filtration is the best condition to decrease the impurity of Cs in the gel. To compare the final concentration of Cs with natural Cs in the environment, water sample was collected in the Loire River Estuary at Mandin station near the city of Saint-Nazaire, France, the results showed that the total level of Cs impurity in agarose gel is 2.2 times lower than the value of natural Cs in the environment. This procedure shows that there is now a possibility to follow the natural Cs in water after treatment of the gel.

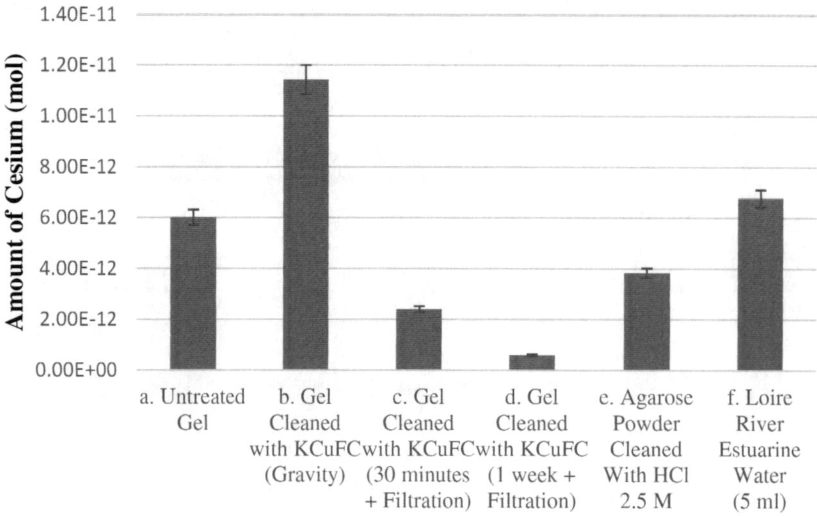

Figure 5. Comparison of various cleaning methods normalized to 2 ml of agarose gel.

The method used for cleaning the agarose gel has several advantages; yet a drawback from using this method is that due to the initial viscosity of the agarose solution. Indeed, whereas the diffusion gel in DGT devices are typically prepared with a ratio of 15 $g_{agarose} \cdot L_{water}^{-1}$, such solutions quickly clog the filter after around 1 ml of filtrate. Instead, a ratio of 5 $g_{agarose} \cdot L_{water}^{-1}$ was used in our research. By using this lower ratio, the filter can be used for up to 5 ml of solution. This results in a gel that is less dense and more fragile, but it may also let Cs ions diffuse easily.

Besides, special care should be done to prevent the formation of air bubbles when the syringe used for filtering the agarose solution is almost empty. The bubbles can cause the filtered gel to sputter, which can lead to air being trapped inside the gel. Air bubbles formed inside the gel may impede Cs diffusion towards the resin. This impedance will lower the amount of Cs in the resin and will lead to an inaccurate calculation of Cs concentration in the water.

5. Conclusions

Agarose gel is a good diffusive medium with a diffusive coefficient similar to polyacrylamide gel. A good thing about agarose is its simplicity in its manipulation and its non-toxicity compared to polyacrylamide. Taking all these results as a whole, it appears that acceptable Cs levels (i.e. with respect to ambient Cs concentration in natural waters) are achievable by the removal of preexisting Cs in the agarose powder before utilization.

In order to reach an even higher degree of purity (in the range of a few tens of 10^{-12} $mol \cdot g^{-1}$), a series of cleaning stages is maybe required. As an example, the agarose powder could undergo an initial cleaning by adding 0.1 $mol \cdot L^{-1}$ HCl solution and mixing at room temperature for at least 1 hour. After discarding the HCl solution and multiple rinsing with ultrapure water, a polishing of the acid cleaned agarose may be undertaken. This would take place by mixing the agarose powder with KCuFC resin and ultra-pure water. By following these steps, the final Cs content of the agarose gel would be 10-50×10^{-12} $mol \cdot L^{-1}$. This is below the detection limit of our ICP-MS device as well as stable Cs concentrations of estuarine and riverine waters (i.e. typically less than 5,000 and 100×10^{-12} $mol \cdot L^{-1}$).

6. References

[1] Tercier ML and Buffle J 1993 *Electroanalysis* **5** 187
[2] Zhang H and Davison W1995 *Analytical Chemistry* **67** 3391

[3] Beaugelin-Seiller K, Roussel-Debet S and Germain P 2005 *Fiche Radionucléide Césium 137 et environnement*. IRSN
[4] Staunton S and Roubaud M 1997 *Clay and Clay Minerals* **45** 251–260
[5] Toxicological Profile for Cesium, in US Department of Health and Human Services, April 2004
[6] Nakanishi T and Tanoi K. (eds) 2013 Agricultural Implications of the Fukushima Nuclear Accident
[7] Kitajima A, Ogawa H, Kobayashi T, Kawasaki T, Kawatsu Y, Kawamoto T and Tanaka H 2014 *Environmental Science: Processes Impacts* **16**, 28
[8] National Research Council 1986 Workshop on Marine Algae Biotechnology: Summary Report, Jakarta, Indonesia, December 11-13, 1985. The National Academies Press, Washington, DC
[9] Johnson CK 1965 *Report ORNL-3794*. Oak Ridge National Laboratory, Oak Ridge, Tennessee
[10] Chang L-Y, Davison W, Zhang H, and Kelly M 1998 *Analytica Chimica Acta,* **368** 243-253
[11] Murdock C, Kelly M, Chang L-Y, Davison W and Zhang H 2001 *Environment Science Technology* **35** 4530-4535
[12] Li W, Wang F, Zhang W and Evans D 2009 *Analytical Chemistry* **81** 5889-5895

Acknowledgment
This research was supported by the Region Pays de la Loire with the funding of the Observatoire des Sciences de l'Univers Nantes Atlantique (OSUNA, France). The authors gratefully acknowledge Dr. Caroline Simonucci and Dr. Damien Tournieux from IRSN (France), Dr. Catherine Landesman from SUBATECH laboratory (France) and Dr. Takumi Saito from Tokyo University (Japan) as well as to Dr. Edouard Metzger from LPG-Angers (France) for their constructive advices and fruitful discussions.

Microwave-Assisted Cationic Polymerization of Turpentine: Change in Infrared and Ultraviolet Spectra

Muhamad Farid[1], Bambang Soegijono[1a], Zainal Alim Mas'ud[2] dan Surya Herlambang[2]

[1] Department of Physics, Faculty of Mathematics and Natural Science, University of Indonesia, Depok 16424, Indonesia
[2] Department of Chemistry, Faculty of Mathematics and Natural Science, Bogor Agricultural University, Bogor 16680, Indonesia

[a]Email: bambangsg11@yahoo.com

Abstract. Indonesia is a turpentine producer country. So far, only one factory is carrying out turpentine fractionation to increase its selling price. Chemical conversion is necessary to produce more valuable derivatives of this renewable biomass-based material. In this experiment, turpentine as the raw material and the four products: α-pinene, β-pinene, limonene, and δ-carene were reacted successively with sulfuric acid, hydrochloric acid, nitric acid, and ethereal boron trifluoride, respectively. The reaction products were further irradiated using a commercial microwave oven. The raw material composition, functional group, and ultraviolet absorption were analyzed using a gas chromatograph, Fourier transforms infrared spectrophotometer, and UV-Vis spectrophotometer, respectively. Half of the reaction products were neutralized with sodium bicarbonate whether or not it irradiated. All products which viscous colored liquid was examined on their ultraviolet and infrared spectrum. The turpentine reacted with sulfuric acid and boron trifluoride ethereal produced a brownish viscous liquid, but did not react with hydrochloric acid and nitric acid. The reaction products with sulfuric acid and boron trifluoride ethereal and microwave irradiation processes showed some shifts on the infrared spectra, particularly in C=C band due to to the polymerization and C=C reforming reaction. The products also demonstrate the presence of OH group in the infrared spectra.

1. Introduction

Indonesia produces about 12000 tons of turpentine [1] and is the second largest producer in the world after China [2]. The main components of turpentine are mono and bicyclic monoterpenes ($C_{10}H_{16}$; Mr 136.2) as α-pinene, β-pinene, δ-carene, and limonene [3]. However, since 2014, Perum Perhutani began to produce turpentine derivatives such as α-pinene, β-pinene, δ-carene, and δ-limonene to increase the value of the products [4].

Based on the turpentine functional group, the transformation is due to a reaction of the unsaturated C=C bond. The transformation includes isomerization, hydration [5], condensation, hydrogenation, cyclization, oxidation [6,7], hydroformylation, rearrangement, ring contraction/enlargement [6], as well as epoxidation [6]. In addition, the C=C group may also undergo polymerization reactions. The cationic polymerization was facilitated by the presence of acid [8]. Cationic polymerization of α-pinene [9-14], β-pinene [15-19], α- and β-pinene [20], α-, β-pinene and limonene [21], limonene [22-24] have been studied, but not for δ-carene. Microwave technique has been used in facilitating the synthesis process, including polymer synthesis [25-33] and has never been used for turpentine and their derivatives.

In this paper, we employed four acidic agents (H_2SO_4, HCl, HNO_3, and boron trifluoride ethereal) as a catalyst for microwave- assisted cationic polymerization of turpentine, α-pinene, β-pinene, δ-carene, and δ-limonene. The resulted products were analyzed by ultraviolet and Fourier transform infrared spectrophotometry and compared with the raw material.

2. Material and methods

2.1 Materials

The raw materials, namely turpentine, α-pinene, β-pinene, δ-carene, and δ-limonene were provided by Perhutani Pine Chemical Industry (PPCI), Pemalang. All raw materials were directly used without purification. Analytical grade purchased from Merck: hexane, H_2SO_4 95%-97% (d=1.84), HCl 32% (d=1,16), HNO_3 65% (d=1.39), and $NaHCO_3$. Boron trifluoride ethyl ether complex ($BF_3.C_4H_{10}O$) was acquired from TCI.

2.2 Instrumentation

Turpentine as the raw material was analyzed for its composition using Shimadzu GCMS QP-2010. Ultraviolet spectra were taken using Shimadzu Pharmaspec UV-1700. The infrared spectra were observed using PerkinElmer Spectro One on a demountable liquid cell module. The reactions were carried out in 100 ml Erlenmeyer inserted in threaded PTFE reactors (workshop made, $Ø_{ID}$ = 60 mm ID, $Ø_{OD}$ = 100 mm, h = 85 mm) (Fig. 1). The microwave oven used was Samsung MG23H3185PK (230 V, 50 Hz, 2450 MHz; input 1400 W output 800 W).

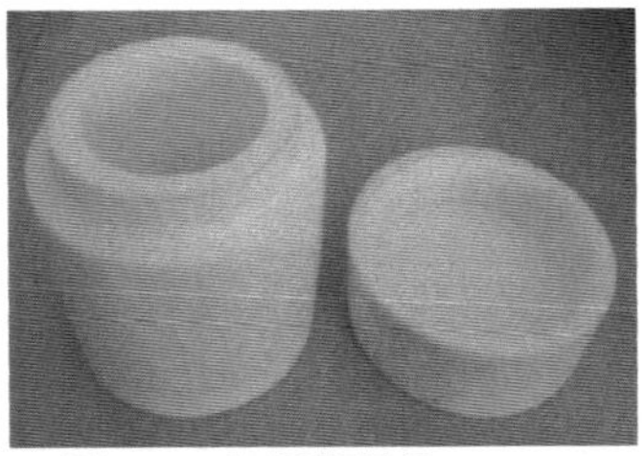

Figure 1. PTFE Reactor.

2.3 Instrumentation

Turpentine as the raw material was analyzed for its composition using Shimadzu GCMS QP-2010. Ultraviolet spectra were taken using Shimadzu Pharmaspec UV-1700. The infrared spectra were observed using PerkinElmer Spectro One on a demountable liquid cell module.

2.4 Method

The amount of raw material uses and time of reaction we use, are based on our initial study in the lab. Reaction with HCl and HNO_3: 60 g of the raw material in 100 mL Erlenmeyer was reacted with 3 g of reagent and stirred for 1 h.

Reaction with H_2SO_4: 60 g of the raw material in 100 mL Erlenmeyer was reacted with 3 g of reagent and stirred in an ice bath for 30 min. we using ice bath because the addition of H_2SO_4 into turpentine is exothermic and it's is can cause the raw material spurt out. The stirring was continued without the ice bath for 30 min because after the several minutes of stirred the temperature did not increase anymore.

Reaction with BF_3-ethereal: 15 g raw material in a 100 mL Erlenmeyer was stirred in an ice bath for 20 min, and carefully reacted dropwise with 0.75 g of BF_3-ethereal. Stirring was continued in the ice bath for 20 min and continued for 40 min without the ice bath. This experiment was repeated 4 times, and all products were mixed.

The polymerization process divided into 2 parts ("A" and "B"). The "A" parts were divided into 2 portions ("A_1" and "A_2"). The "A_1" portions were stored directly in vials, while the "A_2" portions were

IOP Conf. Series: Materials Science and Engineering **599** (2019) 012028 doi:10.1088/1757-899X/599/1/012028

neutralized with $NaHCO_3$. The "B" parts were irradiated with the microwaves under 800W of power for 10 min. The irradiated products were divided into 2 portions ("B_1" and "B_2"). The "B_1" portions were stored into vials, while the "B_2" portions were neutralized first with $NaHCO_3$. The products showing thick brownish liquid were picked up for observation using ultraviolet and infrared spectrophotometers.

3. Result and discussion

3.1 Turpentine composition

The GCMS analytical results showed that turpentine was composed of 81.6% α-pinene, 10.94% δ-carene, 2.92% β-pinene, and 0.93% δ-limonene. Since α-pinene is the main component of turpentine, it is expected that the reaction behavior is analogous to α-pinene.

3.2 Effect of microwave irradiation on raw materials

Microwave irradiation treatments against turpentine, α-pinene, β-pinene, δ-carene, and δ-limonene do not visually change the color and the viscosity. However, all UV spectra pattern differ, showing an increase in absorption. The pattern change of the turpentine UV spectrum is not like that of the α-pinene. The IR spectrum intensity of C=C band, $\upsilon = 1600\text{-}1680\ cm^{-1}$ [34] do not change due to irradiation. The pattern of IR spectrum change is random and cannot be generalized.

3.3 The reaction products of the raw material and the acid and the effects of microwave irradiation

Formation of polymers was characterized by the appearance of thick brownish red liquid and the reaction underwent exothermically. H_2SO_4 (Fig. 2(b) before and 2(c) after irradiation), HCl (Fig. 2(d), before and Fig. 2(e) after irradiation) and HNO_3 (Fig. 2(f) before and Fig. 2(g) after irradiation) reagents do not produce the desired product with all the raw materials. and BF_3-ethereal reagents (Fig. 2(h) before and (i) after irradiation) produced thick brownish-colored liquid; all reactions with all raw materials were exothermic.

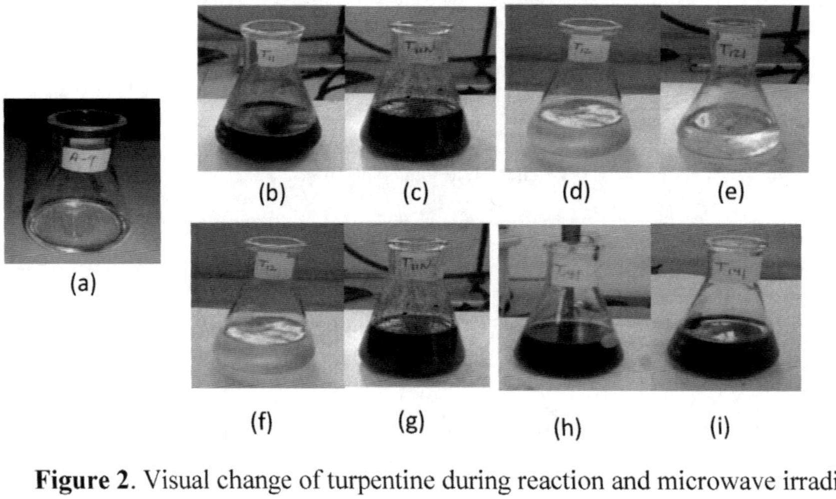

Figure 2. Visual change of turpentine during reaction and microwave irradiation.

3.4 Infrared spectra changes caused by treatments

IR spectra changes due to H_2SO_4 and BF_3-ethereal treatments, microwave irradiation, and neutralization are shown in Fig. 3. The H_2SO_4 and BF_3-ethereal treatments increase the transmittance (%T) (decrease the absorbance) in the range of $\upsilon = 1600\text{–}1680\ cm^{-1}$, due to the transformation of the C=C bond to the C–C bond as the consequence of polymerization. However, the microwave irradiation treatment decreases the transmittance in the same range, presumably due to the re-forming of C=C bonds. Surprisingly, the appearance of absorption around 3500 cm^{-1} wavenumber is strong and wide and unchanged due to the irradiation and the neutralization. This band corresponds to the OH group.

IOP Conf. Series: Materials Science and Engineering **599** (2019) 012028 doi:10.1088/1757-899X/599/1/012028

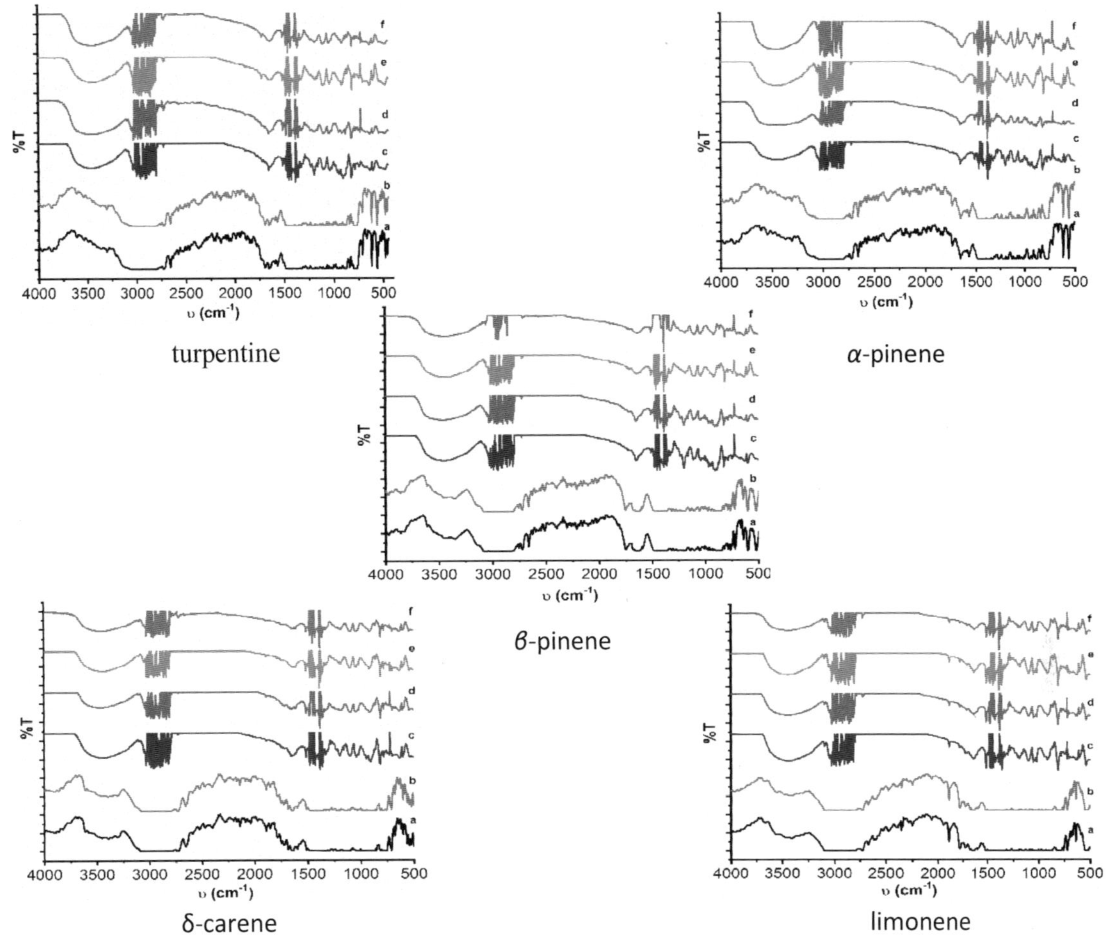

Figure 3. Changes in infrared spectrum of turpentine, *α*-pinene, *β*-pinene, δ-carene and limonene without catalyst (b) and with acid catalysts BF_3-etheral (c), HCl (d), HNO_3 (e), H_2SO_4 (f) towards control (a).

3.5 Ultraviolet spectra changes caused by treatments

UV spectra changes due to H_2SO_4 and BF_3-ethereal treatment, microwave irradiation, and neutralization are shown in Fig. 4. The H_2SO_4 treatments increase UV absorbance (except that for δ-carene). Microwave irradiation treatment varies, decreasing the absorbance (α-pinene, β-pinene, δ-limonene) and increasing (turpentine and δ-carene). Neutralization processes increase the absorbance.

The BF_3-ethereal treatments increased the UV absorbance (except for β-pinene). The effect of microwave irradiation treatment varies, decreasing the absorbance (α-pinene, and δ-limonene) and increasing (turpentine, β-pinene, δ-carene). Neutralization increases the absorbance only for δ-limonene. It is interesting since this treatment changes the color from dark to light without changing the viscosity.

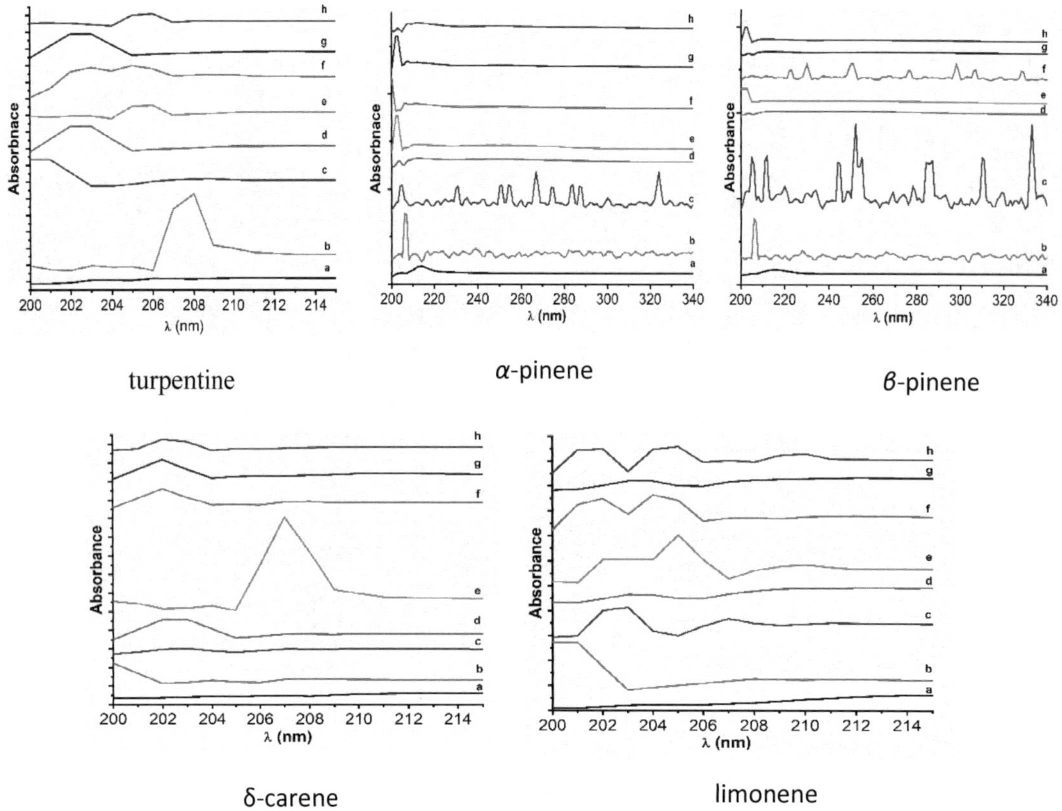

Figure 4. Changes in ultraviolet spectrum of turpentine, *α*-pinene, *β*-pinene, δ-carene and limonene without catalyst (b) and with acid catalysts BF_3-etheral (c), HCl (d), HNO_3 (f), H_2SO_4 (h), and acid catalyst neutralization HCl (e) and HNO_3 (g) towards control (a).

4. Conclusions

Turpentine, α-pinene, β-pinene, limonene, and δ-carene react with sulfuric acid and ethereal boron trifluoride to produce polymers which visually forming viscous liquids and reduced intensity of C=C infrared absorption bands. In addition to polymer formation, these reactions also produce hydroxyl groups.

5. References

[1] Perum Perhutani 2015 Statistik Perum Perhutani Tahun 2010–2014
[2] Ukkonen K 2015 A Pine Chemicals-Global view
[3] Gscheidmeier M and Fleig H *Turpentines* 2012 In B Elvers (Eds) *Ullmann's Encyclopedia of Industrial Chemistry, Vol 37* Wiley-VCH Verlag GmbH & Co KGaA, Weinheim
[4] Perum Perhutani 2015 Profil Perum Perhutani
[5] Robert W J and Day A R 1950 *J. Am. Chem. Soc.* **72** 1226–1230
[6] Monteiro J L F and Veloso C O 2004 *Topics Catal.* **27** 169–180
[7] Schwab W Fuchs C and Huang F C 2013 *Eur. J. Lipid Sci. Technol.* **115** 3–8
[8] Faust R and Shaffe T D 1997 Cationic Polymerization-Fundamentals and Applications *ACS Symp. Ser.* **665** 226
[9] Higashimura T, Lu J, Kamigaito M, and Sawamoto M 1992 *Makromol. Chem.* **193** 2311–2321
[10] Higashimura T, Lu J, Kamigaito M, and Sawamoto M 1993 *Makromol. Chem.* **194** 3441–3453
[11] Higashimura T, Lu J, Kamigaito M, and Sawamoto M 1993 *Makromol. Chem.* **194** 3455–3465
[12] Shiwei L, Congxia X, Shitao Y, and Fusheng L 2009 *Catal. Commun.* **10** 986–988

[13] Zuguang L, Taishun Z, Wei Z, Hualong Z, and Xinnan A 2011 *React. Kinet. Mech. Cat* **104** 125–137
[14] Shiwei L, Lin Z, Shitao Y, Congxia X, Fusheng L, Zhanqian S 2013 *Biomass and Bioenergy*, **57** 238–242
[15] Pietila H, Sivola A, and Sheffer H 1970 *J Polymer Sci Part A: Polymer Chem.* **8** 727–737
[16] Martinez F 1984 J Polymer Sci Part A: Polymer Chem. **22** 673–677
[17] Jian L, Kamigaito M, Sawamoto M, Hicashimura T, and Yun-Xiang D 1996 *J. Appl. Polymer Sci.* **61** 1011–1016
[18] Kukhta NA, Vasilenko IV and Kostju SV 2011 *Green Chem* **13** 2362–2364
[19] Karasawa Y, Kimura M, Kanazawa A, Kanaoka S and Aoshima S 2014 *Polymer J.* **47** 1–6
[20] Corma A, Iborra S, and Velty A 2007 *Chem. Rev.* **107** 2411–2502
[21] Lu J, Liang H, Zhang R and Deng YX 1998 *Acta Polymerica Sinica* **68** 698–703
[22] Norström E 2011 *Terpenes as renewable monomers for biobased-materials* KTH Chemical Science and Engineering Thesis p 125
[23] Firdaus M 2013 Terpenes as Renewable Resources for Organic and Macromolecular Chemistry. Karlsruher Institut für Technologie Dissertation p 193
[24] Zhang Y J 2014 *Copolymerization of Limonene* Department of Chemical Engineering Faculty of Engineering University of Ottawa Thesis p 91
[25] Zong L, Zhou S, Sgriccia N, Hawley M C and Kempel L C 2003 *J. Microw. Power Electromagn. Energy* **38** 49–74
[26] Wiesbrock F, Hoogenboom R and Schubert U S 2004 *Macromol. Rapid Commun.* **25** 1739–1764
[27] Evalueserve 2005 *Developments in Microwave Chemistry* Royal Society of Chemistry Burlington House London p 52
[28] Koopmans C, Iannelli M, Kerep P, Klink M, Schmitz S, Sinnwell S and Ritter H 2006 *Tetrahedron* **62** 4709–4714
[29] Hoogenboom R and Schubert U S 2007 *Macromol. Rapid Commun.* **28** 368–386
[30] Sinnwell S and Ritter H 2007 *Aust. J. Chem.* **60** 729–743
[31] Mallakpour S and Rafiee Z 2008 *Iranian Polym. J.* **17** 907–935
[32] Sosnik A, Gotelli G and Abraham G A 2011 *Prog. Polym. Sci.* **36** 1050–1078
[33] Mishra A, Vats T and Clark J H 2016 *Microwave-Assisted Polymerization* The Royal Society of Chemistry Milton Road Cambridge p 183
[34] Pavia D L, Lampman G M, Kriz G S, and Vyvyan J R 2009 *Introduction to Spectroscopy (4th ed)* Belmont: Brooks/Cole, Cengage Learning

Preparation and Characterization of Local Indonesian Chitosan-graft-Maleic Anhydride as Drug Carrier

Daniel Timotius[1], Rochmadi[1], Yuni Kusumastuti[1,*]

[1]Department of Chemical Engineering, Universitas Gadjah Mada, Yogyakarta 55281, Indonesia

[*]yuni_kusumastuti@ugm.ac.id

Abstract. The modification of local Indonesian chitosan as a drug carrier is conducted by grafting with maleic anhydride (MA). The grafted film resulting in a negative charge of carboxyl group attach on chitosan backbone. Several ratio of MA/chitosan is conducted to evaluate the effect of MA on the drug release behavior. Chitosan and MA was dissolved and stirred in 1% (v/v) acetic acid for 24 hours. Tween 80 and curcumin are added to the solution as surfactant and drug model respectively. The homogeneous solution is casted into polyethylene petri dish and dried under room temperature. Chitosan-graft-MA is characterized by Fourier Transform Infra-Red (FTIR) in addition to gravimetric titration to characterize the existing of carboxyl groups of the films. The swelling and degradation ability in phosphate buffered saline (PBS) solution is showed an increasing trend as the addition of MA. The kinetic study of drug release is performed in Peppas and first order model. An optimum condition of grafting is achieved at 150% (w/w).

1. Introduction

Chitosan, (1-4)-2-amino-2-deoxy-β-D-glucan, is the second abundant polysaccharide after cellulose [1] and can be obtained by deacetylation of chitin from exoskeleton of crustacean [2]. The properties of chitosan such as biodegradation, biocompatible and non-toxic make it to be a good materials in biomedical application [3], [4]. Chitosan (≥50 % degree of deacetylation) is soluble in dilute acid solutions [5]. The solubility occur due to protonation of $-NH_2$ which lead to formation of polycationic polymer [6]. It is also known that chitosan has a good film forming properties [5], [7]–[11].

In order to improve on some properties of chitosan film (e.g. mechanical properties, swell-ability, pH sensitivity) it is commonly use crosslinking method [12]. There are at least 2 classification of crosslinking method on chitosan, chemical crosslinking and physical crosslinking [6]. Chemical crosslinking provide better mechanical strength than physical crosslinking [13]. However, some chemical crosslinker are tended to be toxic [12]. Some researchers have begun to find a cross-linker with low toxicity [7], [8], [10], [14]. MA becomes one of the candidates as crosslinker. A group of researchers [15] have reported that modification hyaluronic acid by MA tend to be non-toxic and showing a good drug release behavior on pilocarpine hydrochloride drug model. In another study [16], addition of MA on chitosan shows a good swelling behavior and antimicrobial properties.

The grafting of chitosan by MA will lead to appearance of carboxyl group attached on chitosan backbone. The carboxyl group will donate negative charges on the film. The double bond from MA also can be used for further modification. In this research, we develop a local Indonesian chitosan film grafted by MA as a potential patch drug delivery system and wound dressing application.

2. Materials and Methods

2.1. Materials

A local Indonesian chitosan (>90% degree of deacetylation, 10-500 cps viscosity, <1.5% ash content, 0.5% protein content) was obtained from PT Biotech Surindo (Cirebon, Indonesia). Acetic acid (100%) was purchased from Merck, Germany. Tween 80 was supplied by Merck, France. Curcumin was obtained from Bio Basic Inc, Canada.

2.2. Film preparation

Preparation of chitosan grafted MA solution was similar to a study [17] with some modification. Chitosan and MA were weighed about 0.6 g and 0.3 g respectively, then dissolved and stirred in 30 mL of acetic acid (1% v/v) for 24 hours. Tween 80 and curcumin were added to the solution followed by stirring to form homogeneous solution. About 5 mL of homogeneous solution was casted on petri dish and dried under ambient temperature to form a thin film. This method was repeated with several ratios of MA as presented in Table 1. All ratios in Table 1 were based on chitosan weight.

Table 1. Composition of each sample

Samples	MA (% w/w)	Tween 80 (% w/w)	Curcumin (% w/w)
CM0	0	70	20
CM50	50	70	20
CM100	100	70	20
CM150	150	70	20

2.3. Fourier Transform Infrared (FTIR) Spectroscopy

The functional groups of samples was measured by FTIR spectroscopy. The FTIR study was performed by SHIMADZU IRPrestige-21 in the range of 400 – 4000 cm^{-1}. All the samples were scanned with a resolution of 4 cm^{-1} and 10 number of scans.

2.4. Carboxyl Content Measurement

In order to measure the amount of carboxyl group in modified chitosan film, back titration method was performed [15]. About 20 mg of each sample was dissolved in 20mL of NaOH solution (0.005 N) and followed by titration using HCl (0.005 N). The data were calculated by Equation (1) to obtain the acidic contents for each sample.

$$Acidic\ content = \frac{C_{NaOH} \times V_{NaOH} - C_{HCl} \times V_{HCl}}{m_{sample}} \tag{1}$$

2.5. Swelling Ratio and Degradation Study

The swelling and degradation properties were measured by immersing the sample in PBS solution. The swelling method was similar to a study [7]. Swelling ratio was done by immersing Ø15mm of film for 30 minutes, while for degradation study was continued for 24 hours. The immersing of film in PBS was done under normal physiological temperature (about 37 ^{0}C). The film was weighed before and after immersing (wet weight for swelling ratio and dried weight for degradation study). The obtained data were calculated by Equations (2) and (3) for swelling ratio and degradation study respectively.

$$Swelling\ Ratio(\%) = \frac{W_t}{W_0} \times 100\% \tag{2}$$

$$Degradation(\%) = \frac{W_D - W_0}{W_0} \times 100\% \tag{3}$$

2.6. Drug Release Kinetics

The drug release kinetics was measured by immersing Ø10 mm of film in 20 mL PBS solution under constant stirring rate. Concentration of curcumin was analysed at periodic interval with

IOP Conf. Series: Materials Science and Engineering **599** (2019) 012029 doi:10.1088/1757-899X/599/1/012029

spectrophotometer VIS. Optimum wave length for curcumin was attained at 420 nm. The concentration was calculated by calibration curve. The drug release model was followed first-order and Peppas model. It was commonly used to express non-constant release from any reservoir [18]. The model was presented in Equation (4) and (5) for first order and Peppas model respectively:

$$\frac{M_t}{M_\infty} = 100(1 - e^{-kt}) \tag{4}$$

$$\frac{M_t}{M_\infty} = Kt^n \tag{5}$$

3. Result and discussion

3.1. FTIR Spectroscopy

A study on FTIR spectroscopy was conducted to analyze the functional group of modified chitosan. The result of FTIR spectroscopy study is presented in figure 1. This result is similar from other study [16]. The scanning result for chitosan show peaks at 1651.07 cm^{-1} and 1381.03 are belong to amide I and amide II functional group. While the broad band around 3425.58 cm^{-1} is signed as –NH and –OH stretching.

The appearance peak around 1712.79 cm^{-1} in CM50 is belong to carboxyl group that attach to the chitosan backbone. This result was similar to other study that use MA to modify hyaluronic acid [15]. In another study, the researcher modify cellulose nanocrystals by MA [19] and show the appearance of carboxyl group around 1724 cm^{-1}.

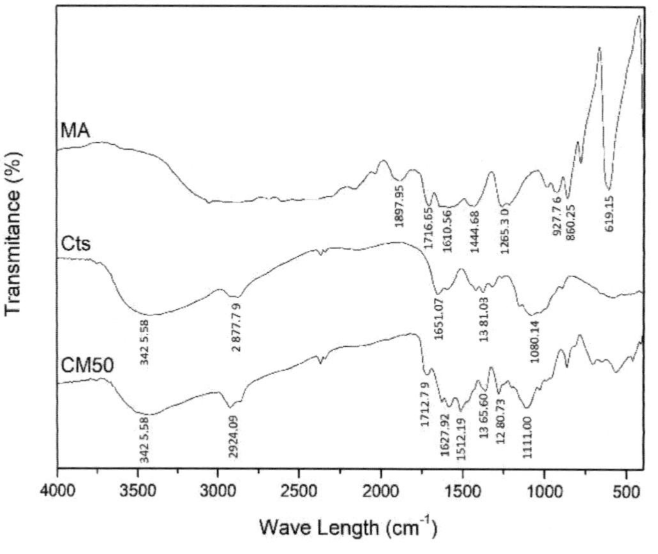

Figure 1. FTIR Spectrum for MA, Cts, and CM50.

3.2. Carboxyl Content Measurement

Back titration method was effective to be used in measuring the acidic content of the films. The result is presented in figure 2. An increasing trend of the acidic content in each sample confirm that the grafting method was successfully done. The carboxyl content can be obtained by calculating the difference on acidic content of CM50, CM100, and CM150 compared CM0. The amount of carboxyl group in CM50, CM100, and CM150 are 2.15±0.08; 3.70±0.02; and 4.35±0.41 respectively. The increasing values of acidic contents occur due to the variation of ratio used in this study. This result is similar to a study that using MA to modify hyaluronic acid in form of gels [15].

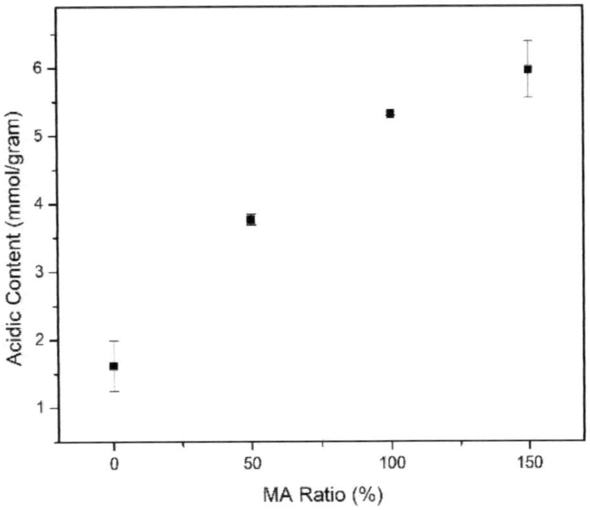

Figure 2. Acidic Content of CM0, CM50, CM100, and CM150.

3.3. Swelling Ratio and Degradation Study

The swelling ability of film is important, since the drugs have to diffuse through film matrix. It also can demonstrate the ability of exudates absorption for some specific wound cases. The result for swelling behavior for each samples are shown in figure 3. The swelling of samples tend to increase as the addition of MA. This may occur because of the appearance of new hydrophilic carboxyl group on the chitosan backbone [7]. The value of swelling ratio for CM0, CM50, CM100, and CM 150 are 176.67±2.89%; 221.93±16.32%; 309.65±17.48%; and 239±25.63% respectively.

While the degradation study was done to determine the degradation ability of each sample. Figure 4 shows the degradation of film after 24 hours of immersing. As it is shown in figure 4, the addition of MA lead to increase the degradation of film.

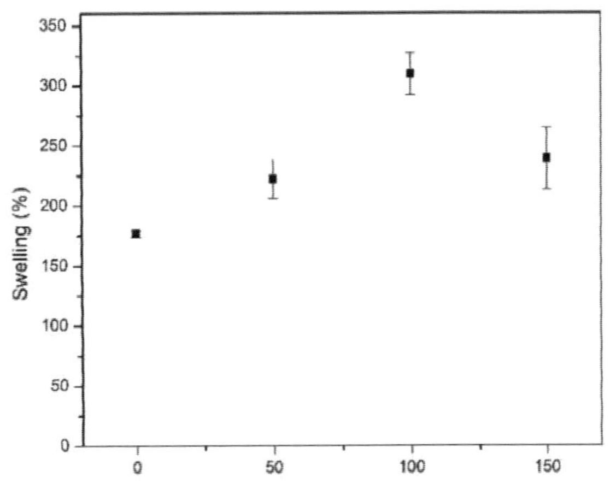

Figure 3. Swelling Ratio of CM0, CM50, CM100, and CM150.

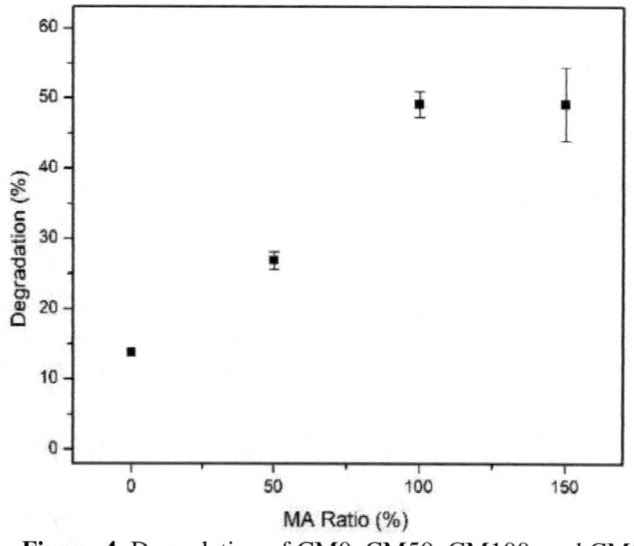

Figure 4. Degradation of CM0, CM50, CM100, and CM150.

3.4. Drug Release Kinetics

Curcumin concentration along the time for each samples are presented in figure 5 for First Order Model and figure 6 for Peppas Model. It can be seen that a burst release occur at the beginning (around the first 30 min). It may be caused by the dissolving of drugs at the surface of film. The concentration of drugs are not significantly increasing after 30 minutes. It means that the release rate is getting slower as time goes by. The concentration properties of each samples also confirm that chitosan grafted with MA will decrease the drug release rate. The constant values of CM0, CM50, CM100, and CM150 are presented in Table 2.

The values of n in Peppas Model indicate the diffusion mechanism of drug. The results show that the release of curcumin in film is following the Fickian diffusion mechanism ($n \leq 0.45$) [20]. While first order kinetic model explain that the release of drug is following first order reaction kinetic. The reaction is assumed as the change of drug phase from solid to liquid. The values of k are the kinetic reaction constant of each samples.

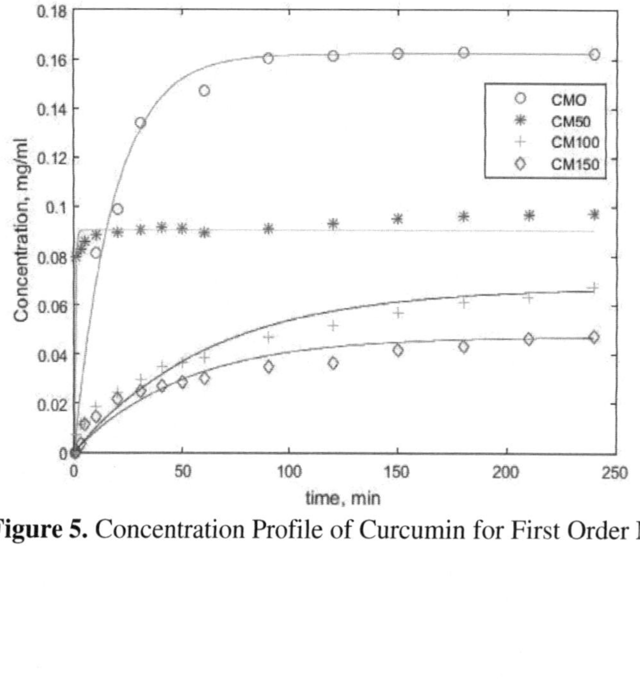

Figure 5. Concentration Profile of Curcumin for First Order Model.

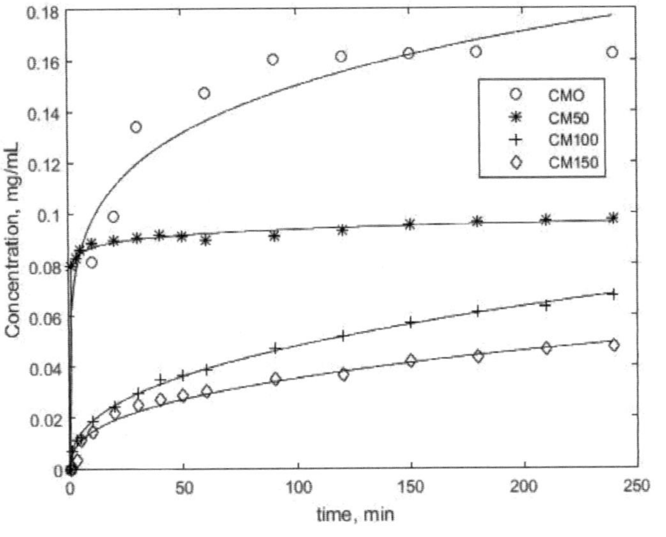

Figure 6. Concentration Profile of Curcumin for Peppas model.

Table 2. Constant Values of Each Samples and Models.

Samples	k	n	K (1/min)
CM0	38.6037	0.1896	0.0553
CM50	88.7728	0.0339	2.0732
CM100	11.0325	0.4042	0.0159
CM150	13.1170	0.3739	0.0178

4. Conclusions

Indonesian chitosan film grafted by MA has been successfully produced. The FTIR analysis showed an appearance of carboxyl functional group in modified chitosan film. The increasing of acidic content also confirm that carboxyl groups are found in the modified film. Swelling ratio and degradation ability of film also increased as the ratio of MA was increased. The addition of MA lead to decrease the release rate of drug in PBS solution. The drug release mechanism for curcumin in film is following Fickian diffusion mechanism. While the optimum ratio of MA/chitosan is 150% (w/w).

5. References

[1] Ali A and Ahmed S 2018 A review on chitosan and its nanocomposites in drug delivery *Int. J.Biol. Macromol.* **109** 273–286

[2] Sinha V R, Singla A K, Wadhawan S, Kaushik R, Kumria R, Bansal K and Dhawan S 2004 Chitosan microspheres as a potential carrier for drugs *Int. J. Pharm.* **274** 1–33

[3] Kean T and Thanou M 2010 Biodegradation, biodistribution and toxicity of chitosan *Adv. Drug Deliv. Rev.* **62** 3–11

[4] Mengatto L, Helbling I and Luna J 2012 Recent advances in chitosan films for controlled release of drugs *Recent patents drug Deliv. formulation* **6** 156–170

[5] Liu Y, Shen X, Zhou H, Wang Y and Deng L 2016 Chemical modification of chitosan film via surface grafting of citric acid molecular to promote the biomineralization *Appl. Surf. Sci.* **370** 270–278

[6] Lawrie G, Keen I, Drew B, Chandler-Temple A, Rintoul L, Fredericks P and Grøndahl L 2007 Interactions between alginate and chitosan biopolymers characterized using FTIR and XPS *Biomacromolecules* **8** 2533–41

[7] Escárcega-Galaz A A, Sánchez-Machado D I, López-Cervantes J, Sanches-Silva A, Madera-

Santana T J and Paseiro-Losada P 2018 Mechanical, structural and physical aspects of chitosan-based films as antimicrobial dressings *Int. J. Biol. Macromol.* **116** 472–481

[8] Costa-Júnior E S, Barbosa-Stancioli E F, Mansur A A P, Vasconcelos W L and Mansur H S 2009 Preparation and characterization of chitosan/poly(vinyl alcohol) chemically crosslinked blends for biomedical applications *Carbohydr. Polym.* **76** 472–481

[9] Pratt D Y, Wilson L D and Kozinski J A 2013 Preparation and sorption studies of glutaraldehyde cross-linked chitosan copolymers *J. Colloid Interface Sci.* **395** 205–211

[10] Hamedi H, Moradi S, Hudson S M and Tonelli A E 2018 Chitosan based hydrogels and their applications for drug delivery in wound dressings: A review *Carbohydr. Polym.* **199** 445–460

[11] Pellá M C G, Lima-Tenório M K, Tenório-Neto E T, Guilherme M R, Muniz E C and Rubira A F 2018 Chitosan-based hydrogels: From preparation to biomedical applications *Carbohydr. Polym.* **196** 233–245

[12] Berger J, Reist M, Mayer J M, Felt O, Peppas N A and Gurny R 2004 Structure and interactions in covalently and ionically crosslinked chitosan hydrogels for biomedical applications *Eur. J. Pharm. Biopharm.* **57** 19–34

[13] Hasipoglu H N, Yilmaz E, Yilmaz O and Caner H 2005 Preparation and characterization of maleic acid grafted chitosan *Int. J. Polym. Anal. Charact.* **10** 313–327

[14] Yucel S, Ozdemir Z O, Kesgin C, Terzioglu P, Unlu S and Erdogan Y 2013 Swelling Behavior and Cytotoxicity of Maleic Acid Grafted Chitosan *Int. J.* **7** 232–235

[15] Hernández-Moreno D, de la Casa Resino I and Soler-Rodríguez F 2014 Maleic Anhydride *Encycl. Toxicol. Third Ed.* **3** 138–141

[16] Vasi A M, Popa M I, Butnaru M, Dodi G and Verestiuc L 2014 Chemical functionalization of hyaluronic acid for drug delivery applications *Mater. Sci. Eng. Mater. Biol Appl. C* **38** 177–185

[17] Lavanya R, Gomathi T, Vijayalakshmi K, Saranya M, Sudha P N and Anil A 2017 Adsorptive removal of copper (II) and lead (II) using chitosan-g-maleic anhydride-g-methacrylic acid copolymer *Int. J. Biol. Macromol.* **104** 1495–1508

[18] Coutts-Lendon C A, Wright N A, Mieso E V and Koenig J L 2003 The use of FT-IR imaging as an analytical tool for the characterization of drug delivery systems *J. Control. Release* **93** 223–248

[19] Zhou L, He H, Li M, Huang S, Mei C and Wu Q 2018 Enhancing mechanical properties of poly (lactic acid) through its in-situ crosslinking with maleic anhydride-modified cellulose nanocrystals from cottonseed hulls *Ind. Crop. Prod.* **112** 449–459

Acknowledgments
This research was supported by Ministry of Education and Culture of Indonesia through PDUPT scheme with contract number No. 122/UN1/DITLIT/DIT-LIT/LT/2018.

IOP Conf. Series: Materials Science and Engineering **599** (2019) 012030 doi:10.1088/1757-899X/599/1/012030

An Optimization Method to Determine Optimum Carbonization Temperature of Banana Stems Based Activated Carbon for Supercapacitors

Erman Taer[1,*], Agrandi Purnama[1], Apriwandi[1], Agustino[1], Rika Taslim[2] and Widya Sinta Mustika[1,3]

[1]Department of Physics, University of Riau, 28293 Simpang Baru, Riau, Indonesia
[2]Departement of Industrial Engineering, State Islamic University of Sultan Syarif Kasim, 28293 Simpang Baru, Riau, Indonesia
[3]Department of Physics, Institute Technology of Bandung, Bandung, West Java, Indonesia

*E-mail: erman_taer@yahoo.com

Abstract. The optimization of carbonization temperature was obtained by experimental design, assisted by using a 3rd-order polynomial equation, for use in supercapacitor application. Activated carbon monoliths are produced based on the carbonization temperatures of 500 °C, 550 °C, 600 °C and 650 °C. The monolithic activated carbon samples were characterized based on their physical properties, including density, degree of crystallinity and surface morphology, and their electrochemical properties, including specific capacitance, energy and power. The optimum conditions, which yielded a minimum electrode density and steak height as well as a maximum combination of energy and power, were found to be at a temperature of approximately 550 °C. In addition, the exact minimum and maximum carbonization temperatures that correlated with density, steak height and a combination of energy and power were found by a simulated mathematical model to be 543 °C, 544 °C, and 553 °C, respectively. The surface morphology was analyzed and was used as supporting data. In conclusion, by experimental and mathematical simulation, the optimum carbonization condition was obtained more accurately.

1. Introduction

Supercapacitors have attracted the interest of many researchers in recent years because they have several advantages, such as a high energy density, maximum power, long life cycle and short charge-discharge time. Supercapacitors are composed of electrodes, electrolytes, separators and current collectors. Studies on supercapacitors have been widely undertaken, with one of the main focuses being modifications to the electrodes. Electrodes can be produced from biomass materials such as rubber wood [1], oil palm empty fruit bunches [2], banana peels [3], rice husks [4] and tea leaves [5]. Biomass materials are selected as the raw materials for the manufacture of supercapacitor electrodes for several reasons, such as ease of production at low carbonization temperature, easy activation, low production cost and abundant availability [6]. Pyrolysis, otherwise known as carbonization, is a common method used in the production of supercapacitor electrodes from biomass materials. Several factors that affect pyrolysis are the type of inert gas, the gas flow rate and the maximum temperature of carbonization, which in turn affects the kinetic energy of the inert gas. A high temperature can lead to high collision energies between the inert gas molecules and the atoms of the biomass material, resulting in the breaking of bonds between carbon and non-carbon atoms so that organic carbon content from biomass can be converted into porous carbon materials. The optimum carbonization temperatures produce a maximum percentage of carbon [7, 8]. The percentage of carbon on the electrode affects the number of ion pairs and electrons formed in the supercapacitor cell. The electron-ion pair occurs at the interface of the electrode and the electrolyte. The number of charge pairs formed determines the capacitive behavior of the supercapacitor cell. A high carbon content also increases the probability of forming charge pairs. The optimum transformation of biomass materials into carbon materials occurs at the optiumum temperature of carbonization, with each type of biomass having a unique optimum carbonization temperature. This condition needs to be considered in choosing the temperature of carbonization. Ma, et. al (2015) [9] produced supercapacitor cell electrodes by carbonizing potato waste residue (PWR) at a temperature range of 600-900 ˚C and obtained a specific capacitance of 255 $F.g^{-1}$ at 700 ˚C. The production of electrodes from ramie in a carbonization temperature range of 400-750 ˚C resulted in a maximum capacitance of 287 $F.g^{-1}$ at a carbonization temperature of 650 ˚C [10]. Meanwhile, carbon electrodes from algae produce a specific capacitance of 234 F.g-1 at a carbonization temperature of 700 ˚C, wherein carbonization is observed at a temperature range of 700-900 ˚C [11]. The objective of this study was to review the maximum carbonization temperature for the production of supercapacitor cell electrodes from banana stem biomass residues. This study was conducted by combining experimental methods and mathematical modeling with polynomial equations. A combination of experimental methods and mathematical modeling to obtain the optimum activation temperature in the production of activated carbon from luscar char [12], coconut husk [13] and palm shell [14] have been performed using quadratic equations based on central composite design (CCD). Faramarzi (2015) [15] and Musabbikhah (2017) [16] optimized the production process of activated carbon from pistachio shell and rice husk through a mathematical model approach and through Taguchi's experimental design. Taguchi's experimental design requires many variables, such as particle size, carbonization temperature, heating rate, holding time, and the N_2 gas flow rate in determining the optimum conditions for the activation of carbon. This approach may get more accurate results but may take a relatively longer time. This study will carry out an optimization of the carbonization temperature in the production process of activated carbon monoliths from banana stems with a mathematical approach using a 3rd order polynomial equation. The mathematical equation used is simpler than the methods previously reported, whereas the more extensive discussion involves the density, steak height, energy and power of a supercapacitor cell.

2. Eperimental Method

2.1. The Electrode preparation and cell fabrication

Activated carbon monolith electrodes were produced from banana stems by previously reported methods [17]. Banana stems were first cut into square shapes ± 2 cm. Furthermore, the banana stems were dried and carbonized at 250 ˚C for 2.5 hours to obtain a green carbonized banana stem that had adhesive properties. The next step involved smoothing using ball milling for 20 hours and sieving to obtain a particle size of 38-53 μm. The powder sample was then activated with 0.4 M KOH. The activated powder sample was molded into pellets to produce Green Monolith Banana Stem (GMBS) electrodes [17] that had a diameter of ± 2 cm and a thickness of ± 2 mm. The GMBS pellet was modified into a carbon pellet electrode through a stage of integrated carbonization and physical activation using a furnace [18]. This stage began with carbonization carried out under N_2 gas at a temperature that varied from room temperature to a maximum temperature, followed by physical activation using CO_2 [19] at a temperature of 850 ˚C with a 2-hour holding time. This process produced an activated carbon monolith (ACM). The carbonization temperatures were 500 ˚C, 550 ˚C, 600 ˚C and 650 ˚C; each ACM sample was thus labeled ACM500, ACM550, ACM600 and ACM650. All the ACM pellets were polished to the desired thickness and were washed with neutral (pH 7) water. The supercapacitor cells were arranged in a sandwich that comprised the insulator body supercapacitor, current collector, electrode, electrolyte and separator, as we have reported in previous research [20][21].

2.2. Physical Characteristics

The physical properties of the ACM analyzed include density, structure, surface morphology and chemical composition. The density was calculated from the mass, diameter and thickness of the electrodes. The ACM structure was analyzed using X-ray Diffraction (XRD). XRD measurements were performed using a Philip X-Pert Pro PW 3060/10 instrument with a K-α ray source at a wavelength of 1.5418 Å and diffraction angles of 2θ = 10-100 degrees. The surface morphology was examined using scanning electron microscopy (SEM) with a SUPRA instrument, S-3400N series. The composition of ACM contents was determined by energy dispersive X-ray (EDX) spectroscopy.

2.3. Electrochemical Characteristics

The electrochemical properties of the supercapacitor cells were analyzed using a Physics CV UR Rad-Er 5841 instrument and were calibrated with Solartron 1280 devices. Electrochemical measurements were carried out at a rate of 1 $mV.s^{-1}$ and a potential of 0 s / d of 500 mV controlled using the Cyclic Voltammetry software CVv6. The specific capacitance, energy and power were calculated by using the standard formula [22] [23].

$$C_{sp} = \frac{I_c - I_d}{sm} \qquad E = \frac{1}{2} C_{sp} \Delta V^2 \qquad P = \frac{E}{\Delta t} \tag{1}$$

3. Result and Discusion

3.1. Physical Properties Analysis

The ACM density data were obtained from the calculation of the mass, diameter and mean thickness by using the standard formula (ρ = m / V). The density of each ACM was reported after the integrated carbonization-activation process and is shown in Figure 1 a. The densities of ACM500, ACM550, ACM600 and ACM650 were 0.808 g cm^{-3}, 0.733 g cm^{-3}, 0.790 g cm^{-3} and 0.813 g cm^{-3}, respectively. ACM550 showed a lower density than did ACM500. The increase in the carbonization temperature for ACM550 led to a greater loss of materials other than carbon, such as moisture content, compared to ACM500, resulting in a significant decrease in mass. The ACM550, ACM600 and ACM650 samples showed a linear trend of increasing density values; an increase in carbonization temperature resulted in an increase in density. In this case, the increase in carbonization temperature resulted in a smaller decrease in mass than a decrease in volume. The mass loss during the carbonization process led to the rearrangement of carbon atoms to form stronger bonds, causing dimension shrinkage.

The density as a function of carbonization temperature was fit using the 3rd-order polynomial shown in Figure 1a. The 3rd-order polynomial equation indicated a minimum density of 0.7329 g cm^{-3} at a carbonization temperature of 543 °C. The polynomial approach supported the experimental results to obtain a more appropriate minimum density. Figures 1b and 1c show micrograph SEM data for the ACM550 and ACM600 samples. These two samples showed differences in the arrangement of carbon microparticles; ACM550 showed more inter-particle pores than the ACM600 sample. The ACM600 sample showed inter-particle pores covered by smaller particles with irregular shapes as well as by fibers. This difference indicated that the increase in carbonization temperature caused the particles to break into finer particles, resulting in the release of fibers. This SEM data supported the relationship between density and carbonization temperature, as shown in Figure 1a, wherein the minimum density occurred at approximately of 550 °C

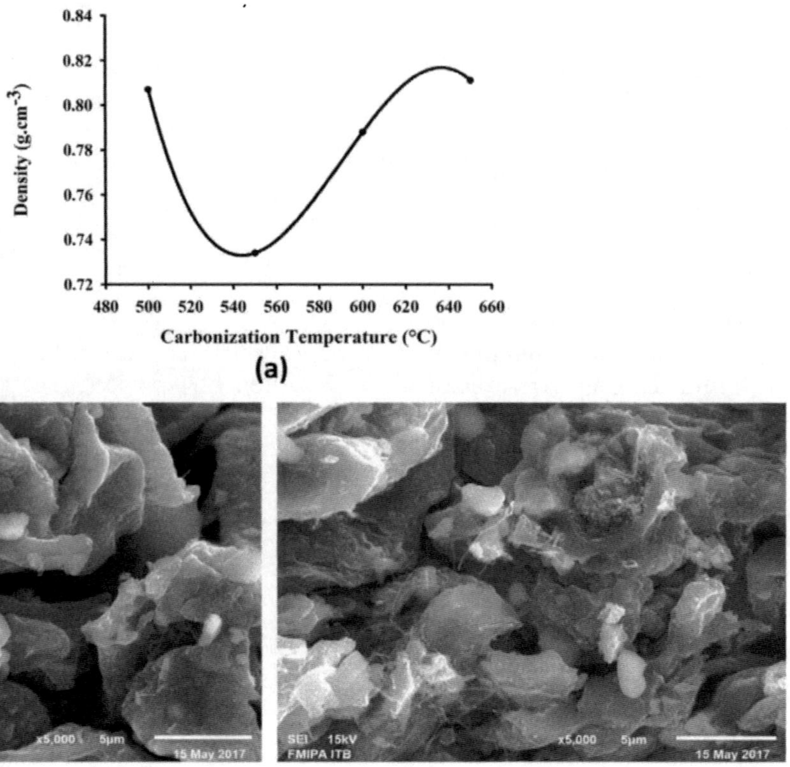

Figure 1. a) Density modeling curve; b) micrograph SEM for ACM550; c) micrograph SEM for ACM600.

The X-Ray diffraction curves for all of the ACM samples is shown in Figure 2. This figure, which plots the X-ray intensity as a function of the diffraction angle, showed no significant difference. In general, Figure 2a shows that the samples of ACM produced two broadening peaks and the same sharp peaks. The broadening peaks appeared at a diffraction angle 2θ of 24.827°- 25.333° and 44.542°-46.505° for reflection planes of 002 and 100, respectively [24]. The presence of broadening peaks in the diffractogram showed that the ACM sample had an amorphous carbon structure [25]. In addition, the sharp peaks, due to the presence of impurities other than carbon material, had a crystalline structure. This condition was caused by the heating process in the modification of carbon from biomass, which produced other materials such as silica [2], potassium and magnesium. The chemical activation process also produced impurities in the sample caused by the imperfect washing process.

The variations of carbonization temperature used in the production of ACM resulted in microcrystallite dimensions. The microcrystallite dimension, such as steak height (L_c), was calculated using the Debye-Scherrer equation [26 - 28] and $L_c = 0.89\ \lambda/\beta\ \cos\theta_{(002)}$, where λ is the wavelength (Å), β is the bandwidth (degrees), $\theta_{(100)}$is the diffraction angle in the hkl plane 100 (degrees) and $\theta_{(002)}$is the diffraction angle in the hkl plane 002 (degrees) [29]. Differences in the temperature of carbonization used also resulted different values of microcrystallite dimensions.

The relationship between the L_c and the carbonization temperature was expressed in the 3rd-order polynomial shown in Figure 2b. The smallest L_c was obtained at a carbonization temperature slightly lower than 550 °C. The minimum L_c was obtained by using the inserted 3rd-order polynomial in Figure 2b; the minimum L_c obtained was 11, 5617 (Å) at a carbonization temperature of 544 °C. The relationship between electrode density and L_c can be observed in Figures 1a and 2b, where the minimum density occurred when L_c was also the minimum. In other words, the minimal density and L_c minimum occurred at almost the same carbonization temperature. These results indicate that density changes in the biomass-based carbon electrode, as a function of the carbonization temperature, were influenced by the arrangement of carbon particles in the height direction, or L_c.

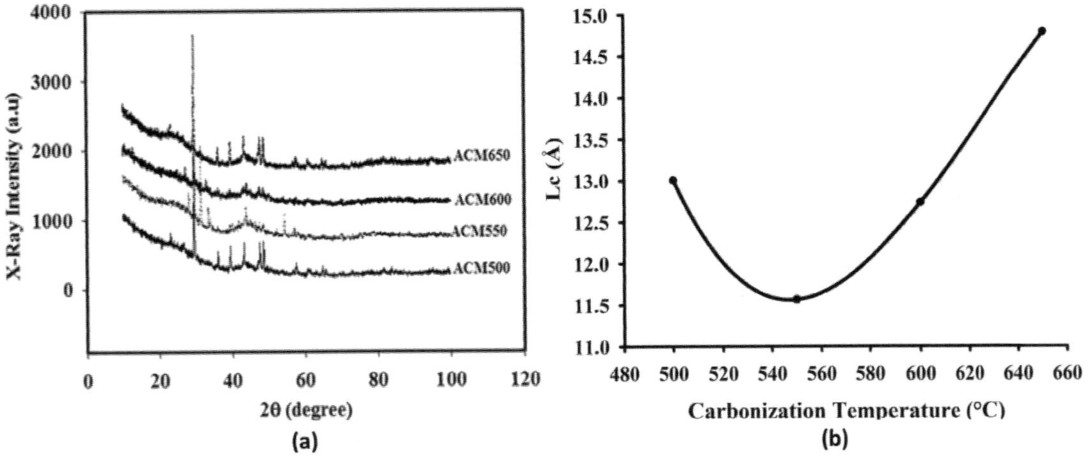

Figure 2. a) Diffractogram curves for ACM; b) polynomial curve of L_c and carbonization temperature.

3.2 Electrochemical Properties Analysis

The cyclic voltammetry (CV) measurement data for all the ACM samples are shown in Figure 3a. The current density was plotted as a function of voltage, at a scan rate of 1 $mV.s^{-1}$. All the samples showed an almost identical curve. The current charging process for each ACM sample started at a potential of 0.0 V, while the current discharge process began at a potential of 0.5 V [30]. The charging and discharging current rates indicate the shape and pore size of the carbon electrode. The difference in the rate of current at a potential slightly larger than 0.0 V indicates the difference in the number of pores on the carbon electrode. The process of ion diffusion from the electrolyte into the pores that exist on the surface of the carbon electrode affects the number of electron pairs formed on the carbon electrode, thereby increasing the current and increasing the specific capacitance of the supercapacitor cell. Figure 3a also shows the area between different charge and discharge currents, at a potential of 0.25 V. ACM500, ACM550 and ACM600 produced a curve with a minimum increase after the maximum current charge from a potential that was slightly larger than 0.0 V to a maximum potential window of 0.5 V. The presence of an increase in slope indicated the presence of a dominant amount of micropores in the electrode samples. ACM550 produced a wider gap between the charge and discharge curves than that of ACM500, which corresponded to an increase in the number of micropores carbon electrode. The increase in carbonization temperature for ACM600 produced a lower curve than that of ACM550. An increase in carbonization temperature of up to 600 °C reduced the number of micropores on the carbon electrode [7]. However, the ACM650 produced a flatter curve shape after the maximum current charge at a potential of 0.0 V. The flatter curve indicated that the charging process, with ions moving into the micropore, had finished when the potential was larger than 0.2 V.

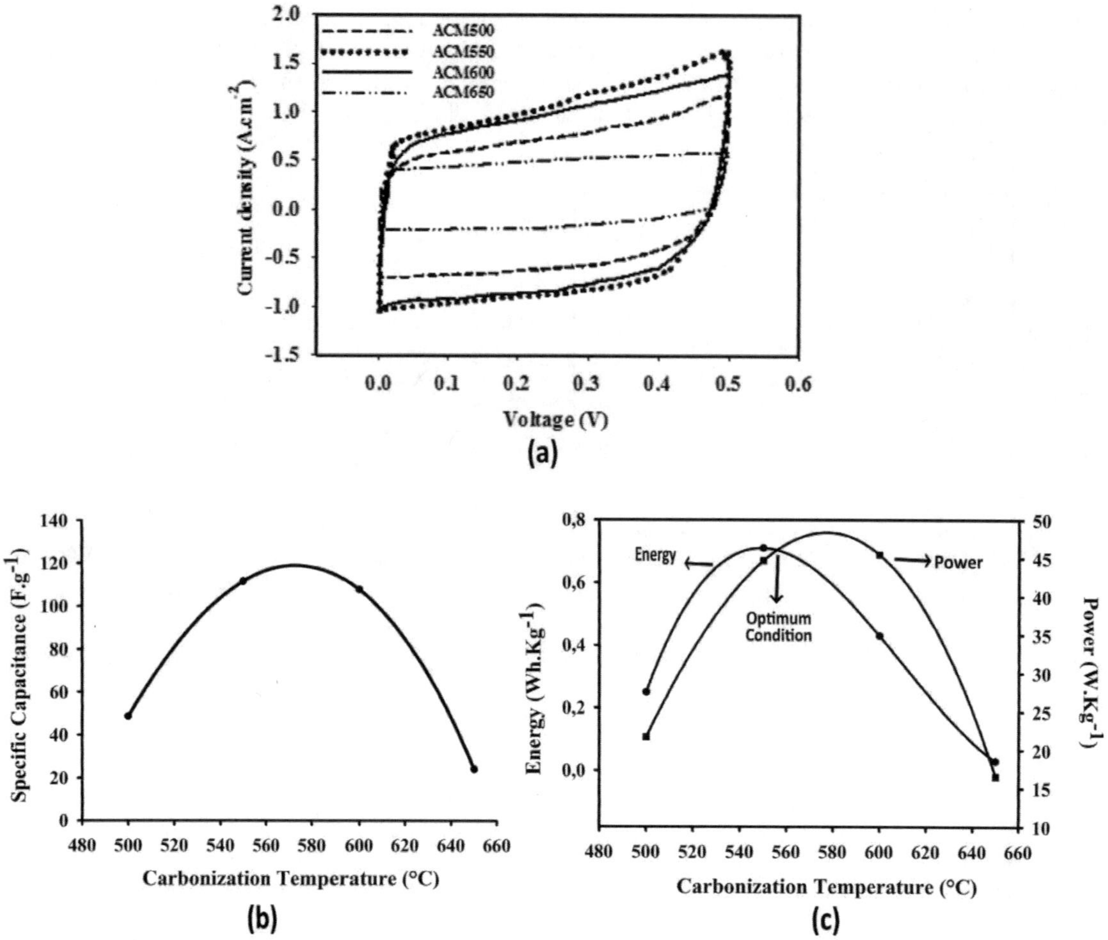

Figure 3. a) CV curve for ACM; b) polynomial curve of Csp and carbonization temperature; c) polynomial curves of energy and power with carbonization temperature.

The CV measurement resulted in the charging current (Ic) and discharging current (Id); by using the same formulas [2, 22], one can calculate the specific capacitance (Csp) of the electrode as well as the specific energy (E) and power (P) of the supercapacitor cell. The Csp initially increased with increasing carbonization temperature. At a carbonization temperature of 500 ˚C, the Csp was 48.58 F g^{-1}; when the carbonization temperature was increased to 550 ˚C, the Csp was as high as 111.56 F g^{-1}. However, as temperatures continued to increase to 600 ˚C and 650 ˚C, the Csp decreased as the carbonization temperature increased; the Csp values were 107.80 F g^{-1} and 24.15 F g^{-1}, respectively. It can be concluded that the optimum carbonization temperature was between 550 ˚C and 600 ˚C. The optimum carbonization temperature, corresponding to the optimum Csp, can be calculated using polynomial equation fitting. Figure 5 was obtained by fitting the Csp, as a function of carbonization temperature, to a 3rd-order polynomial. The fit showed that the best carbonization temperature for the ACMs from banana stems was in the range 550-600 ˚C, exactly 572 ˚C, with an optimum Csp as high as 119 F g^{-1}. In addition, the resulting capacitive electrode is almost the same as the other capacitive electrode with a different biomass such as durian shell as high as 88.39 F g^{-1} [31], banana stem 170 F g^{-1} [32] and corn stalk as high as 140 F g^{-1} [33].

The energy and power obtained from the experimental data were also fit using a 3rd-order polynomial. These data are shown in Figures 3c. The maximum energy of 0.7101 $Wh.kg^{-1}$ was obtained at a carbonization temperature of 548 ˚C, while a maximum power of 48.3531 $W.kg^{-1}$ was obtained at a carbonization temperature of 577 ˚C. These results indicate that there was little

IOP Conf. Series: Materials Science and Engineering **599** (2019) 012030 doi:10.1088/1757-899X/599/1/012030

difference in optimum carbonization temperature conditions for the Csp, energy and power produced by the supercapacitor cells. The optimum condition of a supercapacitor cell is an optimum combination of energy and power. Figure 3c shows that the energy and power curves intersected at a certain point; this intersection point indicated the optimum condition of energy and power. This intersection point appeared at a carbonization temperature of 553 °C. This point is indicated by arrows in Figure 3c. This temperature was selected as the optimum carbonization temperature for the production of supercapacitor electrodes from banana stem waste material. This optimum carbonization temperature is almost equal to the optimum temperature of carbonization in other material samples, such as rubber wood sawdust [20] and bamboo waste [34].

Table 1. Coefficient parameters of the 3rd-order polynomial equation

Parameters	Coefficients			
	y0	*a*	*b*	*c*
Density (g cm^{-3})	43.2820	-0.2188	0.0004	-2,1067E-007
Steak Height (Å)	549.4660	-2.6532	0.0043	-2.2947E-006
Csp (F g^{-1})	-1358.9200	-0.5927	0.0156	-1.7533E-005
Energy (Wh kg^{-1})	-181.4500	0.9127	-0.0015	8.2667E-007
Power (W kg^{-1})	230.2600	-3.9788	0.0122	-1.0053E-005

The mathematical equation used in analyzing the experimental results, to determine the optimum value of density, steak height, Csp, energy and power, is

$$f = y0 + ax + bx^2 + cx^3 \quad (2)$$

which is known as a 3rd-order polynomial equation. This 3rd-order polynomial is a cubic polynomial with one variable. The mathematical equations obtained had a correlation coefficient (R) equal to 1.00, which means that the equations used had a perfect correlation. Coefficients y0, a, b and c are real coefficients obtained using Ms.excel, if all coefficients are real numbers then 3rd-order polynomial solutions have more than one real root or critical point, but in the data shown in Figure 3a and 3b only one critical point appears due to the limited of data. This critical point signify the optimum conditions of each parameters being modeled. This simple modeling helps researchers to determining optimum conditions faster and easier without the use of complicated software. The coefficients of the 3rd-order polynomial equations (such as y0, a, b and c) for density, steak height, Csp, energy and power can be seen in Table 1. The density and steak height modeling showed curves with a minimum value, whereas the modeling data for Csp, energy and power showed curves with a maximum value.

In general, the results of the analysis of the physical and electrochemical properties as a function of the carbonization temperature of the carbon electrode supercapacitor showed a connection among density, surface morphology, steak height and the combination of energy and power. The minimum density and minimum L_c occurred at nearly equal carbonization temperatures, and these conditions also led to an optimum combination of energy and power. It can be concluded that the experimental approach and the associated 3rd-order polynomial equation have been successfully used to obtain the optimum temperature conditions for the combination of energy and power. Finally, the experimental approach and the 3rd-order polynomial equation shown in this paper can be considered as one way of determining the best temperature variables in the process of preparing supercapacitor electrodes, such as the temperatures of carbonization and activation.

4. Conclusion

The optimum physical and electrochemical properties of carbon electrodes have been successfully found experimentally, assisted by mathematical modeling using a 3rd-order polynomial equation. Varying the carbonization temperature in the preparation process resulted in different physical and electrochemical properties. The experimental data were fitted by using a 3rd-order polynomial equation to find the optimum conditions. The information in the mathematical models was analyzed, and the corrected optimum condition was found, for the experimental data. Finally, the mathematical modeling approach that has been shown can determine the optimum conditions of carbonization temperature more accurately.

5. References

[1] Taer E, Taslim R and Deraman M 2016 *AIP Conf. Proc.* **1712** 050011-1
[2] Farma R, Deraman M, Awitdrus A, Talib I A, Taer E, Basri N H, Manjunatha J G, Ishak M M, Dollah B N M and Hashmi S A 2013 *Bioresource Technol.,* **132** 254
[3] Taer E, Taslim R, Aini Z, Hartati S D and Mustika W S 2017 *AIP Conf. Proc.* **1801** 040004
[4] Teo E Y L, Muniandy L Ng E, Adam F, Mohamed A B, Jose R and Chong K F 2016 *Electrochim. Acta* **192** 110
[5] Peng C, Yan X, Wang R, Lang J, Ou Y and Xue O 2013 *Electrochim. Acta* **87** 401
[6] Abioye D M and Ani F N 2015 *Renewable and Sustainable Energy Reviews* **52** 1282
[7] Daud W M A W, Ali W S W and Sulaiman M Z 2000 *Carbon* **38** 1925
[8] Li W, Yang K, Peng J, Zhang L, Guo S and Xia H 2008 *Industrial crops and products* **28** 190
[9] Ma G, Yang Q, Sun K, Peng H, Ran F, Zhao X and Lei Z 2015 *Bioresource Technol.* **197** 137
[10] Du X, Zhao W, Wang Y, Wang C, Chen M, Qi T, Hua C and Ma M 2013 *Bioresource Technol.* **149** 31
[11] Tian Z, Xiang M, Zhou J, Hu L and Cai J 2016 *Direct Carbonization and Excellent Electrochemical Properties* **211** 225
[12] Azargohar R and Dalai A K 2005 *Microporous and Mesoporous Mater* **85** 219
[13] Tan I A W, Ahmad A L and Hameed B H 2008 *Chemical Engineering Journal* **137** 462
[14] Arami-Niya A, Daud W M A W, Mjalli F S, Shafeeyan M S and Abnisa F 2012 *Chemical engineering research and design,* **90** 776
[15] Faramarzi A H, Kaghazchi T, Ebrahim H A and Ebrahimi A A 2015 *J. Anal. Appl. Pyrolysis*
[16] Musabbikhah, Saptoadi H, Subarmono and Wibisono M A 2017 *Int. J. Renew. Energy Research* **7** 1219
[17] Farma R, Deraman M, Omar R, Awitdrus, Ishak M M, Taer E, and Talib I A 2011 *AIP Conf. Proc.* **1415** 180.
[18] Taer E, Apriwandi, Yusriwandi, Mustika W S, Zulkifli, Taslim R, Sugianto, Kurniasih B, Agustino and Dewi P 2018 *AIP Conf. Proc.* **1927** 030036-1
[19] Deraman M, Ishak M M,Farma R, Awitdrus, Taer E, Talib I A, Omar R 2011 AIP Conf. Proc. **1415** 175
[20] Taer E, Deraman M, Taslim R and Iwantono 2013 *AIP Conf. Proc.* **1554** 33
[21] Taer E, Sugianto, Sumantre M A, Taslim R, Iwantono, Dahlan D, Deraman M 2014 *Adv. Mater. Res.* **896** 66
[22] Li L, Liu E, Li J, Yang Y, Shen H, Huang Z, Xiang X and Li W 2010 *J. Power Sources* **195** 1516
[23] Zhao Y-Q, Lu M, Tao P-Y, Zhang Y-J, Gong X-T, Zhang G-Q, Li H-L,Yang Z 2017 *J.Power Sources* **307** 391
[24] Nabais J M V, Teixera J G and Almeida I 2011 *Bioresource Technol.* **102** 2781
[25] Deraman M, Omar R, Zakaria S, Mustapa I R and Talib M 2002 *J. Mater. Science* **37** 3329
[26] Cullity B D 2001 Elements of X-Ray Diffraction, Ed. 3, Amazon Prentice Hall
[27] Awitdrus, Deraman M, Talib I A, Omar R, Jumali M H, Taer E and Saman M H 2010 *Sains Malaysiana* **39** 83
[28] Nabais J M V, Teixeira J G and Almeida I 2010 *Bioresource Technol.* **102** 2781
[29] Carrott P J M, Nabais J M V, Carrott M M L R and Pajares J A 2001 *Carbon* **39** 1543
[30] Inagaki M, Konno H and Tanaike O 2010 *J. Power Sources* **195** 7880
[31] Taer E, Dewi P, Sugianto, Syech R, Taslim R, Salomo, Susanti Y, Purnama A, Apriwandi, Agustino and Setiadi R N 2018 *AIP Conf, Proc,* **1927** 030026-1

[32] Taer E, Taslim R, Mustika W S, Kurniasih B, Agustino, Afrianda A and Apriwandi 2018 *Int. J. Electrochem, Sci,* **13** 8428
[33] Yu K, Zhu H, Qi H and Liang C 2018 *Diamond & Related Mater.* **88** 18.
[34] Zhang Y-J, Xing Z-J, Duan Z-K, Li M and Wang Y 2014 *Applied Surface Science* **315** 279

Acknowledgements
The author would like to thank the DRPM Kemenristek-Dikti through the second year Project of PDUPT with the title "Potential of Urban Solid Waste Utilization as a Supercapacitor Electrode" with contract number: 360/UN.19.5.1.3/PP/2018. The author also thanks the SEM FMIPA ITB Laboratory, which has assisted in obtaining the SEM and EDX data.

A Study of Physical and Mechanical Properties of Nata de Coco in the Market

Handika Dany Rahmayanti [1,2,a], Nadya Amalia[1], Rahmawati Munir[1], Elfi Yuliza[1], Fisca Dian Utami[1], Euis Sustini[1] and Mikrajuddin Abdullah[1,b]

[1]Departement of Physics, Faculty of Mathematics and Natural Science, Bandung Institute of Technology, Jl. Ganesha 10 Bandung 40132, Indonesia
[2]Politeknik Negeri Media Kreatif, Jl. Srengseng Sawah Jakarta Selatan, Indonesia

E-mail: [a]handikadany44@gmail.com, [b]din@fi.itb.ac.id

Abstract. Nata de coco is a bacterial cellulose produced by Acetobacter xylinum in the process of fermenting coconut water. It is renowned for its high fiber content. The chewiness of nata de coco is proportional to the fiber content. There are many brands of nata de coco available in the market. Six different brands are included in this study. The mechanical strengths have been evaluated using a Universal Testing Machine. Based on the tensile test results, the lowest Young's modulus is 3 MPa, attributed to the brand labeled as F as the most elastic nata de coco among the six brands. The degree of elasticity of nata de coco also corresponds to the sugar content. Further, the optical properties have been observed using a UV/Vis Spectrometer. Clearer nata de coco yields higher transmittance value. Meanwhile, the turbidity commensurate the nitrogen content.

Keywords: nata de coco, elasticity and transmittance

1. Introduction

One of the popular products from coconut water is nata de coco. Nata de coco is produced from the fermentation process of Acetobacter xylinum using coconut water as a source of micronutrients called bacterial cellulose [1,2]. Nata de coco is easy to make, easy to process and easy to obtain with inexpensive production costs. At present, nata de coco industry is growing rapidly. Many brands of nata de coco are sold in the market. The price offered varies for each brand of nata de coco. If we observe with the naked eye, nata de coco available in the market show generally yellowish white, white to clear color variations. The texture of each brand of nata de coco also varies, some are rather hard or even very soft/chewy. On the other hand, there has been no published research on the characterization of various brands of nata de coco in the market. The study on the characterization of several brands of nata the coco available in the market may seem unimportant, as many people assume it in a normal way and apparently can not benefit from it. However, we will show here that it is necessary. Information about the characteristics of each market-traded nata de coco is needed to determine whether the nata de coco sold in the market has the same standarization. Indeed, there are many common phenomena around us that could be interesting topics of research that have not been considered as such by many people [3-5]. The study of the characteristics of each nata de coco is also important for the development of research on nata de coco. This material is interesting since it is composed (network) of strong nano fibers. Some researchers have also examined the application of Nata de coco for different purposes. These findings have highlighted the benefits of bacterial cellulose composite for material of future application [6-11].

2. Methods

In this study, six brands of nata de coco in the market were tested (Figure 1). Several characterizations have been employed such as physical properties, optical properties and mechanical properties. Mechanical strength of the sample was tested using a Universal Testing Machine (Micro Control Systems model MCS-UTE60 equipped with a MCSUTE STDW2KXP), as shown in Figure 2(a). Spectrometer UV-Vis (Nano Calc 2000-UV-VIS Serial No. I2J8280) was used to examine the transmittance of all nata de coco is shown Figure 2(b).

Figure 1. Six brands of nata de coco available in the market and are labeled as (A) brand A, (B) brand B, (C) brand C, (D) brand D, (E) brand E and (F) brand F.

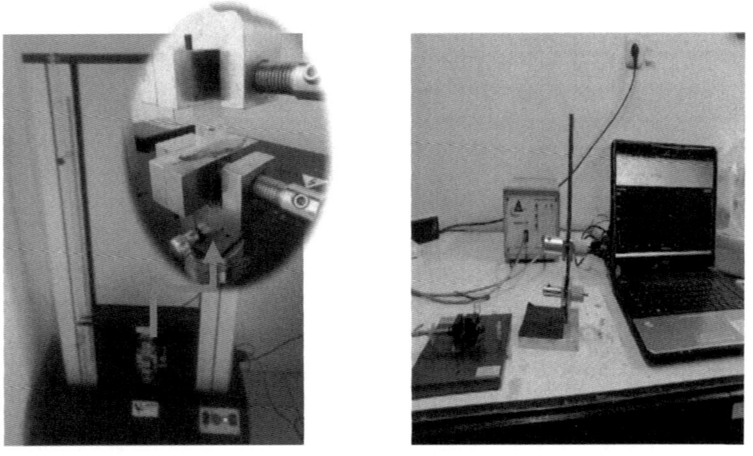

(A) (B)

Figure 2. (A) Tensile strength tester dan (B) UV/Vis Spectrometry.

3. Results and Discussion

Nata de coco is a fermentation product derived from coconut water with the help of the bacterium Acetobacter xylinum. Coconut water can stimulate bacterial growth because of its rich and relatively complete nutritional content, so that it can be a natural medium suitable for microbial growth. Bacteria also need other additives for fermenting nata de coco, including carbon and nitrogen sources. Maltose, sucrose, lactose, fructose and mannose are carbonaceous compounds necessary for the fermentation of nata derived from monosaccharides and disaccharides. The type and level of sugar added will affect the thickness and nature of the Nata formed. Sucrose is often used as a carbon source to produce nata a thick and hard nata. Sucrose levels of 5 to 10% in a fermented medium will produce a thick and hard nata [12]. In this study, to measure the degree of elasticity of nata de coco, we used tensile strength test. The results of the characterization of tensile strength on commercial nata de coco samples are presented in Table 1. In this study, Universal Tensile Machine (UTM) were used to investigate tensile behavior of all samples. The Young's modulus also can be estimated. The Young's modulus value of all samples can be determined from the gradient value of the stress versus strain curve, as shown in Figure 3.

Table 1. Results of tensile strength test.

	$\Delta l_m\,(mm)$	$F_m\,(N)$	$\sigma_m\,(MPa)$
Brand A	3.30	18.42	0.41
Brand B	3.06	7.36	0.35
Brand C	11.71	43.04	0.31
Brand D	11.12	57.96	0.21
Brand E	12.15	46.99	0.14
Brand F	6.42	11.42	0.12

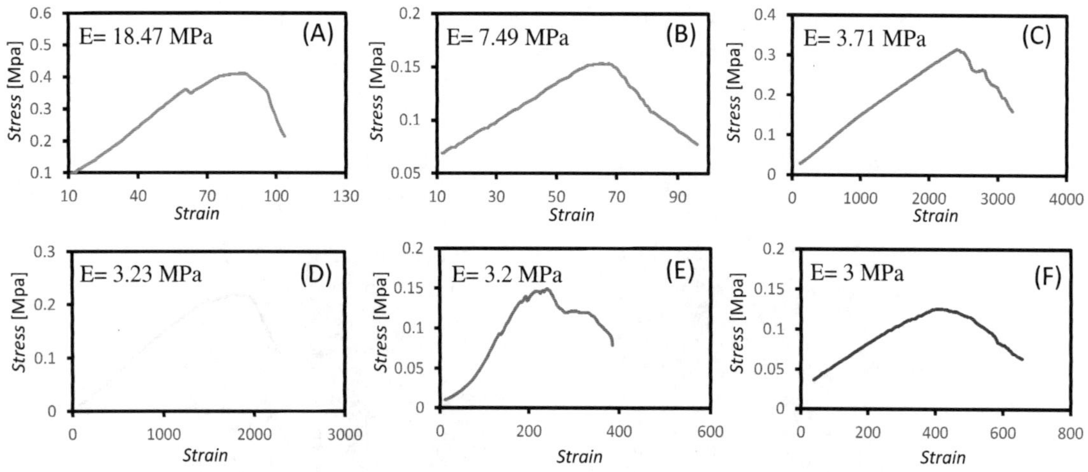

Figure 3. Curve of stress vs strain by (A) brand A, (B) brand B, (C) brand C, (D) brand D, (E) brand E and (F) brand F.

From tensile strength test, data of increase of length Δl_m were obtained. The value of maximum tensile strength F_m and maximum voltage σ_m can also be estimated. We observe from Table 1 that the highest maximum voltage was obtained by the sample of nata de coco brand A, 0.41 MPa. The higher value of maximum voltage correlates with higher tensile strength. Physically, the sample of nata de coco brand A has texture had such a hard texture. Based on the result of the tensile strength test, the Young's modulus can be estimated. Figure 3 shows that the lowest Young's modulus was obtained from the sample of nata de coco brand F, which is equal to 3 MPa. Meanwhile, the sample of nata de coco brand A has the highest Young's modulus value, 18.47 MPa. It also in line with the measurement graph of tensile strength test, in which the line of elongation at break of the sample nata de coco brand F is higher and longer than any other method. The line of elongation at break in the tensile strength test determines the elasticity of the material. It can therefore be concluded that the sample of nata de coco brand F was the most chewy nata. With increasing levels of sugar in the medium, the hardness of the nata will be lower and the elasticity will increase. This is because high sugar levels will cause looseness of the bonds between the fibers and, therefore, most of the formed gel will be filled with water and little solid. The elasticity of nata is influenced by a lot of lack of fibers [13]. The higher the fiber content, the more elastic the texture of nata.

Other supporting factors for the growth of bacterial activity in nata de coco are sources of nitrogen which can come from organic nitrogen and inorganic nitrogen. Sources of organic nitrogen include protein and yeast extract, peptone and tripton, while inorganic nitrogen such as ammonium phosphate, urea, potassium nitrate, and ZA. In the manufacture of nata de coco, it generally uses inorganic nitrogen, such as ammonium phosphate because it is very economical and its function is not inferior when compared to organic nitrogen sources. In addition, ammonium sulfate has advantages such as cheap, soluble, and selective to other microorganisms [14]. In the market, Nata de coco is generally yellowish white, white to clear or transparent, as shown in Figure 1. The clearer nata de coco indicates even greater transmittance. Based on results of UV/Vis spectrometer (Figure 4) for samples of nata de coco brand A, brand B, brand C, brand D, brand E and brand F, 75%, 57%, 50%, 48%, 42 and 38% were obtained, respectively. Meanwhile, the sample of nata de coco brand A attained the highest transparency value. The higher the level of ammonium sulfate added to the nata, the more yellow it will be. In fact, the hydrolysis ions of ammonium sulphate react with sugar, so that the results give a darker color [15].

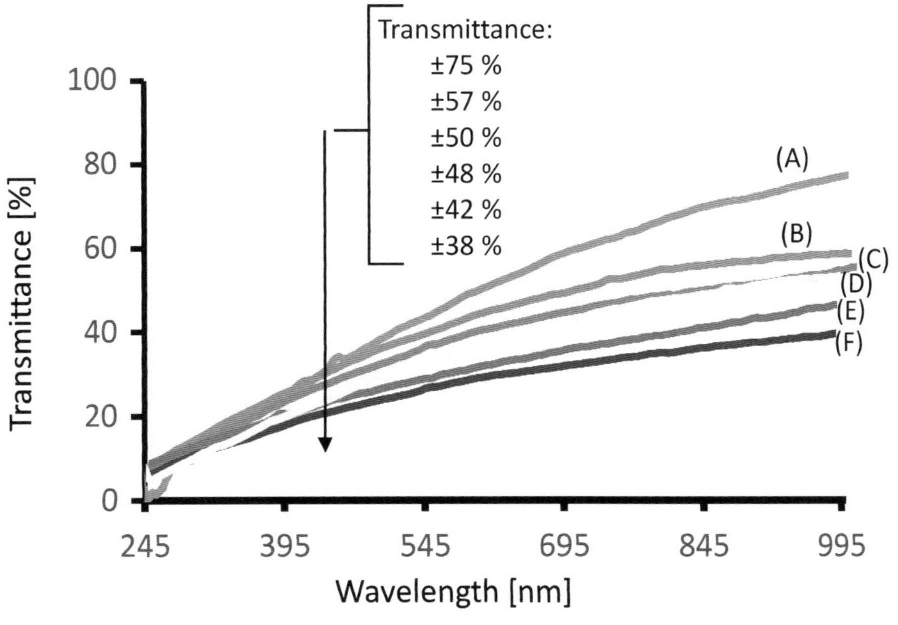

Figure 4. Characterization results of UV/Vis Spectrometer for (A) brand A, (B) brand B, (C) brand C, (D) brand D, (E) brand E and (F) brand F.

4. Conclusion

Based on the UV/Vis test, the highest transmittance was obtained by sample of nata de coco brand A, which visually showed the clearest color. If the color of nata de coco is more yellowish white, it indicates higher nitrogen content in nata de coco. The tensile strength test shows that nata de coco brand F is the most elastic nata de coco of the others. The more elastic the nata de coco is, more sugar content in nata de coco.

5. References

[1] Sasithorn K 2008 *Appl. Biochem. Biotech.* **148** 245 – 256
[2] Byrom D 1991 *Journal of Biomaterials* 263-284
[3] Rahmayanti H D, Munir R, Sustini E and Abdullah M 2019 *Journal of Statistical Mechanics: Theory and Experiment* **1** 013401
[4] Rahmayanti H D, Utami F D and Abdullah M 2016 *European Journal of Physics* **37**(6) 065806
[5] Mayer H C and Krechetnikov R 2012 *Phys. Rev. E* **85** 046117
[6] Rahmayanti H D, Amalia N, Dewi Y C, Sustini E and Abdullah M 2018 *Mater. Res. Express* **5**(5) 054004
[7] Puspitasari T and Radiman C L 2012 *Atom Indonesia* **32** 119-28
[8] Saputra A H, Andini G and Anindita H N 2015 *International Journal of Technology* **6** 1198-04
[9] Yang J, Sun D, Li J, Yang X, Yu J, Hao Q, Liu W, Liu J, Zou Z and Gu J 2009 *Electrochim. Acta* **54** 6300-05
[10] Evans B R, O'Neill H M, Malyvanh V P, Lee I and Woodward J 2003 *Bisensors and Bioelectronics* **18** 917–23
[11] Onggo D, Putri O K and Aminah M 2015 *IOP Conference Series: Materials Science and Engineering* **79** 012021
[12] Suratiningsih S 1994 *Duta Farming* **29**
[13] Rossi E, Pato U and Damanik SR 2008 *J.SAGU* **7**(2) 30-36
[14] Ajaban C A 1962 *Studies on Optimum Conditions for Nata de Coco Bacterium or Formulation in Coconut Water* Phillipine: Phillipine Agricultural
[15] Mashudi 1993 *Mempelajari pengaruh penambahan sumber nitrogen dengan berbagai konsentrasi pada pembuatan nata de coco* Bogor: Institut Pertanian Bogor

Acknowledgement

Thank you for the PMDSU (Program Magister Doktor Sarjana Unggul) research grant from the Ministry of Research and Higher Education, Republic of Indonesia No. 535C/I1.C01/PL/2018 for HDR.